T0331822

OPTICAL COATINGS AND THERMAL NOISE IN PRECISION MEASUREMENT

Thermal noise from optical coatings is a growing area of concern, and overcoming limits to the sensitivity of high-precision measurements by thermal noise is one of the greatest challenges faced by experimental physicists.

In this timely book, internationally renowned scientists and engineers examine our current theoretical and experimental understanding. Beginning with the theory of thermal noise in mirrors and substrates, subsequent chapters discuss the technology of depositing coatings and state-of-the-art dielectric coating techniques used in precision measurement. Applications and remedies for noise reduction are also covered.

Individual chapters are dedicated to specific fields where coating thermal noise is a particular concern, including the areas of quantum optics/optomechanics, gravitational wave detection, precision timing, high-precision laser stabilization via optical cavities, and cavity quantum electrodynamics. While providing full mathematical detail, the text avoids field-specific jargon, making it a valuable resource for readers with varied backgrounds in modern optics.

GREGORY HARRY has worked in the field of gravitational wave detection for over 15 years and is currently the Optics Chair and Coating Cognizant Scientist for the Laser Interferometer Gravitational Wave Observatory (LIGO), and Professor at American University, Washington DC. He is amongst the pioneers of coating thermal noise research.

TIMOTHY P. BODIYA is a graduate student at the Physics Department of Massachusetts Institute of Technology. He is conducting research in the field of gravitational wave physics and quantum optomechanics with the goal of measuring quantum effects on everyday-sized objects (gram to kilogram size).

RICCARDO DESALVO is Professor at the University of Sannio in Benevento, Italy. Previously he has held the positions of Senior Staff Scientist at LIGO, Caltech, Passadena, and that of Staff Scientist at INFN in Pisa, Italy. He is a member of ASME, APS, and SIF and has authored more than 200 refereed papers.

OPTICAL COATINGS AND THERMAL NOISE IN PRECISION MEASUREMENT

Edited by

GREGORY HARRY
American University, Washington DC

TIMOTHY P. BODIYA
Massachusetts Institute of Technology

and

RICCARDO DESALVO
Universitá degli Studi del Sannio, Benevento, Italy

Shaftesbury Road, Cambridge CB2 8EA, United Kingdom

One Liberty Plaza, 20th Floor, New York, NY 10006, USA

477 Williamstown Road, Port Melbourne, VIC 3207, Australia

314–321, 3rd Floor, Plot 3, Splendor Forum, Jasola District Centre, New Delhi – 110025, India

103 Penang Road, #05–06/07, Visioncrest Commercial, Singapore 238467

Cambridge University Press is part of Cambridge University Press & Assessment,
a department of the University of Cambridge.

We share the University's mission to contribute to society through the pursuit of
education, learning and research at the highest international levels of excellence.

www.cambridge.org
Information on this title: www.cambridge.org/9781107003385

© Cambridge University Press & Assessment 2012

First published 2012

A catalogue record for this publication is available from the British Library

Library of Congress Cataloging-in-Publication data
Optical coatings and thermal noise in precision measurement / edited by Gregory M. Harry,
Timothy Bodiya and Riccardo DeSalvo.
p. cm.
Includes bibliographical references.
ISBN 978-1-107-00338-5 (hardback)
1. Optical coatings. 2. Quantum optics. 3. Light – Scattering. 4. Electromagnetic waves – Scattering.
I. Harry, Gregory M., 1967– II. Bodiya, Timothy P. III. DeSalvo, Riccardo.
TS517.2.O64 2012
621.36 – dc23 2011039977

ISBN 978-1-107-00338-5 Hardback

Contents

Contributors

Markus Aspelmeyer
University of Vienna, Faculty of Physics, Boltzmanngasse 5, VIENNA 1090, Austria

Stefan Ballmer
Syracuse University, New York, Department of Physics, SYRACUSE, NY 13244, USA

Peter Beyersdorf
San Jose State University, Department of Physics and Astronomy, 1, Washington Square, SAN JOSE, CA 95192-0160, USA

Vladimir B. Braginsky
M. V. Lomonosov Moscow State University, Faculty of Physics, Leninskie Gory, MOSCOW 119991, Russia

Shiuh Chao
National Tsing Hua University, Institute of Photonics Technologies, 101 Kuangfu Rd, Sec. 2, HSINCHU, Taiwan

Garrett D. Cole
University of Vienna, Faculty of Physics, Boltzmanngasse 5, VIENNA 1090, Austria

Riccardo DeSalvo
Universitá degli Studi del Sannio, Via Port'Arsa, 11, 82100 Benevento, Italy

Matthew Evans
Massachusetts Institute of Technology, LIGO Laboratory MIT, NW22-295, CAMBRIDGE, MA 02139, USA

Andreas Freise
University of Birmingham, School of Physics and Astronomy, Edgbaston, BIRMINGHAM B15 2TT, UK

Michael L. Gorodetsky
M. V. Lomonosov, Moscow State University, Faculty of Physics, Leninskie Gory,
MOSCOW 119991, Russia

Yuri Levin
Leiden Observatory, Niels Bohrweg 2, LEIDEN 2300 RA, The Netherlands

Iain Martin
University of Glasgow, Department of Physics and Astronomy, GLASGOW G12 8QQ,
UK

Michael J. Martin
University of Colorado, Boulder, JILA 440 UCB, BOULDER, CO 80309-0440, USA

Tracy E. Northup
Universität Innsbruck, Institut für Experimentalphysik, Technikerstrasse 25/4,
INNSBRUCK 6020, Austria

Kenji Numata
NASA-Goddard Space Flight Center Code 663, 8800 Greenbelt Rd, GREENBELT, MD
20771, USA

Greg Ogin
California Institute of Technology, LIGO Laboratory, MS 100-36, Room 252B W. Bridge,
PASADENA, CA 91125, USA

David J. Ottaway
University of Adelaide, School of Chemistry and Physics, ADELAIDE, SOUTH
AUSTRALIA 5005, Australia

Steven D. Penn
Hobart and William Smith Colleges, Department of Physics, 300 Pulteney Street,
GENEVA, NY 14456, USA

Innocenzo M. Pinto
University of Sannio, Department of Engineering, Corso Garibaldi 107, Pal. dell'Aquila
Bosco-Lucarelli, BENEVENTO I-82100, Italy

Maria Principe
Department of Engineering, Corso Garibaldi 107, Pal. dell'Aquila Bosco-Lucarelli,
BENEVENTO I-82100, Italy

Stuart Reid
University of Glasgow, Department of Physics and Astronomy, GLASGOW G12 8QQ,
UK

Sheila Rowan
University of Glasgow, School of Physics and Astronomy, GLASGOW G12 8QQ, UK

Joshua R. Smith
Cal State Fullerton, Department of Physics, 800 N State College Blvd, FULLERTON, CA
92831, USA

Kentaro Somiya
Waseda Institute for Advanced Study, 1-6-1 Nishiwaseda, Shinjuku, TOKYO 169-8050,
Japan

Sergey P. Vyatchanin
M. V. Lomonosov, Moscow State University, Faculty of Physics, Leninskie Gory,
MOSCOW 119991, Russia

Phil Willems
California Institute of Technology, LIGO Laboratory, MS 18-34, PASADENA, CA
91125, USA

Kazuhiro Yamamoto
Leibniz Universitaet Hannover, Max-Planck-Institut fuer Gravitationsphysik,
Callinstrasse 38, HANNOVER D-30167, Germany

Jun Ye
University of Colorado, Boulder, JILA 440 UCB, BOULDER, CO 80309-0440, USA

Michael E. Zucker
Massachusetts Institute of Technology, LIGO Laboratory, NW22-295, 185 Albany Street,
CAMBRIDGE, MA 02139, USA

Foreword

As Lord Kelvin was renowned for saying – "to measure is to know" – and indeed precision measurement is one of the most challenging and fundamentally important areas of experimental physics.

Over the past century technology has advanced to a level where limitations to precision measurement systems due to thermal and quantum effects are becoming increasingly important. We see this in experiments to test aspects of relativity, the development of more precise clocks, the measurement of the Gravitational Constant, experiments to set limits on the polarisation of the vacuum, and the ground based instruments developed to search for gravitational radiation.

Many of these experimental areas use laser interferometry with resonant optical cavities as short term length or frequency references, and thermal fluctuations of cavity length present a real limitation to performance. This has received particular attention from the community working on the upgrades to the long baseline gravitational wave detectors, LIGO, Virgo and GEO 600, the signals from all likely sources being at a level where very high strain sensitivity – of the order of one part in 10^{23} over relevant timescales – is required to allow a full range of observations. Research towards achieving such levels of strain measurement has shown that the thermal fluctuations in the length of a well designed resonant cavity are currently dominated by those due to mechanical losses in the dielectric materials used to form the multi-layer mirror coating used, with the fluctuations of the mirror substrate materials also playing an important part.

Now that the importance of thermal noise in coatings and substrates is of clear importance in a range of precision experiments using optical cavities, it is very timely that a book be dedicated to these issues, and that the theoretical and experimental physicists at the forefront of their field from many laboratories around the world, have collaborated together in writing this.

The book is unique in that it ranges from discussions of the theoretical basis of thermal noise in mirrors and substrates, through the technology of depositing coatings and the techniques for measuring mechanical loss and thermal noise to the importance of this noise source in a range of applications. The real challenge of bringing this about will become very clear to the reader as will the rewards to be gained in areas such as precision timing and gravitational wave detection, these areas being well described by another quotation of Lord Kelvin – "When you are face to face with a difficulty, you are up against a discovery."

Professor James Hough, Kelvin Professor of Natural Philosophy,
University of Glasgow, January 2011

Preface

Dedicated to Robert Kirk Burrows

In 1999, I was a young postdoc moving to Syracuse University to work on LIGO, which had been a dream of mine since I was first introduced to gravitational wave detection as an undergraduate in Kip Thorne's class at Caltech. I had done my PhD in gravitational wave detection, but using the older technology of resonant masses rather than LIGO's laser inteferometry. I was concerned that my background would not prove appropriate. I soon found a common issue, thermal noise, that I was able to focus on. Beyond just a good fit for me, thermal noise was actually a topic in flux within LIGO at the time. A talented young theorist at Caltech named Yuri Levin had just shown that the optical coatings on the LIGO mirrors could well contribute much more thermal noise than anyone had anticipated. What was missing were realistic numbers to plug into Yuri's formulas to see just how big of an impact coating thermal noise might have. This became one of my principal roles in LIGO, as part of a group of experimentalists interested in this question at Stanford, Glasgow, as well as Syracuse and other collaborating institutions.

Since then, we in LIGO have found that coating thermal noise is a very important limit to sensitivity, and we have engaged in over a decade of theoretical, experimental, and modeling work to better understand and reduce it. One of the key difficulties was that we had to engage coating thermal noise within the strict limits of optical performance, as LIGO coatings also have to satisfy some of the strictest specifications on optical absorption, scatter, uniformity on a large scale, and other more conventional optics concerns. In the last few years, I started to see that other precision measurement fields were also hitting the same coating thermal noise limit.

Collaboration between fields on coating thermal noise started with a discussion in a bar in Harvard Square between myself and Markus Aspelmeyer, having been introduced by Professor Nergis Mavalvala whose research interests overlap with both of ours. I saw that a workshop on coating thermal noise involving researchers from many precision measurements fields as well as coating technologists, optical engineers, and others could be mutually beneficial. We held this workshop in March of 2008, and it is still accessible on the web at http://www.ligo.mit.edu/~gharry/workshop/workshop.html.

Finally, this book came out of late night conversations with my graduate school roommate and friend, Kirk Burrows, during our annual vacations on North Carolina's Outer Banks. He would always encourage me to write a book, so after the workshop had proved a success and the opportunity with Cambridge University Press presented itself, I decided the time was right. All of us who have worked to make this book happen hope it proves valuable to both those currently in the trenches battling coating thermal noise and all the other coating issues discussed herein, but also to researchers in new fields just coming up to these limitations.

This book, like any book, is the result of many people's hard work, inspiration, dedication, and collaboration. The editors and authors would like to especially thank Matt Abernathy, Juri Agresti, Warren Anderson, Craig Benko, Eric Black, Birgit Brandstätter, Aidan Brooks, Gianpietro Cagnoli, Christof Comtet, Rand Danenberg, Carly Donahue, Raffaele Flaminio, Ray Frey and the entire LIGO Scientific Collaboration Publication and Presentation Committee, Daniel Friedrich, Peter Fritschel, Eric Gustafson, Ramin Lalezari, Yige Lin, Jean-Marie Mackowski, Andrew McClung, John Miller, Nazario Morgado, Mark Notcutt, Laurent Pinard, Takakazu Shintomi, David Shoemaker, Matthew Swallows, Toshikazu Suzuki, Takashi Uchiyama, Akira Villar, Stephen Webster, Valerie Williams, Dal Wilson, Hiro Yamamoto, and the post-graduate class in Advanced Electromagnetics at the University of Sannio for useful input and feedback on chapter drafts. Some material in this book is based upon work supported by the United States National Science Foundation under grants 0757058 and 0970147. Any opinions, findings, and conclusions expressed in this material are those of the authors and do not necessarily reflect the views of the National Science Foundation. We would also like to thank the Italian National Institute for Nuclear Physics (INFN) for financial support.

1

Theory of thermal noise in optical mirrors

YURI LEVIN

1.1 Introduction

Mechanical and optical thermal noises play an important role in many precise optomechanical experiments, in which positions of test bodies are monitored by laser beams.
Much of the initial experimental and theoretical research in this area was driven by the
physics of gravitational-wave interferometers, where thermal fluctuations are expected to
be the dominant source of noise in the frequency band between about 10 and 100 Hz
(Harry *et al.*, 2002) (see Chapter 14). Recently, it has become clear that controlling thermal noise will be key in several other fields, notably in designing laser cavities with
higher frequency stability (Numata *et al.*, 2004) (see Chapter 15), in reaching the quantum
limit in macroscopic opto-mechanical experiments (Kippenberg and Vahala, 2008) (see
Chapter 16), and in cavity QED experiments (Miller *et al.*, 2005) (see Chapter 17). In this
chapter we review the statistical-mechanics formalism which is used to theoretically calculate mechanical and optical thermal noise. For completeness, we also add a discussion of
another important limitation in mechanical measurements, the so-called Standard Quantum
Limit.

1.2 Theory of mechanical thermal noise

The theory of time-dependent thermodynamical fluctuations has been extensively developed for the past century. One of the fundamental results in this field is the Fluctuation–
Dissipation Theorem, which was originally formulated by Callen and Welton (1951).
Callen and Welton's insight was that the intensity of random fluctuation in some *macroscopic* degree of freedom $\hat{x}$ of the thermodynamic system was proportional to the strength
with which $\hat{x}$ was coupled to the *microscopic* degrees of freedom of the heat bath. Since
the same coupling is responsible for damping of motion in $\hat{x}$, one obtains a proportionality
relation between the microscopic thermal fluctuations of the quantity $\hat{x}$ and the damping
coefficient for the macroscopic motion when $\hat{x}$ is driven externally. For optomechanical

Optical Coatings and Thermal Noise in Precision Measurement, eds. Gregory M. Harry, Timothy Bodiya and Riccardo DeSalvo.
Published by Cambridge University Press. © Cambridge University Press 2012.

experiments, $\hat{x}$ is typically an integrated quantity over a mirror surface. For example, in experiments with a test mass position readout by a laser, $\hat{x}$, is given by

$$\hat{x} = \int f(\vec{r})x(\vec{r}, t)d^2r. \tag{1.1}$$

Here $\vec{r}$ is the location of a point on the test-mass' mirror surface, and $x(r, t)$ is the displacement of the mirror along the direction of the laser beam at point $\vec{r}$ and time t. The form factor $f(\vec{r})$ depends on the laser beam profile and is proportional to the laser light intensity at the point $\vec{r}$ (Gillespie and Raab, 1995); it is usually normalized so that $\int f(\vec{r})d^2r = 1$.

The Fluctuation–Dissipation Theorem is then applied to the variable $\hat{x}$ by performing a mental experiment (Levin, 1998) consisting of the following three steps.

- Apply an oscillatory pressure $P(\vec{r}, t) = F_0 \cos(2\pi f t)f(\vec{r})$ to the face of the test mass. This is equivalent to driving the system with an external interaction Hamiltonian $H_{int} = -F_0 \cos(2\pi f t)\hat{x}$.
- Work out the average power, W_{diss}, dissipated in the test mass under the action of this oscillatory pressure.
- Compute the spectral density of fluctuations in $\hat{x}$ from

$$S_{\hat{x}}(f) = \frac{2k_B T}{\pi^2 f^2} \frac{W_{diss}}{F_0^2}, \tag{1.2}$$

where k_B and T are the Boltzmann's constant and the temperature of the mirror respectively, and f is the frequency at which the spectral density is evaluated.

This method is straightforward to use. Once the dissipative processes are understood, the calculation reduces to a problem in elasticity theory (Levin, 1998; Bondu et al., 1998) and, in the case of thermoelastic noise (see Section 1.3 and Chapter 9), time-dependent heat flow (Braginsky et al., 1999; Liu and Thorne, 2000). For the important case of so-called structural damping (Saulson, 1990),

$$W_{diss} = 2\pi f U_{max}\phi(f), \tag{1.3}$$

where U_{max} is the energy of elastic deformation at a moment when the test mass is maximally contracted or extended under the action of the oscillatory pressure, and $\phi(f)$ is the loss angle characterising the dissipation. If the frequency band being measured is well below the normal modes of the test mass (as is often the case in gravitational wave detectors, see Chapter 14, frequency stabilization, see Chapter 15, and cavity QED experiments, see Chapter 17) one can assume constant, non-oscillating pressure $P(\vec{r}) = F_0 f(\vec{r})$ when evaluating U_{max}. On the other hand, in many microscopic opto-mechanical experiments the detection frequencies are comparable to mechanical resonance frequencies, such as discussed in Chapter 16, and one needs to solve a time-dependent elasticity problem in order to find U_{max}.

For the case of structural damping, and for frequencies much smaller than the mechanical resonant frequencies of the system, the thermal noise is approximately given by (Harry *et al.*, 2002)

$$S_{\hat{x}} = \frac{2k_B T}{\sqrt{\pi^3} f} \frac{1 - \sigma^2}{Y w_m} \left[\phi_{\text{substrate}}(f) + \frac{1}{\sqrt{\pi}} \frac{d}{w_m} A\phi_{\text{coating}}(f) \right], \tag{1.4}$$

where

$$A = \frac{Y'^2(1 + \sigma)^2(1 - 2\sigma)^2 + Y^2(1 + \sigma')^2(1 - 2\sigma')}{YY'(1 - \sigma'^2)(1 - \sigma^2)}. \tag{1.5}$$

Here Y and Y' are the Young's moduli of the substrate and coating, respectively, σ and σ' are their Poisson's ratios, d is the coating thickness, w_m is the radius where the field amplitude of the Gaussian laser beam is $1/e$, and $\phi_{\text{substrate}}$ and ϕ_{coating} are the loss angles of the substrate and the coating, respectively. This expression is valid when the beam size w_m is much smaller than the size of the mirror. When the latter approximation breaks down, one can find a series solution (Bondu *et al.*, 1998; Liu and Thorne, 2000) or an analytical solution for the coating thermal noise contribution (Somiya and Yamamoto, 2009) for axisymmetric configurations and use direct finite-element methods to calculate U_{max} for cases without the axial symmetry (Numata, 2003). See also Chapter 4.

Of the two contributions on the right-hand side of Equation 1.4, the one from the coating is projected to be the greater in most precision experiments. There are two essential reasons why the contribution from the coating is so large. The first one is geometrical: the sources of thermodynamically fluctuating random stress are spread out throughout the substrate and the coating, but the ones near the coating are closer to the surface and thus have greater effect on its displacement. This geometrical effect explains why the coating noise scales as $1/w_m^2$ while the structural substrate noise scales as $1/w_m$ (Levin, 1998). The second reason is that coating losses per volume are typically orders of magnitude greater than that of the substrates (see Chapters 4 and 7). Several practical proposals on how to reduce the unfavorable geometric factor have been investigated. The basic idea behind these proposals is to either increase the effective beam size by reconfiguring its shape towards a flat-top geometry or by working with higher-order cavity modes, see Chapter 13. Decreasing the mechanical loss of the coating is a great challenge in thermal noise research, and is discussed in Chapter 4. Other ideas for improving coating thermal noise are discussed in Chapters 6 and 8.

1.3 Theory of optical thermal noise

So far we have described mechanical thermal noise which appears due to small thermally driven random changes of the mirror's shape and results in random displacement of the mirror's surface. However, this description is not complete. Upon striking the mirror surface, the light penetrates several wavelengths into the coating, before being completely reflected.

The temperature within this skin layer of the coating is not constant, but fluctuates thermodynamically. These temperature fluctuations lead to random changes in the phase of the reflected light via two physical effects.

- The mirror surface shifts randomly due to the coating's thermal expansion. This is known as the coating thermoelastic noise (Braginsky and Vyatchanin, 2003a; Fejer *et al.*, 2004), and
- The optical pathlength inside the coating changes randomly, due to the temperature dependence of the coating's index of refraction. This is known as the thermorefractive noise (Braginsky *et al.*, 2000).

The thermorefractive and thermoelastic noises are typically anti-correlated, which reduces their impact on noise (Evans *et al.*, 2008; Gorodetsky, 2008) and are discussed in detail in Chapter 9 (see also Section 6.6). The theoretical evaluation of both of these noises involves calculating the spectra of thermal fluctuations of temperature-dependent quantities of the form

$$\delta \hat{T}(t) = \int q(\vec{r}) \delta T(\vec{r}, t) d^3 r. \tag{1.6}$$

Here $\delta T(\vec{r}, t)$ is the local fluctuation in temperature and $q(\vec{r})$ is the form factor proportional to the local intensity of light and the thermo-refractive coefficient $\partial n / \partial T$. A variation of the Fluctuation–Dissipation Theorem has been devised that allows one to calculate directly the spectral density $S_{\hat{T}}(f)$ (Levin, 2008). The calculation proceeds via a mental experiment similar to the one in the previous section. It consists of the following three steps.

- Periodically inject entropy into the medium, with the volume density of the entropy injection given by

$$\frac{\delta s(\vec{r})}{dV} = F_0 \cos(2\pi f t) q(\vec{r}), \tag{1.7}$$

  where F_0 is an arbitrarily small constant.
- Track all thermal relaxation processes in the system (e.g. the heat exchange between different parts of the system) which occur as a result of the periodic entropy injection. Calculate the total entropy production rate and hence the total dissipated power W_{diss} which occurs as a result of the thermal relaxation.
- Evaluate the spectral density of fluctuations in $\delta \hat{T}$ from

$$S_{\hat{x}}(f) = \frac{2k_B T}{\pi^2 f^2} \frac{W_{\text{diss}}}{F_0^2}. \tag{1.8}$$

This formalism was instrumental in the calculation of total thermo-optical coating noise (Evans *et al.*, 2008). It is likely to be useful for computing thermorefractive noise in experiments with non-trivial optical geometry, see Benthem and Levin (2009) for example.

1.4 Standard quantum limit

There is another source of noise which places an important limitation on opto-mechanical experiments. From the early days of quantum mechanics, it was clear that by precise measurement of a test mass' position one inevitably, and randomly, perturbs its momentum in accordance with the Heisenberg uncertainty relation $\Delta p \geq \hbar/(2\Delta x)$. Thus no matter which measurement device one uses, one inevitably introduces a *back-action* noise into the system; the higher the intrinsic precision of the measuring device, the greater the back-action noise. One can express this mathematically via the uncertainty relation (Braginsky and Khalili (1992))

$$S_x(f)S_F(f) - |S_{xF}(f)|^2 \geq \hbar^2/4, \qquad (1.9)$$

where $S_x(f)$ is the spectral density of the intrinsic measurement noise, $S_F(f)$ is the spectral density of the back-action noise, and $S_{xF}(f)$ is the spectral density of the correlation between the measurement error and the back-action perturbation. For a vast majority of measuring devices the cross-correlation term S_{xF} is zero. Then the intrinsic and back-action noises, added in quadrature, enforce a limit on how precisely the position of a test-mass can be monitored. This is known as the standard quantum limit (SQL) (Braginsky and Khalili (1992) and references therein). For a free test body of mass m, the SQL is given by

$$S_x^{SQL}(f) = \frac{\hbar}{m(2\pi f)^2}. \qquad (1.10)$$

As the coating noise is reduced due to technological progress, precision optical experiments will reach the SQL. However, the SQL is not a fundamental limit and can be overcome by using techniques of quantum optics which are capable of introducing correlations between a measurement error and a back-action perturbation, see Section 11.3 for discussion. Practical proposals exist on how to reach sensitivities below the SQL, see Kimble *et al.* (2001); Buonanno *et al.* (2001).

2
Coating technology

SHIUH CHAO

2.1 Introduction

The preparation and deposition of coatings can determine many of their basic properties. This chapter discusses the major technologies for creating thin film coatings, with an emphasis on those technologies most useful to precision measurement applications, as a baseline for the other chapters to expand on. There are many types of optical coatings, but we will be concerned with stacks of multi-layer thin films of dielectric materials with different refractive indices and thicknesses. Through optical interference effects, various optical functions can be achieved by properly selecting the materials and designing the layer thicknesses. Traditionally, the major application of coatings has been in imaging systems, including coatings on lenses, windows, and filters for purposes such as anti-reflection, band passing, polarization selection, etc. (Macleod, 2010; Baumeister, 2004a). With the advent of the laser and its diverse applications, high quality coatings for laser optics have become in high demand. Dielectric mirror coatings, which are often used in active or passive optical cavities, are particularly important.

The dielectric mirror is composed of a stack of thin films with pairs of alternating high and low refractive index materials. Conventionally, each layer has a quarter wave of optical thickness. Optical thickness is defined as $d_{\text{opt}} = dn$, where d is the physical thickness of the layer, and n is the refractive index of the coating material. Given this, a quarter wave layer has $d_{\text{opt}}/\lambda = 1/4$, where λ is the wavelength of light for which the coating is designed. Reflectance of the mirror increases with the increasing number of pairs and the increasing refractive index difference between the pair materials in general. For a quarter wave stack, the reflectivity in air at normal incidence can be found from

$$r = \frac{1 - n_s \left(n_1/n_2\right)^{2p}}{1 + n_s \left(n_1/n_2\right)^{2p}}, \tag{2.1}$$

where r is the reflected field amplitude, n_s is the refractive index of the substrate, n_1 and n_2 are the refractive indices of the two coating materials, and p is the number of pairs of a high and low index material in the coating.

Optical Coatings and Thermal Noise in Precision Measurement, eds. Gregory M. Harry, Timothy Bodiya and Riccardo DeSalvo.
Published by Cambridge University Press. © Cambridge University Press 2012.

High-end applications for the mirror coatings all require that the coatings have low optical losses, i.e. low absorption (see Chapter 10) and low scattering (see Chapter 11). In a ring laser gyroscope, for example, extremely low backscattering (typically less than a few ppm) in the mirror is required in order to avoid frequency lock-in so that low rotation rates can be detected (Kalb, 1986). Furthermore, some additional applications now require low thermal noise (see Chapter 5, 14–17). Beginning in the 1970s, researchers also intensively studied laser induced damage stemming from optical coating losses in high energy laser applications, such as laser ignition fusion (Stolz and Taylor, 1992).

In addition to low optical losses, narrow-band thin film filters composed of multiples of half-wavelength thick films in between quarter wave stacks are required to have a narrow pass-band width as low as sub-nm and to be environmentally stable while operating in the field. These types of filters are critical for dense wavelength division multiplexing (DWDM) technology in optical communication. These criteria impose high thickness control and environmental stability requirements on the coatings (Takashashi, 1995). The stringent requirements of high-end applications drove rapid progress in optical coating technology since the 1970s. This chapter introduces coating methods, coating processes, and thin film materials used in high-end coating technologies with the purpose of stimulating further research and development activities on coatings for precision measurement.

2.2 Coating methods

Various coating methods have been developed to provide films with desired qualities such as good optical characteristics, uniformity, precise thickness control, absence of stress and defects, strong adhesion, ease of fabrication, large throughput, and low fabrication cost. Among the many factors that affect the film qualities, one of fundamental importance is the kinetic energy of the coating material atoms when impinging on the substrate prior to condensation. Upon arriving at the substrate surface, the atoms need a sufficient amount of energy to overcome the activation energy of various mechanisms to reach proper sites for nucleation and growth. This will allow for a close-packed structure and good adhesion to the substrate and the neighboring layer (Neugebauer, 1970; Ohring, 2002). In the following sections, we shall introduce different coating methods, with emphasis on energetics.

2.2.1 Thermal evaporation

Thermal evaporation is the most commonly used coating method. Source material is heated by resistance heating or by electron beam bombardment to either the sublimation or melting point. The evaporants then condense on the substrate to form a thin film. Most compounds do not evaporate to form films that have the same composition as the source material, and various means for reactive evaporation or multiple single element source co-evaporations need to be implemented to insure the correct stoichiometry for the films. The average kinetic energy of the evaporant when impinging the substrate is the thermal energy at the melting point, typically of the order of 10^{-1} eV. This is a relatively low energy and therefore the atoms have difficulty migrating on the substrate surface and forming a dense, less porous film. The film is therefore susceptible to moisture when exposed to the atmosphere, which can lead to weak adhesion to the substrate and a low refractive index.

There are many methods to increase the energy of the evaporant (Vossen and Kern, 1978); substrate heating, DC or RF biasing of the substrate, and more recently employed, ion beam assisted deposition (IBAD). IBAD uses a low energy and high current broad ion beam to bombard the substrate, assisting the film formation with denser packing and enhanced oxidation/nitridation when the ion beam contains oxygen/nitrogen ions for deposition of oxide/nitride films (Green *et al.*, 1989). Since the evaporation process is performed in a high vacuum, typically 10^{-6} Torr, the mean free path of the evaporant is a few tens of meters. The evaporant distribution is nearly Lambertian; the evaporant has a large field of view. In addition, a high evaporation rate is easy to achieve by increasing the temperature of the melt with electron beam bombardment. Therefore, high throughput deposition can be realized in a large box coater that can accommodate a large quantity of substrates positioned in planetary rotation dome-shaped holders. Currently, electron beam evaporation with the IBAD method is the primary coating technique for large quantities of fairly good quality batched optical coatings.

2.2.2 Glow discharge sputtering

The glow discharge sputter deposition technique dates back to the nineteenth century, when deposits of cathode material were observed in DC glow discharge environments. In a straightforward planar DC glow discharge, the target serves as the cathode and the substrate serves as the anode. At the state of "abnormal glow" for sputtering, most of the discharge voltage falls in the Crookes dark space adjacent to the cathode. The positive ions in the neighboring negative glow region gain kinetic energy, typically a few hundreds to 1000 eV, from the cathode fall and bombard the cathode to sputter off the cathode atoms. The sputtered atoms have kinetic energy of a few tens of eV (see Section 2.2.3). Initiating voltage for the glow discharge varies with the product of the gas pressure and the electrode separation with a deep minimum according to Paschen's law, which limits the operating gas pressure and electrode separation. A typical value for electrode separation is a few cm and for gas pressure is in the range of 10^{-2} Torr. This is a relatively high pressure and the mean free path of the sputtered atoms is therefore short. The consequence is that the kinetic energy of the sputtered atoms tends to be thermalized through multiple collisions on the way to the substrate. A typical value for the kinetic energy of the sputtered atoms when impinging the substrate is on the order of 1 eV, which is lower than that in ion beam sputtering (see Section 2.2.3) but about ten times higher than that of the evaporation process. Therefore, film qualities of adhesion, density, refractive index, and moisture susceptibility are generally better than that of the films deposited by thermal evaporation. Nevertheless, since the substrate is immersed in plasma during deposition, the films may be subjected to UV damage and re-sputter.

Both triode sputter, in which a thermionic electrode is added to the diode, and magnetron sputter, in which a magnetic field is applied to confine the plasma, can increase the collision probability between the electron and the gas atoms to enhance the plasma generation efficiency. This allows lower gas pressure, lower power or larger electrode distance to be accommodated. When the sputter target is an electrical insulator, a radio frequency (13.56 MHz) source is applied. The heavier ion is less responsive to the high frequency RF field than the electrons in the plasma, establishing a self-bias at the target for sputtering. One advantage of glow discharge sputtering is that very large targets can be used, such that large substrates, e.g. architectural window glass, solar panels, display panels, plastic rolls, etc, can be coated in a load-lock conveyer-fed coater or a web-coater for uninterrupted continuous large volume coating operations.

2.2.3 Ion beam sputter deposition (IBSD)

Most of the optical coatings for high-end applications mentioned in Section 2.1 are fabricated by the ion beam sputter deposition method (IBSD). The ion beam source was originally used as a spacecraft thruster, only later was it applied to ion etching and thin film deposition. In recent years, the focused ion beam (FIB) technique has been developed for semiconductor device fabrication, diagnosis, and nano-patterning (Joe *et al.*, 2009). Wei and Louderback (1979) first used the IBSD technique to sputter deposit mirrors for a ring laser gyroscope, obtaining unprecedented quality. The technique then became the major coating method for high quality optical coatings.

The schematic of an IBSD apparatus with a conventional Kaufman type ion source, i.e. hot filament with extracting grids, is shown in Figure 2.1. High density plasma is generated in the discharge chamber and ion beamlets are extracted, accelerated through the apertures in the grids to form a broad ion beam, and hit the target. The target atoms are, then, sputtered off and condense on the substrate to form the films. Several targets can be attached to the rotatable target holder for coating multi-layers. Planetary rotation fixtures can be used to obtain films with uniform thickness distribution.

Ion source

The conventional ion beam generation method is the use of the Kaufman type ion source. Referring to Figure 2.1, electrons emitted from the hot cathode through thermionic emission collide with gas atoms, typically argon gas, to produce positive ions and electrons. The ionization energy for argon is 15.76 eV, and around a 40 V voltage difference between the cathode and the anode with a few mTorr of gas pressure is sufficient to sustain the plasma. The anode is typically held at a voltage from 500–1000 V above ground, and the plasma potential is nearly the same value. The target is held at the ground potential. The average kinetic energy of the ions is therefore 500–1000 eV when hitting the target. The screen grid is held to roughly the anode potential and the accelerator grid is negatively biased to about -100 V. Both grids are precisely aligned to each other so that ion beamlets are extracted

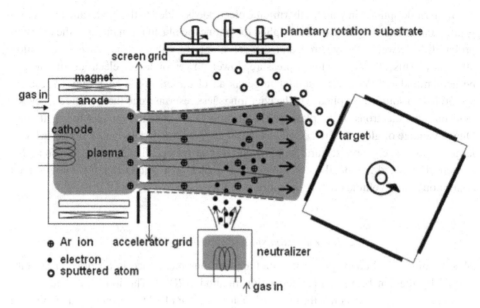

Figure 2.1 Schematics of an ion beam sputter setup.

and accelerated through the apertures. Factors such as grid separation, aperture size, grid curvature, applied voltage, plasma density, etc, affect the beam shape, beam divergence and the current density. Thorough reviews for the physics and characteristics of the Kaufman ion source are given in Kaufman *et al.* (1982); Harper *et al.* (1982). The grids are made of graphite or molybdenum, which have higher resistance to ion bombardment within the aperture. A plasma bridge neutralizer, usually a thermionic cathode or a hollow cathode which emits electrons to neutralize the ion beam for sputtering the insulator targets, is positioned aside the beam path. For depositing oxide films, oxygen gas with pressure around 10^{-4}–10^{-5} Torr is fed into the sputter chamber to oxidize the films during deposition. A second ion source is sometimes used to bombard the substrate for ion beam assisted sputter deposition. The second ion source is generally low energy and high current so that the film will not be re-sputtered and yet sufficient energy can be added to assist film growth. The second ion beam could be an oxygen ion beam or a mixture of oxygen and argon to enhance the oxidation for the oxide film.

A major drawback to the hot filament ion source is the frequent filament maintenance required. The filament might break down during long-term continuous deposition (e.g. coating for DWDM filters, the filament is subjected to ion bombardment and the ion beam may become contaminated). Two other advanced plasma generation methods have been developed to avoid hot filament issues; radio frequency (RF) and electron cyclotron resonance (ECR) plasma generation. RF power, with a frequency of 13.56 MHz, is fed into the discharge chamber, usually by inductive coupling, i.e. through a solenoid coil or a flat spiral coil that is embedded in a dielectric shield. Electrons oscillate in the RF field and collide with the gas atoms to form plasma. In ECR plasma generation, a microwave of 2.45 GHz is fed into the discharge chamber through an antenna shielded in a quartz container. A static magnetic field with a strength of 856 Gauss, satisfying the electron cyclotron resonance condition $f = eB/(2\pi m)$, is provided perpendicular to the electric field. Electrons cycle the magnetic field lines and pick up energy resonantly from the microwaves and make multiple collisions with the gas atoms along the way. Highly efficient plasma generation can be achieved in a smaller discharge volume and with lower gas pressure. Both RF and ECR plasma generation methods use the same grid configuration as the Kaufman source to extract, accelerate, and focus the ion beam. Each ion source designer typically has a unique design to achieve uniform plasma distribution, high current density, proper energy range, good focus quality, large beam size, and low beam contamination. The 13.56 MHz RF and 2.54 GHz microwave generations have become industry standards, and the generators and tuning networks are readily available. However, among the different types of ion sources, the hot filament ion source is still the least expensive choice.

Sputtering process

For depositing metal oxides or nitrides (e.g. tantala (Ta_2O_5), titania (TiO_2), alumina (Al_2O_3), aluminum nitride (AlN), disilicon trinitride (Si_2N_3)), metal targets are usually used for their better thermal conductivity and oxygen or nitrogen gas is introduced into the sputter chamber or discharge chamber to react with the sputtered atoms. Sputter yield, angular distribution of the sputtered atom, and energy distribution of the sputtered atoms in the IBSD process are major factors that affect the deposition rate, thickness uniformity, and film quality. The following discussion is focused on polycrystalline targets as these are the most commonly used metal targets in the IBSD process.

The ion energy for IBSD is in the range of 500–1000 eV when hitting the target. Within this range, the ion penetration depth is only a few atomic layers (Harper, 1984). The sputter yield, defined as the number of sputtered atoms per incident ion, is about unity for heavy inert ions of argon, krypton, and xenon, and about 0.5 for neon and 0.01 for helium (Vossen and Kern, 1978). Argon is therefore the most commonly used gas for IBSD due to its high sputter yield and lower cost. The sputter yield is a function of the angle of incidence of the ion beam. It increases with the angle of incidence to a maximum around 40–70° depending on the target material, and down to zero for grazing incidence (Melliar-Smith and Mogab, 1978). This gives a guideline for orienting the target relative to the direction of the ion beam in the IBSD chamber. The angular distribution of the sputtered atoms is generally "under-cosine" for a normal incident ion beam, i.e. less than the cosine distribution around the normal direction, but it is preferentially in the forward direction for oblique ion incidence (Wehner and Rosenberg, 1960). Therefore, in a conventional IBSD chamber, the normal of the substrate surface is oriented perpendicularly to the direction of the ion beam and the angle of incidence for the ion beam is about 45° to the target.

The kinetic energy of the IBSD sputtered atoms has been systematically investigated only in limited situations. It was found that the kinetic energy of the sputtered atoms from a copper target bombarded by krypton ions with ion energy in the range 80–1200 eV follows the Maxwell distribution with a maximum at a few eV and the tail of the distribution extending to 40 eV (Stuart and Wehner, 1964). Similar results were found for silver, palladium, rhodium, and zirconium targets bombarded by mercury and krypton ions (Wehner and Anderson, 1970). The average kinetic energy of the IBSD sputtered atoms is generally of the order of 10 eV (Harper *et al.*, 1982). Since the gas pressure in the sputter chamber of IBSD is in the range of 10^{-5}–10^{-4} Torr, the mean free path for the sputtered atoms is longer than the target–substrate separation. Therefore, unlike the situation in the glow discharge sputtering (see Section 2.2.2), the sputtered atoms in IBSD maintain their kinetic energy upon arriving at the substrate surface. This energy is about 100 times larger than that of the evaporation process and about 10 times larger than that of the glow discharge sputter. Upon arriving at the substrate surface, the atoms have sufficient energy to migrate and reside on the proper sites to form films with close-packed structures, high refractive indexes and stronger adhesion than the films deposited by other means.

Advantages of IBSD can be summarized as follows.

- The ion energy and current density can be independently controlled.
- The target and the substrate are separated from the plasma generation environment.
- The angle of incidence to the target can be varied.
- Layers of different materials can be deposited by simply flipping the target around.
- The pressure of the deposition environment is low.
- The kinetic energy of the sputtered atom is high.

A disadvantage for IBSD is that the deposition rate is low, typically of the order of a few tens of Ångstroms per minute. It normally takes about 10 minutes to deposit a quarter wave film for a visible wavelength, several hours for a mirror stack, and more than 20 h for a DWDM thin film filter. Therefore, highly reliable automation and stable coating processes are important.

The coating methods introduced so far are in the category of physical vapor deposition (PVD). Chemical vapor deposition (CVD), in which the reaction of chemical substances is used to grow thin film material on the substrate epitaxially, i.e. lattice matching between the neighboring materials, either in a plasma enhanced environment (PECVD), in a liquid phase environment (LPE), with a metal-organic precursor (MOCVD), or other variations such as molecular beam epitaxial (MBE), are major deposition methods, particularly for semiconductor devices (Dobkin and Zuraw, 2003). Quarter wave stacks with high and low index films in semiconductor optoelectronic devices can be deposited by alternately changing the composition of the materials, e.g. $Al_x Ga_{1-x} As$ (see Section 16.2.3).

2.3 Substrates

Substrates for high-end mirror applications are required to satisfy some or all of the following criteria; chemical and mechanical stability to endure exotic operation environments or fabrication processes, low thermal expansion for displacement-sensitive precision measurements or large optics, absence of bubbles and inclusions to avoid scattering and absorption in high energy applications, optical homogeneity to avoid wave front distortion in precision measurements, and low mechanical loss where low thermal noise is required. See Chapter 7 for a full discussion of thermal noise in substrate materials.

Fused silica is the most commonly used material that satisfies all of these criteria to some extent. Fused silica has a wide transparency range from deep UV to IR (Neuroth, 1995). Synthetic fused silica is 100% SiO_2 in amorphous form and is fabricated by using a flame hydrolysis process from silicon halides such as $SiCl_4$. Different manufacturers use different trade names and numbering systems for different grades of fused silica. Other materials, for example, Zerodur® (Schott) have been used for ring laser gyroscopes, Pyrex® (Corning) for reflective telescope mirrors, and ULE® (Corning) specifically for the 98 inch diameter primary mirror of the Hubble Space Telescope. A concise introduction on optical glass is given in Bach and Neuroth (1995). Grinding and polishing the substrate to meet the stringent requirements for surface roughness and curvature are critical; a detailed and thorough introduction for grinding and polishing is given in Karow (2004). See also Chapter 11 for the role of surface roughness in scattering.

It is crucial that the substrate surface be free from contaminations to ensure good adhesion, durability and lifetime for the coatings (Pulker, 1984a). Possible contaminants include residues from polishing, oil and grease, metal ions, and dust particles. The cleaning procedure is something of an art depending on the application and the experience of the operator. In general, basic cleaning procedures include cleaning in an aqueous solution of demineralized water with detergents, dilute acids or bases, and in organic solvents such as isopropyl alcohol and acetone, often accompanied with heating and/or ultrasonic agitation. Bonding strength between the contaminant particle and the surface can be very strong as the particle size is reduced. Rubbing with lens tissue in solvent is often used. Stripping an adhesive or a lacquer coating can be effective to remove small particles. Prior to thin film deposition, effective cleaning can be performed in the coating chamber by plasma cleaning with DC or RF biasing of the substrate, and more recently, with ion beam bombarding of the substrate. A good introduction to the cleaning of optical glasses can be found in Pulker (1984b) and Brown (1970).

2.4 Coating uniformity

Thickness uniformity of the film is crucial for coating large optics or a large quantity of small pieces in a batch. The substrate holder is usually rotated with single axis rotation or planetary rotation during the deposition. A static mask with a certain shape is often used to partially shade the substrate in order to balance the flux distribution. The angular distribution for the evaporant or the sputtered atoms is obtained by measuring the thickness of the distributed test pieces in practice, then geometrical calculations can be performed to design the proper mask shape (Pulker, 1984c; Baumeister, 2004b; Villa *et al.*, 2000;

Arkwright, 2006). Uniformity control with precision less than 0.7% over 500 mm diameter has been reported (Sassolas *et al.*, 2009).

It is important that the angular distribution of the evaporant and the sputtered atoms be constant from coating run to coating run. A new target needs to be pre-sputtered to change the surface from polycrystalline to a larger crystalline structure, which is a stable structure resulting from annealing during prolonged sputtering. The angular distribution of the sputtered atoms will then be constant till noticeable indentation appears from long term sputtering by an ion beam with an uneven beam profile (nearly Gaussian). For evaporation processes, it takes a skillful operator to pre-melt the source material in the crucible each time in order to form a constant molten surface for the deposition.

2.5 Thickness control

Over the past decade, the needs of the DWDM thin film filter have driven the development of thickness control. For the narrow band-pass filter in DWDM applications, several thin film cavities are between the quarter wave high reflectors to give a bandwidth of a few tenths of a nanometer with a flat top pass-band. The shape of the passband will be out of specification for only a few tenths of a nanometer random deviation of the optical thickness of the films. It is therefore of ultimate importance to control the refractive index and the thickness of the films precisely and constantly.

There are, in general, three categories of thickness control. For less critical cases, control through deposition time with a stable deposition process is sufficient. Control can also be done by a crystal monitor in which the change of resonance frequency of a quartz crystal as a layer of thin film is deposited is monitored. The thickness of the deposited film can be deduced once the acoustical constants of the film material, which can be difficult to obtain accurately, are known. However, crystal monitoring is indirect, and further calibration between the thickness on the crystal and on the substrate needs to be performed. Optical monitoring is the most effective means for optical coatings (Macleod, 1981). A laser source for single wavelength monitoring or a wideband light source for spectroscopic monitoring are used either in reflection or transmission. For quarter wave stacks, the reflectance/transmittance undergoes a minimum or a maximum, depending on the relative refractive indices of the neighboring layers, when the layer's optical thickness reaches quarter wave. Terminating the deposition at the turning point of the reflectance/transmittance yields a layer with quarter wave optical thickness. Sophisticated techniques of sensitive optical monitoring include monitoring the non-quarter wave thickness, multi-wavelength monitoring, spectroscopic monitoring, and thickness error compensation (Sullivan and Dobrowski, 1992a,b). The substrate can be directly monitored as well. An up-to-date review of optical monitoring is given by Buzea and Robbie (2005). The optical monitoring system is often integrated into the coating system as a major accessory for coating process automation.

2.6 Coating materials

Inorganic optical thin film materials are mostly halides (primarily fluorides), metal oxides, nitrides, and chalcogenides. Fluorides have a transparency range down to 0.11 μm (LiF, MgF), with refractive indices in the range of 1.3–1.6, and are mostly used for applications in the UV. Chalcogenides, e.g. ZnS, CdS, ZnSe, ZnTe, are semiconductor materials for use mostly in the IR range (up to 15 μm) together with silicon (1–9 μm) and germanium (2–23 μm), and the refractive indices range from 2.2 to 4.5. Si_3N_4 is the most commonly used nitride material with a transparency range from 0.25–9 μm and a refractive index around 1.9 in the visible. Metal oxides are the optical thin film material that are most widely used, with a transparency range going from UV (0.16 μm for SiO_2) up to IR (8 μm) and a refractive index range from 1.45 (SiO_2) to 2.4 (TiO_2) in the visible. These materials are mostly used in the visible and near IR ranges.

Physical and optical properties of thin films vary a great deal depending on coating methods and coating process parameters. Because of this, caution should be used regarding optical constants from the literature. There are good reviews with detailed tables for most of the optical thin film materials deposited by different methods, with the early reference work of Pulker (1984c) and more recently Bange (1997a); Friz and Waibel (2003); Baumeister (2004c) as comprehensive sources when searching for desirable optical thin film materials.

For high-end optical coating applications, the major requirements for the thin film materials include, but are not limited to, (1) low absorption and scattering loss (see Chapters 10 and 11), (2) mechanical and environmental stability, (3) low thermal noise for some applications (see Chapter 4). Absorption loss mainly comes from non-stoichiometrical composition, impurities, and defects in the film. For oxides and nitrides, introducing O_2/N_2 in the thin film formation process such as an O_2/N_2-containing ion beam or background pressure, or post-deposition annealing in the O_2/N_2 environment are common practices used to enhance the oxidation/nitridation of the film. Co-sputtering from multi-elemental targets or co-evaporation from multi-elemental sources to adjust the stoichiometry of the film from independent targets and sources are also useful. Scattering loss in the films mainly comes from the grain boundaries and rough polycrystalline film structures. An amorphous structure in which the film is in a non-crystalline state is therefore desirable not only for reducing scattering loss but also for simplifying the design and the analysis of the thin film stack. This is especially true when the crystalline structure of the film material is optically anisotropic. Any microscopic inhomogeneity such as defects, voids, intrusions in the film or inhomogeneities in the composition or thickness of the film can also contribute to scattering loss.

For most applications in this book and many other high-end applications mentioned in Section 2.1, tantala (Ta_2O_5), titania (TiO_2) (for the high index layer) and silica (SiO_2) (for the low index layer) are commonly used thin film materials in the dielectric stack. In the following, a summary of some of the known properties for these materials and their composites deposited by IBSD is introduced. Properties by other deposition methods can

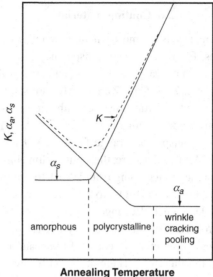

Annealing Temperature

Figure 2.2 Qualitative illustration of K, α_a, and α_s for TiO$_2$ versus annealing temperature. Annealing time was 24 h (Wang and Chao, 1998).

be found in Anderson *et al.* (1997) for TiO$_2$, Anderson and Ottermann (1997) for SiO$_2$ and Bange (1997b) for Ta$_2$O$_5$.

TiO$_2$. Titanium dioxide films have the largest refractive index in the visible among all the oxides. It is the most-used high index material for optical thin film stacks. However, its structural, physical and chemical properties vary too widely, depending on the preparation methods, for it to lend itself to diversified applications. These may include photo-catalyst and dye-sensitized solar cells in which a loosely packed film is required, and optical coatings in which a densely packed film is required. Optical constants of TiO$_2$ films have a wide range, depending on the deposition methods and parameters (Bennett *et al.*, 1989). Stress in TiO$_2$ films vary a great deal from tensile to compressive, heavily depending on the coating method and process parameters (Anderson *et al.*, 1997). The refractive index of TiO$_2$ films deposited by IBSD is 2.54 at 550 nm with an extinction coefficient in the lower 10^{-4} range (Wang and Chao, 1998), practically the largest refractive index among all the PVD deposition methods with TiO$_2$. The structure is amorphous when deposition is at an ambient temperature. Figure 2.2 shows qualitatively the dependence of the extinction coefficient K, absorption coefficient α_a and scattering coefficient α_s on the annealing temperature deduced from experimental data (Wang and Chao, 1998). The absorption coefficient decreases as annealing temperature increases, indicating oxidation towards complete stoichiometry for the film. The scattering coefficient begins to increase at around 225 °C, associated with the appearance of the crystalline anatase phase.

Ta₂O₅. Tantalum pentoxide films are extensively used as dielectric layers for microelectronic devices, and various methods including both PVD and CVD are used to deposit the films. Its structural and electrical properties have been extensively studied and reviewed (Chaneliere *et al.*, 1998; Bange, 1997b). It can be readily deposited into an amorphous structure by IBSD, and the refractive index can be as high as 2.18 at 550 nm, practically the largest value obtained among the various PVD deposition methods for Ta₂O₅ (Demiryont *et al.*, 1985). A low extinction coefficient of 3×10^{-6} has been reported by Binh *et al.* (1985). Absorption loss, scattering loss, and the laser induced damage threshold of Ta_2O_5–SiO_2 high reflector mirrors are all good, with surprising immunity to metal contaminants that can be sputtered from the stainless steel of the coater walls and to a high content of argon in the film (Becker and Scheuer, 1990). Currently, it is common practice to perform post-deposition annealing at 400–500 °C for several hours. This can be done without crystallization, bringing the total loss of the Ta_2O_5-high reflector down to the ppm range or lower.

SiO₂. Silicon dioxide is a glass-forming material; it is readily deposited into the amorphous form by almost all deposition methods. Refractive indices of the films that were deposited by various PVD methods and subjected to different treatments ranged from 1.45 to 1.60 in the visible (Martin and Netterfield, 1989). The differences in refractive indices is mainly attributed to the differences in porosity, and hence density, of the films. Substrate heating, UV radiation, and ion bombardment during deposition and post-deposition annealing increase the density of the film and the effect can be attributed to changing of the bond angle distribution of the Si–O–Si bonds (Anderson and Ottermann, 1997). Stress in the films also varies from tensile to compressive depending on the coating process. The extinction coefficient of SiO_2 films is as low as 5×10^{-6} (Kalb *et al.*, 1986). A transparent range close to bulk fused silica can be obtained. SiO_2 films are the most commonly used low index material for optical coatings from UV to IR.

Composites. The mixing of two or more different materials forms composite films with properties that cannot be obtained from a single component material. Film stress can be reduced by mixing, e.g. SiO_2/Al_2O_3 (Selhofer and Müller, 1999), SiO_2/ZrO_2 (Pond *et al.*, 1989), germanium with fluorides and chalcogenides (Sankur *et al.*, 1988), TiO_2/Ta_2O_5 (Lee and Tang, 2006). A refractive index that is not obtainable from a single component material can be obtained by mixing a high and a low refractive index material in the proper proportion, e.g. SiO_2/TiO_2 (Demiryont, 1985; Chen *et al.*, 1996). Continuously changing the proportion during deposition gives a graded index film for use as a rugate filter (Lee *et al.*, 2006). Adding SiO_2 into TiO_2 films increases the crystallization temperature of the TiO_2 so that it can sustain a higher annealing temperature. Higher annealing temperatures reduce absorption loss and yet maintain the amorphous structure for low scattering loss (Chen *et al.*, 1996; Chao *et al.*, 1999). The total loss of a high reflector mirror with a TiO_2/SiO_2 mixed film as the high index layer can be reduced (Chao *et al.*, 2001). Similar results were recently found in ZrO_2/SiO_2 and Nb_2O_5/SiO_2 mixed systems, and the laser induced damage threshold for the mirror was increased (Melninkaitis *et al.*, 2011). Mixed films for UV coatings were also recently reported (Stenzel *et al.*, 2011). Significant progress in reducing the thermal noise of mirror coatings was achieved by using Ta_2O_5/TiO_2 mixed film to replace the Ta_2O_5 layer in the high reflector stack (Cimma *et al.*, 2006) (see also Chapter 4). Mixed films can be deposited by co-evaporation (Chen *et al.*, 1996), co-sputter (Chao *et al.*, 1991), and mosaic targets in IBSD (Chao *et al.*, 1999; Lee and Tang, 2006; Melninkaitis *et al.*, 2011).

2.7 Special coatings for Mesa beams

Mesa beams are a particular shape of beam designed to reduce thermal noise. A complete discussion of Mesa beams and thermal noise is given in Section 13.3.2. Mesa beams require a specially shaped mirror, which can be accomplished by coating onto a substrate. The special surface contour of the Mesa beam mirror is required to match the wave front of the Mesa beam in a "Mexican Hat" profile (see Figure 14.3). Starting with a flat or spherical fused silica substrate, a shaping layer of SiO_2 film with a thickness distribution forming the Mexican Hat profile is deposited on the substrate, and a multi-layer high reflector stack is then deposited to form the surface contour of the mirror.

The construction method for the shaping layer is given in Agresti *et al.* (2006) and is accomplished in two steps; a rough shape coating followed by a corrective coating. In the first step, a static mask is put between the sputter target and the rotating substrate. The mask profile is calculated according to the sputter atom distribution and the Mexican Hat distribution (see Section 2.4) so that the distribution of the accumulated sputtered atom flux on the substrate is equal to the required Mexican Hat distribution. This coating step gives the Mexican Hat a general shape with a precision of 60 nm. A correction map is then generated by comparing the thickness profile of the rough shaping layer measured interferometrically with the theoretical Mexican Hat profile. A corrective coating is then accomplished by adding thickness to the substrate according to the correction map with a pencil-like sputtered atom beam. This beam is generated by putting a small orifice mask between the sputter target and the substrate. The effectiveness of the two-step coating is dominated by the spatial resolutions of the interferometric measurement and the pencil-like beam spot size. Therefore, it is easier to coat a Mexican Hat profile on a larger substrate where the rate of thickness variation along the radius is lower than that on a smaller substrate. Alignment and eccentricity of the rotation axis for the substrate is critical to assure a symmetrical Mexican Hat profile, especially on a small substrate. The masks need to be positioned as close to the substrate as possible to avoid down-grading the spatial resolution of the Mexican Hat profile from the diffusion of the sputtered atom behind the mask. Low energy sputtering of the mask material by the sputtered ions might cause contamination to the coatings.

It is possible that the rough shaping step for the Mexican Hat profile could be replaced by a magnetorheological finishing (MRF) technique. In this technique, magnetorheological abrasive fluid is conveyed through the gap between the work piece and the rotating spindle wall, with the magnetic field and the position of the work piece computer-controlled for preferential polishing with a high degree of precision. Using this method, surface accuracy in tens of nm can be achieved in a relatively fast and reliable way (Kordonski and Golini, 2000).

2.8 Conclusion

Optical coating technology for applications in precision measurement is in its initial stage. With existing coating technologies developed during the past 30 years for other high-end applications, foundations for coating techniques, materials, characterization methods, and understanding of the underlying physics are ready and can benefit the coating developments for precision measurement. Fast turnaround methods for reliable thermal noise measurements on the coatings are needed for optimizing the coating process parameters as well as tailoring the coating materials and their structures, which are crucial to reducing their thermal noise.

3

Compendium of thermal noises in optical mirrors

VLADIMIR B. BRAGINSKY, MICHAEL L. GORODETSKY,
AND SERGEY P. VYATCHANIN

Phase noise and shot noise are often the fundamental limiting factors of sensitivity in precision optical systems. These noises determine the so-called standard quantum limit (Braginsky *et al.*, 2003), see Section 1.4. At the same time the fundamental frequency stability in high-finesse optical resonators may also be determined by other fundamental effects originating in mechanical, thermodynamical, and quantum properties of solid boundaries (Braginsky *et al.*, 1979). Many of these effects were initially identified and calculated on the forefront of laser gravitational wave antenna research (see Chapter 14) but are becoming increasingly important in other optical systems (Numata *et al.*, 2004; Webster *et al.*, 2008; Matsko *et al.*, 2007; Savchenkov *et al.*, 2007) (see Chapters 15, 16, and 17).

Excess optical phase noise is added to a probe optical wave reflected from mirrors forming an optical cavity due to variation of boundary conditions produced by fluctuations of the surface and optical thickness of the multilayer coating. We characterize below different effects leading to phase noise starting from fluctuations originating in the mirror's substrate and in its interferometric multi-layer reflective coating. In addition to intrinsic noises produced by internal properties of optical mirrors, there are also extrinsic noises imprinted onto the output phase due to different nonlinear effects in the bulk and coating of the mirrors. As, for example, fluctuations of input power producing fluctuations of thickness and refractive index in the coating due to local heating (the photothermal effect, see Sections 3.4 and 3.10).

In this chapter we briefly review known sources of phase noise produced by the mirrors. Detailed analysis of the most important noises is given elsewhere in the book, primarily Chapters 4, 7, and 9. Most of the noise effects reveal themselves both in the substrate and in the coating of the mirror. Though the thickness of the coating is small and the depth of optical power penetration is even smaller, material parameters of the coatings, primarily mechanical loss, lead to noises quite comparable to the noises of the same origin in the substrate.

Optical Coatings and Thermal Noise in Precision Measurement, eds. Gregory M. Harry, Timothy Bodiya and Riccardo DeSalvo.
Published by Cambridge University Press. © Cambridge University Press 2012.

3.1 Substrate Brownian thermal noise

Historically, the first noise identified as a problem for laser interferometer gravitational wave antennas was intrinsic thermal noise produced by internal friction in the mirror substrate's material (Gillespie and Raab, 1995) (see Chapter 7 for a complete discussion of substrate thermal noises). This noise, noted frequently as "Brownian", from Brown (1828), produces fluctuations of the mirror's surface. It may be calculated using the Fluctuation–Dissipation Theorem (Callen and Welton, 1951; Levin, 1998), see also Chapter 1. The spectral density of surface fluctuations of a mirror averaged over the Gaussian beam radius w_m (determined at $1/e^2$ decay of intensity) is

$$S_{SB}(f) = \frac{2k_B T \phi(f)(1 - \sigma^2)}{\sqrt{\pi^3} Y w_m f}, \tag{3.1}$$

where k_B is Boltzmann's constant, T is the temperature of the mirror, $\phi(f)$ is the mechanical loss angle of the mirror substrate at frequency f, σ is Poisson's ratio, and Y is Young's modulus of the substrate. This expression was obtained in the approximation where the mirror is represented by a half-infinite elastic space. See Section 7.1.2 for corrections due to finite size.

Internal friction in materials was described traditionally using a model of viscous damping where the friction force is proportional to the first derivative of the strain tensor (in the frequency domain $\sim f u_{ii}$) or, equivalently, dissipation power is proportional to the frequency squared ($W_{diss} \sim f^2$). However, the viscous damping model was found to contradict experimental data and Saulson (1990) proposed a model of so-called structural damping, postulating that the dissipated power W_{diss} in a material is proportional to stored elastic energy, U, *constant* loss angle, and frequency: $W_{diss} = 2\pi f \phi U$. Note that structural damping is only a phenomenological model with the main drawback that it can not be derived from first principles. In particular, it is not clear how the friction term may be simply and correctly accounted for in the elasticity equation. However, nowadays the structural damping model has been found to be generally in agreement with experimental data for low loss materials and is used as a first approximation by the scientific community (see also Chapters 4 and 7). Another useful model is described by a Debye peak, as discussed in Section 4.2.6.

3.2 Substrate thermoelastic noise

Braginsky *et al.* (1999) suggested that thermodynamic temperature fluctuations in the bulk of the mirror transformed through thermal expansion to surface fluctuations should produce additional mechanical noise. This noise may be calculated using two approaches. The first approach uses the Fluctuation–Dissipation Theorem (see Chapters 1, 7, and 9). The second way is the Langevin approach for the analysis of the thermal correlation functions analogous to that developed earlier by van Vliet *et al.* (1980); van Vliet and Menta (1981). This approach uses the solution of the coupled thermoconductivity equations for temperature

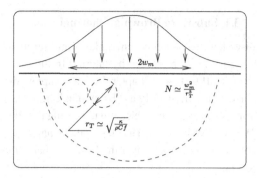

Figure 3.1 Illustration of a semi-qualitative consideration of thermoelastic noise.

with fluctuating sources and elasticity with a final calculation of the mirror's surface fluctuations. These two methods produce identical results and for a half-infinite mirror the corresponding spectral density of averaged surface fluctuations is equal to (Braginsky *et al.*, 1999)

$$S_{\text{STE}} = \frac{4k_{\text{B}}T^2\alpha^2(1+\sigma)^2\kappa}{\sqrt{\pi^5}(C\rho)^2 w_m^3 f^2},$$ (3.2)

where α is the coefficient of linear thermal expansion, C is specific heat capacity per unit mass, κ is thermal conductivity, and ρ is the density of the substrate.

Note that Equation 3.2 may be illustrated using a semi-qualitative consideration, see Figure 3.1. We consider the surface fluctuations averaged over a spot with radius w_m which is larger than the characteristic diffusive heat transfer length r_{T},

$$r_{\text{T}} = \sqrt{\frac{\kappa}{\rho C f}}, \quad r_{\text{T}} \ll w_m.$$ (3.3)

For a frequency of $f \simeq 100\,\text{Hz}$, $w_m \simeq 6\,\text{cm}$ and a fused silica mirror with $r_{\text{T}} \simeq 3.9 \times 10^{-3}$ cm, for example, the condition $r_{\text{T}} \ll w_m$ is fulfilled.

In a volume $\sim r_{\text{T}}^3$, the length variations due to thermodynamic temperature fluctuations ΔT are roughly

$$\Delta x_{\text{T}} = \alpha \Delta T r_{\text{T}} \simeq \alpha \sqrt{\frac{k_{\text{B}}T^2}{\rho C r_{\text{T}}^3}} r_{\text{T}}.$$ (3.4)

Temperature fluctuations in each volume may be considered independent. The number of such volumes that contribute to surface fluctuations is about $N \simeq w_m^3/r_{\text{T}}^3$, and hence the displacement $\overline{X}$ averaged over the spot with radius w_m consists of the sum of the displacements of these independently fluctuating volumes, and is approximately

$$\overline{X}^2 \simeq \frac{(\Delta x_{\text{T}})^2}{N} \simeq \alpha^2 r_{\text{T}}^2 \left(\frac{k_{\text{B}}T^2}{\rho C w_m^3}\right) = \frac{\alpha^2 k_{\text{B}}T^2}{(\rho C)^2 w_m^2 f}.$$ (3.5)

Comparing Equation 3.2 with Equation 3.5 we see that $\overline{X}^2 \simeq S_{TE}^{sub} \Delta f$ (correct to a multiplier of about unity) if one assumes $\Delta f \simeq f$. It confirms that our semi-qualitative consideration is correct.

3.3 Substrate thermorefractive noise

The same fundamental thermodynamic fluctuations of temperatures in a material produce fluctuations of refractive index which, in turn, give rise to phase fluctuations of light waves propagating inside the material. This kind of noise was called thermorefractive nose by Braginsky *et al.* (2000). Initially this noise was calculated using the Langevin approach (Braginsky and Vyatchanin, 2004). Its spectral density in the adiabatic approximation (Equation 3.3) recalculated to effective displacement δz (variation of phase $\delta\varphi = 2k\delta z$) is equal to

$$S_{\text{STR}} \simeq \frac{k_{\text{B}} T^2 \beta^2 \kappa \ell}{\pi^3 (\rho C)^2 w_m^4 f^2}, \tag{3.6}$$

where $\beta = dn/dT$ is the thermorefractive coefficient, n is the refractive index, ℓ is the length of the material, and k is the wavenumber. The same result may be more easily obtained using the Fluctuation–Dissipation Theorem approach developed by Levin (2008). See Chapter 9 for a derivation and discussion using this method on a coating.

Note that Equation 3.6 is valid for a wave *propagating* through the material, whereas in some optics we have the wave propagating through the material twice, once each in opposite directions, i.e. a *standing wave*. Benthem and Levin (2009) stated that for this case the formula for calculation of the fluctuational phase shift, $\delta\varphi$, averaged over the effective volume of the material should contain an additional term of $\sim \sin^2 kz$, with k the wave vector of the light propagating along the z-axis. Calculations presented by Benthem and Levin (2009) show that for relatively small frequencies (< 1 kHz) the result in Equation 3.6 is still valid, however, for larger frequencies (> 1 kHz), Equation 3.6 should be corrected. This type of excess noise was earlier predicted by Wanser (1992) and measured by Knudsen *et al.* (1995) in optical fibers. This noise was also measured in optical microspheres by Gorodetsky and Grudinin (2004) and in microtoroids by Anetsberger *et al.* (2010).

3.4 Substrate photothermoelastic noise

Noise may be produced not only by intrinsic thermal fluctuations but also by fluctuations in absorbed optical power which heats the mirror (Braginsky *et al.*, 1999). As the intensity of the light field decays exponentially in the coating, the optical power is absorbed mostly in a thin layer of thickness

$$d_r = \frac{\lambda(n_L + n_H)}{8 n_L n_H \ln(n_H/n_L)}, \tag{3.7}$$

which is typically of order 0.5 μm for 1.064 nm light. So to calculate the spectral density we may assume only surface absorption. Noise in the absorbed light beam produces fluctuations in temperature which, in turn, cause fluctuations of the surface through thermal expansion. This expansion has a spectral density of

$$S_{\mathrm{SPTE}} = \frac{\alpha^2 (1+\sigma)^2 S_{\mathrm{abs}}}{\pi^4 \rho^2 C^2 w_m^4 f^2},$$

(3.8)

where S_{abs} is the spectral density of the absorbed power.

For the limiting case of shot noise in an absorbed power of W_{abs}, the spectral density becomes $S_{\mathrm{abs}} = 2h\frac{c}{\lambda}W_{\mathrm{abs}}$. The more general case of power fluctuations was analyzed by Wu et al. (2006).

3.5 Substrate cosmic ray noise

During the last few decades substantial progress in data collection and measurement resolution of cosmic ray showers (cascades) has been made (see, e.g., Ryazhskaya (1996)). Based on this, Braginsky et al. (2006b) revised the estimate of the contribution of cosmic rays' impact on mirrors to the noise in precision experiments through three possible mechanisms.

(1) Direct transfer of momentum from the cascade to the mirror.
(2) Distortion of the mirror's surface due to heating by the cascade and subsequent thermal expansion – thermoelastic effect.
(3) Fluctuating component of the Coulomb force between an electrically charged mirror and grounded metal elements located near the mirror's surface.

It was shown that the first two effects are relatively weak and may be vetoed by requiring coincidence between detectors, at least in the case of gravitational wave detection (see Chapter 14) where multiple detectors are expected to be running simultaneously. On the other hand, a veto cannot be considered as an absolute "remedy" for low values of the signal-to-noise ratio.

This conclusion may not be automatically extended to the third effect, however. Negative charge buildup may be high when mirrors spend a long time in vacuum (a year or longer) without removal of the accumulated electrical charge. A second reason is proximity of metal parts, either in the mirror support structure or the surrounding vacuum enclosure. It is necessary to use small areas of all metal elements that are near the mirror's surface and to place these elements as far away as possible from the mirror.

Cosmic rays can potentially cause noise in precision optical experiments other than gravitational wave detectors. However, Braginsky et al. (2006a) showed that in the case of clock frequency stability (see Chapter 15), noise from cosmic rays will be low enough that clock frequency deviations can reach the standard quantum limit (see Section 1.4).

3.6 Substrate thermochemical noise

One more source of thermal noise in substrates is thermochemical noise, first proposed by Benthem and Levin (2009). Fused silica, for example, used for beamsplitters and mirrors, can contain minute quantities of contaminants such as hydroxyl (OH^-), Cl^- ions, and other defects that have an effect on the refractive index depending on their concentration. As these optically active contaminants diffuse up and down the steep gradient of the standing-wave intensity, they cause fluctuations of the overall beam's phase shift. However, estimates show this noise is extremely small for the very pure fused silica used in high-quality mirrors.

3.7 Coating Brownian thermal noise

Brownian thermal noise in a coating has the same physical origin as Brownian noise in a substrate, i.e. it depends on the mechanical losses in each coating layer. Coating Brownian thermal noise is discussed in detail in Chapter 4. Modern technology has made great progress in improving coating quality (see Chapter 2). Unfortunately, mechanical losses in thin layers are still typically much higher than in bulk materials. Even fused silica, used in low refractive index layers, has a ϕ value four orders of magnitude worse than the best achieved value of ϕ in fused silica substrates (Penn *et al.*, 2006). Losses in the high index layers are often even higher.

In the model of independent thin layers on an infinite half space substrate, each layer behaves the same as if it was the only layer. This model has been heavily studied and the solution is known (Harry *et al.*, 2002; Gurkovsky and Vyatchanin, 2010):

$$S_{\text{CB,j}} = \frac{2k_B T \phi'_j d_j}{\pi^2 w_m^2 f} \left[\frac{(1+\sigma'_j)(1-2\sigma'_j)}{Y_j(1-\sigma'_j)} + \frac{Y'_j(1+\sigma)^2(1-2\sigma)^2}{Y^2(1-\sigma'^2_j)} \right], \qquad (3.9)$$

where Y'_j, σ'_j, and ϕ'_j are, correspondingly, the Poisson's ratio, Young's modulus, and mechanical loss angle of a coating layer j, while unprimed values correspond to the substrate.

The first term in brackets of Equation 3.9 corresponds to fluctuations in thickness of a coating layer, and the second one shows fluctuations in the substrate surface induced by losses in the coating. If the losses in the layer responsible for both fluctuations (internal and expansion losses) are equal, which is usually, somewhat arbitrarily, assumed, then these two spectral densities are uncorrelated in each layer. In the opposite case, cross correlation terms should be taken into account. This splitting may be obtained using the approach presented in Gurkovsky and Vyatchanin (2010).

A direct approach to calculating Brownian thermal noise in a multilayer coating suggests simple summation of the spectral densities in Equation 3.9 for each layer. However, this summation ignores the fact that the beam actually penetrates the coating and Brownian expansion of the layers leads to dephasing of interference. This, consequently, causes additional change in the reflected phase. An accurate account of interference decreases coating thermal noise by 2%–3% (Gurkovsky and Vyatchanin, 2010; Kondratiev et al., 2011).

3.8 Coating photoelastic noise

Thermal fluctuations can produce not only surface displacement but, through the photoe-lastic effect, can produce fluctuations in the index of refraction of the coating:

$$\delta n' = -\frac{n'^3}{2} p_{13} \frac{\delta d}{d}, \tag{3.10}$$

where p_{ij} is the photoelastic tensor and indices $i, j \in [1; 6]$. This noise is produced by the same thermodynamical fluctuations in strain as the Brownian noise, and so it should be considered as a coherent correction to the latter one. Unlike Brownian coating noise, variations in refractive index play a role only in the top few layers with total thickness $\sim d_r$. Taking into account that for most materials $p_{13} \sim 0.15 - 0.30$ and that only the part of the Brownian noise leading to fluctuations of coating layer thicknesses is essential, this effect produces a correction of the order of 1%. Accurate calculations confirm this estimate, see Kondratiev et al. (2011).

3.9 Coating thermo-optic noise

Initially, both thermoelastic and thermorefractive noise in coatings were considered inde-pendently until it was shown that they are produced by the same temperature fluctuations and thus should be added coherently (Gorodetsky, 2008; Evans et al., 2008). See Chapter 9 for a detailed discussion of coating thermo-optic noise, as the combined thermoelastic and thermorefractive noise is called. The effect of thermal refraction leads to lengthening of the optical thickness of the coating. This moves the effective surface from which the beam is reflected deeper in the mirror, in the same direction as the incoming beam. At the same time, thermal expansion generally moves the surface of the mirror in the opposite direction, against the incoming beam. This phenomenon, and an observation that the level of coat-ing thermorefractive and thermoelastic noise are typically similar, explain the interference suppression of the combined noise. This suppression may be controlled by tweaking the thickness of the topmost layer (Gorodetsky, 2008).

3.9.1 Coating thermoelastic noise

Thermoelastic noise is produced by the same thermodynamical fluctuations of temperature discussed in Section 3.2, through the thermal expansion of the coating (Braginsky and Vyatchanin, 2003a; Fejer et al., 2004). For the case of a single layer coating with thickness

d, the spectral density of this noise is

$$S_{\text{CTE}}(f) = \frac{8(1+\sigma)^2}{\pi} \frac{\alpha_{\text{eff}}^2 d^2 k_B T^2}{w_m^2 \sqrt{\kappa \rho C f}},$$ (3.11)

where

$$\alpha_{\text{eff}} = \alpha \Lambda,$$ (3.12)

$$\Lambda = -\frac{\rho' C'}{\rho C} + \frac{\alpha'}{2\alpha} \left[\frac{1+\sigma'}{(1-\sigma')(1+\sigma)} + \frac{Y'(1-2\sigma)}{Y(1-\sigma')} \right].$$ (3.13)

Here primes mark the parameters of the coating. Equation 3.13 may be generalized to a multi-layer coating consisting of N alternating sequences of quarter-wavelength dielectric layers with refraction indexes n_H and n_L and optical thickness $d_H = \lambda/4n_H$, $d_L = \lambda/4n_L$ (with λ the wavelength of the light) by substituting

$$\alpha_{\text{eff}} = \frac{\alpha_H d_H \Lambda_H + \alpha_L d_L \Lambda_L}{d_H + d_L},$$ (3.14)

$$d = N(d_H + d_L).$$ (3.15)

Here the factors Λ_H and Λ_L are calculated using Equation 3.13 and the material parameters for high and low refractive layers, correspondingly.

Note that Equation 3.13 for a monolayer may be obtained from a semi-qualitative consideration. Surface fluctuations produced by thermodynamical temperature fluctuations consist of two parts; a "substrate" part that depends on the substrate thermal expansion coefficient α_s (which has been calculated in Braginsky *et al.* (1999)) and a "coating" part due to temperature fluctuations in the layer with thickness d and effective thermal expansion coefficient $\alpha_{\text{eff}} = \alpha' - \alpha$. It is important that the layer thickness d is much smaller than the diffusive heat transfer length r_T (generally, for an optical coating at 1.064 nm, $d \leq 10\ \mu\text{m}$) and overall the following conditions must be valid:

$$w_m \gg r_T \gg d, \quad r_T = \sqrt{\frac{\kappa_s}{\rho_s C_s f}}.$$ (3.16)

Therefore, we may consider temperature fluctuations in our layer to be the same as in a layer with thickness r_T. This means that temperature fluctuations in the layer may be considered as a sum of $N \simeq w_m^2/r_T^2$ independent thermodynamical fluctuations in volumes $\sim r_T^3$ (and do not depend on the thickness d). Hence, we may estimate the fluctuation of an averaged surface position, $\overline{X}_d$, as

$$\overline{X}_d^2 \simeq (\alpha d)^2 \Delta T^2 \simeq (\alpha d)^2 \times \frac{k_B T^2}{\rho_s C_s r_T^3} \times \frac{r_T^2}{w_m^2} = (\alpha d)^2 \frac{k_B T^2}{\rho_s C_s w_m^2 r_T}.$$ (3.17)

Equation 3.17 is in good agreement with Equation 3.11 for the spectral density $S_{\text{TE}}^{\text{coat}}(f)$. Indeed, within a multiplier of about unity we have

$$\overline{X}_d^2 \simeq S_{\text{TE}}^{coat}(f) \Delta f, \quad \text{if } \Delta f \simeq f$$ (3.18)

Equations 3.11 and 3.14 are derived for the model of a half-infinite mirror. A generalization of this result to a finite cylindrical mirror shows that a correction of about several percent is required (Liu and Thorne, 2000). Another correction is required to account for the finite total thickness of the coating as compared to r_T which leads to a frequency dependence of $\alpha_{\text{eff}}(f)$ (Fejer *et al.*, 2004; Somiya and Yamamoto, 2009).

3.9.2 Coating thermorefractive noise

Thermal fluctuations in the mirror due to the thermorefractive factor $\beta_{L,H} = dn_{L,H}/dT$ produce phase fluctuations of the reflected wave and should also be proportional to thermal fluctuations (Braginsky *et al.*, 2000):

$$S_{\text{CTR}} = \frac{2k_B T^2 \beta_{\text{eff}}^2 \lambda^2}{\sqrt{\pi^3} w_m^2 \sqrt{\kappa \rho C f}}. \tag{3.19}$$

The factor β_{eff} depends on the thickness d_c of the topmost low reflection cap layer and, in the limit of a highly reflective $\lambda/4$ coating, is

$$\beta_{\text{eff}} = \frac{1}{2\pi} \frac{\pi n_L^2 (\beta_L + \beta_H) + \beta_L (\varphi_c - \sin \varphi_c)(n_h^2 - n_l^2)}{(n_h^2 - n_l^2)(n_l^2 + 1 + (n_l^2 - 1)\cos \varphi_c)}, \tag{3.20}$$

where $\varphi_c = 4\pi d_c n/\lambda$. Thermorefractive noise is essentially produced by variations in the refractive index of a very thin layer of optical power penetration, d_r.

3.9.3 Combined thermo-optic noise

As thermoelastic and thermorefractive coating noise add coherently, the total thermo-optic noise may be smaller than both contributing noises:

$$S_{\text{CTO}} = \chi(f) S_{\text{CTR}}, \tag{3.21}$$

where $\chi(f) < 1$ is a suppression factor. If the thicknesses of the substrate d and penetration thickness d_r are thin compared to r_T, both the thermoelastic and thermorefractive coating noise will be mostly caused by fluctuations of temperature determined by the substrate. In the adiabatic limit of a thin coating, the suppression factor may be approximated as

$$\chi(f) \simeq \left(\frac{2(1 + \sigma')\alpha_{\text{eff}}(f)d}{\beta_{\text{eff}}\lambda} - 1 \right)^2. \tag{3.22}$$

However, as d and d_r do not coincide, the correlation of the noises can depend on frequency and become zero only at $f = 0$. Nevertheless, significant compensation ($\chi_{min} \sim 0.1$) at low frequencies ($f \sim 100$ Hz) may be attained even for standard coatings with a half wavelength cap layer (see Chapter 12) used on a highly reflective mirror (Gorodetsky, 2008; Evans *et al.*, 2008).

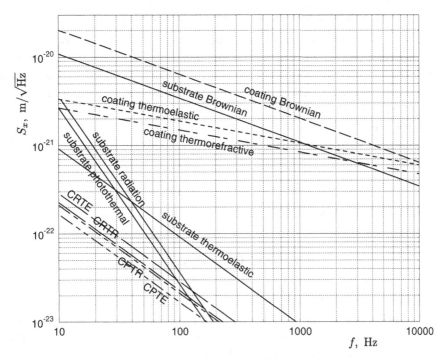

Figure 3.2 Thermal noises in the Advanced LIGO gravitational wave detector mirrors. Abbreviations CPTE, CPTR, CRTE, and CRTR are related to coating thermoelastic and thermorefractive noises produced by photothermal (see Section 3.10) and radiation effects (see Section 3.11). See Figure 14.2 for the full noise curve of Advanced LIGO and Figure 15.9 for thermal noise in a frequency stabilization cavity.

3.10 Coating photothermo-optic noise

Similarly to Sections 3.9.1 and 3.9.2, extrinsic thermal fluctuations produced by absorbed optical power will produce coating noise due to thermal expansion (photothermoelastic noise) and fluctuations of the refractive index in the surface layers (photothermorefractive noise) (Rao, 2003; Gorodetsky, 2008):

$$S_{CPTE} = \frac{4 S_{abs}(1 + \sigma_s)^2 \alpha_f^2 d_N^2}{\pi^3 \rho C \kappa w_m^4 f} \tag{3.23}$$

$$S_{CPTR} = \frac{S_{abs} \beta_{eff}^2 \lambda^2}{\pi^3 \rho C \kappa w_m^4 f}, \tag{3.24}$$

with S_{abs} the absorbed optical power. Combined coating photothermo-optic noise may be suppressed as

$$S_{CPTO} = \chi(f) S_{CPTR}, \tag{3.25}$$

where $\chi(f)$ is the same as in Section 3.9.3.

3.11 Substrate and coating Stefan–Boltzmann radiation noise

While deriving thermal properties of mirrors, Stefan–Boltzmann radiation from the surface is often neglected because thermal conductivity dominates thermal processes. However, thermal radiation as a dissipative process produces additional fluctuations of temperature on the surface. The spectral density of these temperature fluctuations was obtained by van Vliet and Menta (1981) and van Vliet *et al.* (1980). As this noise is applied to the surface nearly in the same way as photothermal noises, we can use the expressions obtained in Sections 3.4 and 3.10 to estimate radiation thermoelastic and radiation thermo-optic noises by simple substitution;

$$S_{abs} \rightarrow S_{SB} = 8\sigma_B k_B T^5 \pi w_m^2, \tag{3.26}$$

where σ_B is the Stefan–Boltzmann constant.

3.12 Conclusion

All the noises discussed above are plotted in Figure 3.2 for the parameters of the Advanced LIGO gravitational wave detector mirrors (Fejer *et al.*, 2004) (see Chapter 14). It is clear that Brownian and coating thermoelastic and thermorefractive noises are the most important to limiting sensitivity. However, the Brownian noises are determined by loss factors which possibly can be improved (see Chapters 4 and 7), as mechanical loss values do not appear to have reached any fundamental limits. This is especially true for the case of coating Brownian noise where the loss factors in thin films are several orders of magnitude higher than in the bulk. It is quite possible that progress in the technology of dielectric film deposition can significantly improve the situation. This and other ways of improving coating thermal noises are discussed in Chapters 4, 6, 8, 12, and 13.

4

Coating thermal noise

IAIN W. MARTIN AND STUART REID

4.1 Introduction

Thermal noise associated with optical coatings is a critical limit to the sensitivity of many high precision measurements which rely on the use of highly reflective mirrors to form laser cavities. These include interferometric gravitational wave detectors (Abramovici *et al.*, 1992; Lück and The GEO 600 Collaboration, 2006; Acernese and The Virgo Collaboration, 2006; Takahashi and the TAMA Collaboration, 2004) (see Chapter 14), highly frequency-stabilized lasers (Rafac *et al.*, 2000; Ludlow *et al.*, 2006; Webster *et al.*, 2004) (see Chapter 15), and fundamental quantum measurements (Schmidt-Kaler *et al.*, 2003; Miller *et al.*, 2005) (see Chapters 16 and 17).

In 1828, Brown reported his observations of the irregular motion of pollen grains and dust suspended in water (Brown, 1828). Brown was not able to fully explain this phenomenon, but later experiments by Gouy (1888) and Exner (1900) showed that the motion was related to the viscosity and temperature of the liquid surrounding the particles, and a full mathematical treatment of Brownian motion was presented by Einstein (1905). As discussed in Chapter 1, any parameter characterizing a dissipative system will exhibit spontaneous thermal fluctuations. The magnitude and frequency spectrum of these fluctuations are related to the dissipative part of the system's impedance (the internal friction), $Z(f)$, by the Fluctuation–Dissipation Theorem (Callen and Welton, 1951; Greene and Callen, 1951; Callen and Greene, 1952). The power spectral density $S_f(f)$ of the fluctuating thermal force on a mechanical system is given by

$$S_f(f) = 4k_\mathrm{B}T\Re\{Z(f)\}, \tag{4.1}$$

where $Z(f)$ is defined as

$$Z(f) \equiv F(f)/v(f), \tag{4.2}$$

for a force $F(f)$ applied to the system resulting in motion with a velocity of amplitude $v(f)$. Alternatively, the theorem can be written in terms of the power spectral density of

Optical Coatings and Thermal Noise in Precision Measurement, eds. Gregory M. Harry, Timothy Bodiya and Riccardo DeSalvo. Published by Cambridge University Press. © Cambridge University Press 2012.

the fluctuating motion of the system $S_x(f)$,

$$S_x(f) = \frac{4k_B T \Re\{Y(f)\}}{4\pi f^2}, \tag{4.3}$$

where $Y(f)$, the mechanical admittance of the system, is equal to $Z^{-1}(f)$. Determining the macroscopic mechanical impedance is therefore the first step in estimating the thermal noise spectrum of a system. The thermal noise is related to the real part of the mechanical impedance; this is the dissipative part of the impedance and comes from the damping coefficient of the system. For the systems considered in this book, external sources of damping are minimized. There are a variety of sources of external damping, such as gas damping due to friction between the critical components and the residual gas molecules, recoil damping where energy may couple into a support structure, electrostatic damping arising from currents induced in nearby structures by residual charge on critical components (Rowan *et al.*, 1997; Mitrofanov *et al.*, 2002; Mortonson *et al.*, 2003), and frictional "slip-stick" damping in the construction joints between different material and components (Quinn *et al.*, 1997). Other sources of damping can arise due to the control schemes adopted within a particular experiment, e.g. where magnets are implemented for position/alignment control or for damping of seismic motion, magnetic hysteresis or eddy current damping can play an important role (Plissi *et al.*, 2004). With these other sources of damping suitably minimized, the relevant damping will typically be the internal friction of the mirror materials. See also Chapter 5 for more on minimizing external damping.

Internal friction arises from the process of anelasticity. When a stress is applied to an anelastic material the strain response is not instantaneous, but develops over a finite relaxation time (Zener, 1948; Nowick and Berry, 1972). An oscillating stress σ applied to the material can be expressed as

$$\sigma = \sigma_0 e^{i2\pi f t}, \tag{4.4}$$

where σ_0 is the stress amplitude and f is the frequency of the oscillation. Since the stress–strain relationship is linear, the resulting strain ϵ will also be periodic with the same angular frequency, but with a phase lag ϕ with respect to the stress,

$$\epsilon = \epsilon_0 e^{i(2\pi f t - \phi)}. \tag{4.5}$$

The phase lag between the stress and strain is known as the mechanical loss angle or the mechanical dissipation factor. Anelasticity, and hence internal friction, can arise from many internal properties of a material which are a function of the internal stress (Nowick and Berry, 1972) such as the density of point defects, dislocations and grain boundaries. If the internal stress in the material is altered, these properties typically require a finite period of time over which to respond to the new state of stress.

The mechanical loss angle can be equivalently defined as a measure of the dissipation of mechanical energy in a material. For a mechanical system oscillating at a resonant frequency f_0, the loss angle can be defined as

$$\phi(f_0) \equiv \frac{E_{\text{lost per cycle}}}{2\pi E_{\text{stored}}} \equiv \frac{\Delta f}{f_0}, \tag{4.6}$$

where E_{stored} is the total energy stored in the oscillating system, $E_{lost\ per\ cycle}$ is the energy dissipated with each cycle of oscillation and Δf is the width of the resonance peak measured at half of its maximum power.

4.1.1 Thermal noise

Early models of the thermal noise of mirrors assumed that the thermally induced motion of each mode was uncorrelated and that the total thermal noise power spectral density could be calculated simply by summing the contribution of each individual mode, calculated via the Fluctuation–Dissipation Theorem (Gillespie and Raab, 1995; Saulson, 1990). However, comparisons between the mode summation technique and a direct application of the Fluctuation–Dissipation Theorem (see Chapter 1) for a simple system of coupled simple harmonic oscillators revealed difficulties with the summation technique (Majorana and Ogawa, 1997). A large number of modes have to be taken into account, with great care taken in the orthogonalization of the modes. Furthermore, the assumption that the modes are uncorrelated can break down if the mechanical dissipation is not homogeneously distributed throughout the mirror. Thus a direct application of the Fluctuation–Dissipation Theorem to calculate the mirror thermal noise was desirable. Nakagawa and collaborators developed a method using a two-point-correlation-function using elastic Greene's functions to calculate the thermal noise (Nakagawa *et al.*, 1997, 2002a). An alternative technique of calculating the thermal noise was pioneered by Levin (1998) and developed by Bondu *et al.* (1998) and Liu and Thorne (2000), discussed in detail in Chapter 1. This approach involves the direct application of the Fluctuation–Dissipation Theorem to the interferometer readout of the position of the mirror face. The thermal noise is calculated by applying a notional pressure, of the same spatial profile as the intensity of the laser beam, to the front face of the mirror and calculating the resulting power dissipated in the mirror. Application of the Fluctuation–Dissipation Theorem then allows the power spectral density of thermal displacement, $S_x(f)$, to be calculated as

$$S_x(f) = \frac{2k_B T}{\pi^2 f^2} \frac{W_{diss}}{F_0^2}, \tag{4.7}$$

where F_0 is the peak amplitude of the notional oscillatory force and W_{diss} is the power dissipated in the mirror. This is Equation 1.2 reproduced here for convenience. For spatially homogeneous loss, W_{diss} can be written as

$$W_{diss} = 2\pi f U_{max} \phi(f), \tag{4.8}$$

where U_{max} is the total energy associated with the peak elastic deformation of the mirror. It should be noted that there are some corrections to the final equation in Levin's paper that are discussed in Liu and Thorne (2000). The final formula for the homogeneous thermal noise, as derived by Bondu *et al.* (1998), is given below in Equation 4.9.

If the laser beam radius is considerably smaller than the radius of the mirror, the mirror can be modeled as being half-infinite and U_{max} can be calculated from elasticity theory (Bondu *et al.*, 1998; Liu and Thorne, 2000). In this case, where mechanical loss is homogeneous, the power spectral density of Brownian thermal noise, $S_x^{\text{ITM}}(f)$, can be shown to be (Bondu *et al.*, 1998)

$$S_x^{\text{ITM}}(f) = \frac{2k_BT}{\sqrt{\pi^3}f} \frac{1-\sigma^2}{Yw_m} \phi_{\text{substrate}}(f), \qquad (4.9)$$

where $\phi_{\text{substrate}}(f)$ is the mechanical loss of the mirror material, Y and σ are the Young's modulus and Poisson's ratio of the material respectively and w_m is the radius of the laser beam where the electric field amplitude has fallen to $1/e$ of the maximum value.[1] See Chapter 7 for more on the role of the mirror substrate in thermal noise.

An expression for the thermal noise power spectral density of a finite sized mirror, $S_x^{\text{FTM}}(f)$, was also derived by Bondu *et al.* (1998), with certain numerical corrections by Liu and Thorne (2000). The finite size (diameter) correction for thermal noise associated with the coating (rather than the mirror substrate) is discussed later in Section 4.1.5.

4.1.2 Inhomogeneous loss and coating thermal noise

A key outcome of Levin's work was the realization that the location of mechanical losses within the mirror is extremely important when calculating thermal noise. Equations 4.7 and 4.8 show that the level of thermal noise is directly related to the power dissipated in the mirror when a notional oscillating pressure is applied to the surface. The power dissipated at any point in the mass is related to both the elastic energy associated with the deformations caused by the pressure and the mechanical loss at that point (Levin, 1998; Liu and Thorne, 2000). Since most of the deformation occurs close to the point where the pressure is applied, it follows that a source of dissipation located close to the location of the reflected laser beam will contribute more to the thermal noise than an identical source of dissipation located further away from this point. In particular, Levin's work led to the realization that the mechanical loss of the multi-layer dielectric coatings, applied to the front face of optics to form highly reflecting mirrors, could be a significant source of thermal noise.

The dissipation of the commonly used silica/tantala (SiO_2/Ta_2O_5) coatings has been found to be in the order of $2-4 \times 10^{-4}$ (Harry *et al.*, 2002; Crooks *et al.*, 2002). This is several orders of magnitude higher than the dissipation of bulk fused silica which may be as low as 1×10^{-9} (Ageev *et al.*, 2004). This higher mechanical loss, plus the above geometrical effect, makes the thermal noise associated with mirror coatings significantly higher than the substrate thermal noise. See Chapter 7 for a discussion of substrate thermal noise.

4.1.3 Calculation of coating thermal noise

Nakagawa *et al.* (2002b) approximated the thermal noise in a coated mirror by modeling a multi-layer reflective coating as a thin surface layer of thickness d and mechanical loss

---

[1] The radius at the point where the *intensity* has fallen to $1/e$ of the maximum value, $r_m = w_m/\sqrt{2}$, is also used in thermal noise literature.

$\phi_{coating}$ with the same material properties as the mirror test-mass substrate. The total power spectral density of the thermal noise associated with such a mirror is given by

$$S_x^{total}(f) = \frac{2k_B T}{\sqrt{\pi^3} f} \frac{1 - \sigma^2}{w_m Y} \left(\phi_{substrate} + \frac{2}{\sqrt{\pi}} \frac{(1 - 2\sigma)}{(1 - \sigma)} \frac{d}{w_m} \phi_{coating} \right), \tag{4.10}$$

where $\phi_{substrate}$ is the mechanical loss of the mirror substrate material.

If the reflective coating is not homogeneous, but is made of alternating layers of two materials (e.g. SiO_2 and Ta_2O_5, see also Chapters 2 and 12), possible anisotropy of the coating mechanical loss factor must be taken into account. The effects of the layer structure of the reflective coating and the differing material properties of the coating and the substrate were investigated as part of a systematic program of research reported by Harry *et al.* (2002). This derivation uses Levin's method of calculating thermal noise by applying the Fluctuation–Dissipation Theorem to a coated mirror to relate the real part of the mechanical admittance $Y(f)$ to the thermal noise power spectral density $S_x(f)$,

$$S_x(f) = \frac{k_B T}{\pi^2 f^2} \Re\{Y(f)\}. \tag{4.11}$$

The mechanical admittance of a mirror is defined as

$$Y(f) \equiv 2\pi i f \frac{x(f)}{F}, \tag{4.12}$$

where $x(f)$ is the response of the mirror to excitation by a cyclic force of amplitude F. This force is taken to have the same profile as the laser beam intensity applied to the face of the mirror. For an interferometer using a Gaussian laser beam to readout the position of the mirror, the resulting pressure distribution is

$$p(r, t) = \frac{2F}{\pi w_m^2} e^{-2r^2/w_m^2} \sin(2\pi f t), \tag{4.13}$$

where $\vec{r}$ is a point on the mirror surface and $|\vec{r}| = r$, f is the frequency, w_m is the field amplitude radius of the laser beam. Since the beam radius is significantly smaller than the radius of the mirror, the mirror can be approximated by a half-infinite space allowing the boundary conditions to be ignored everywhere apart from at the mirror face. Corrections for a finite mirror have been calculated, and will be discussed in Section 4.1.5.

The real part of the admittance can be written as

$$\Re\{Y(f)\} = \frac{4\pi f U(f)}{F^2} \phi_{readout}, \tag{4.14}$$

where $U(f)$ is the elastic energy stored in the mirror when it is under maximum deformation due to the action of the oscillating pressure, and $\phi_{readout}$ is the mechanical loss factor associated with the response of the mirror to the notional Gausian pressure. Here

$$U(f) = \frac{1}{2} F |x(f)|, \tag{4.15}$$

and the response to the pressure is

$$x(f) = |x(f)|\exp(-i\phi_{\text{readout}}) \approx |x(f)|(1 - i\phi_{\text{readout}}). \tag{4.16}$$

In the simple case where the loss in the coating is homogeneous and isotropic, the loss associated with the Gaussian pressure distribution can be separated into contributions from the coating and from the substrate, weighted by the elastic energy stored in the coating, U_{coating}, and the elastic energy stored in the substrate $U_{\text{substrate}}$:

$$\phi_{\text{readout}} = \frac{1}{U}(U_{\text{substrate}}\phi_{\text{substrate}} + U_{\text{coating}}\phi_{\text{coating}}), \tag{4.17}$$

where U is the total elastic energy stored in the mirror under peak deformation from the Gaussian pressure. In the case where the frequencies of interest are far below the first resonant frequency of the mirror, a static pressure distribution $p(r)$ can be used to calculated the elastic energy terms,

$$p(r) = \frac{2F}{\pi w_m^2}\exp\left(\frac{-2r^2}{w_m^2}\right). \tag{4.18}$$

When the coating is very thin compared to the substrate we can make the approximation that $U_{\text{substrate}} \approx U$. Furthermore if the coating is thin in comparison to the width of the pressure distribution then $U_{\text{coating}} \approx \delta U d$ where d is the thickness of the coating and δU is the energy density at the surface of the coating, integrated over the surface. Substituting into Equation 4.17 gives

$$\phi_{\text{readout}} = \phi_{\text{substrate}} + \frac{\delta U d}{U}\phi_{\text{coating}}. \tag{4.19}$$

However, this result only applies if the mechanical loss of the coating is homogeneous and isotropic. Since reflective coatings typically have a stacked layer structure with alternating layers of a high refractive index and low refractive index material (see Chapters 2 and 12), it is likely that the coating mechanical loss may be anisotropic. To account for this the elastic energy density can be expressed in cylindrical co-ordinates with a different loss angle associated with every energy term. Since the individual coating materials are isotropic and amorphous, it can be assumed that the losses associated with strains parallel to the coating plane are all equal. The losses associated with strains perpendicular to the coating plane, however, may be different due to the layer structure of the coating. Harry *et al.* (2002) show that the only loss angles which need to be considered are those associated with the following components of the energy density;

$$\rho'_{v\parallel} = \frac{1}{2}(\epsilon'_{rr}\sigma'_{rr} + \epsilon'_{\theta\theta}\sigma'_{\theta\theta} + \epsilon'_{r\theta}\sigma'_{r\theta}) \tag{4.20}$$

$$\rho'_{v\perp} = \frac{1}{2}\epsilon'_{zz}\sigma'_{zz}. \tag{4.21}$$

Equation 4.17 can now be re-written to include the effect of the layer structure of the coating,

$$\phi_{\text{readout}} = \phi_{\text{substrate}} + \frac{\delta U_{\parallel} d}{U} \phi_{\parallel} + \frac{\delta U_{\perp} d}{U} \phi_{\perp}, \tag{4.22}$$

where

$$\delta U_{\parallel} = \int_S \rho'_{v\parallel} d^2 r \tag{4.23}$$

$$\delta U_{\perp} = \int_S \rho'_{v\perp} d^2 r \tag{4.24}$$

and $\phi_{\parallel}$ and $\phi_{\perp}$ are the loss angles associated with $\rho'_{v\parallel}$ and $\rho'_{v\perp}$, respectively.

The final step of calculating ϕ_{readout} involves finding the elastic energies U, $\delta U_{\parallel}$ and $\delta U_{\perp}$ for the coated mirror, approximated as a half-infinite mass, under the action of the Gaussian pressure distribution $p(r)$. The general solution for the axially symmetric elasticity equations for an un-coated mirror is given by Bondu *et al.* (1998), with corrections by Liu and Thorne (2000). Owing to the thin nature of the coating, the application of axial symmetry and the traction free boundary conditions, it is possible to find expressions for the stresses and strains in the coating in terms of the stresses and strains at the surface of the substrate. This calculation is detailed in the appendix of Harry *et al.* (2002). The solutions give the following expressions for the required elastic energies,

$$U = \frac{F^2(1 - \sigma^2)}{2\sqrt{\pi} w_m Y} \tag{4.25}$$

$$\delta U_{\parallel}/U = \frac{1}{\sqrt{\pi} w_m} \frac{Y'(1 + \sigma)(1 - 2\sigma)^2 + Y\sigma'(1 + \sigma')(1 - 2\sigma)}{Y(1 + \sigma')(1 - \sigma')(1 - \sigma)} \tag{4.26}$$

$$\delta U_{\perp}/U = \frac{1}{\sqrt{\pi} w_m} \frac{Y(1 + \sigma')(1 - 2\sigma') - Y'\sigma'(1 + \sigma)(1 - 2\sigma)}{Y'(1 - \sigma')(1 + \sigma)(1 - \sigma)}. \tag{4.27}$$

By substituting these expressions into Equation 4.17 and Equation 4.14 the power spectral density of the Brownian thermal noise of the coated mirror can be shown to be (Harry *et al.*, 2002)

$$\begin{aligned}
S_x(f) = \frac{2k_B T}{\sqrt{\pi^3} f} \frac{1 - \sigma^2}{w_m Y} \Big\{ &\phi_{\text{substrate}} + \frac{1}{\sqrt{\pi}} \frac{d}{w_m} \frac{1}{YY'(1 - \sigma'^2)(1 - \sigma^2)} \\
&\times [Y'^2(1 + \sigma)^2(1 - 2\sigma)^2 \phi_{\parallel} \\
&+ YY'\sigma'(1 + \sigma)(1 + \sigma')(1 - 2\sigma)(\phi_{\parallel} - \phi_{\perp}) \\
&+ Y^2(1 + \sigma')^2(1 - 2\sigma')\phi_{\perp}] \Big\},
\end{aligned} \tag{4.28}$$

where f is the frequency, T is the temperature, Y and σ are the Young's modulus and Poisson's ratio of the substrate, Y' and σ' are the Young's modulus and Poisson's ratio of the coating, $\phi_{\parallel}$ and $\phi_{\perp}$ are the mechanical loss values for the coating for strains parallel and perpendicular to the coating surface, d is the coating thickness and w_m is the laser beam radius. In the case where $Y' = Y$, $\sigma' = \sigma$ and $\phi_\perp = \phi_\parallel$, this equation agrees with the result of Nakagawa *et al.* (2002b) (Equation 4.10 in this chapter). Note that the coating thermal noise is a function of the Young's modulus of the substrate material, and that the same coating applied to different substrates will therefore result in a different level of coating thermal noise. See Chapter 7 for more on the role of substrates in thermal noise.

Occasionally, it can be useful to re-write Equation 4.28 in terms of the "effective loss", $\phi_{\text{effective}}$, of the coated substrate:

$$S_x(f) = \frac{2k_B T}{\sqrt{\pi^3} f} \frac{1 - \sigma^2}{w_m Y} \phi_{\text{effective}}, \tag{4.29}$$

where

$$
\begin{aligned}
\phi_{\text{effective}} = \phi_{\text{substrate}} + \frac{1}{\sqrt{\pi}} \frac{d}{w_m} \frac{1}{YY'(1 - \sigma'^2)(1 - \sigma^2)} \\
\times [Y'^2(1 + \sigma)^2(1 - 2\sigma)^2 \phi_\parallel \\
+ YY'\sigma'(1 + \sigma)(1 + \sigma')(1 - 2\sigma)(\phi_\parallel - \phi_\perp) \\
+ Y^2(1 + \sigma')^2(1 - 2\sigma')\phi_\perp].
\end{aligned}
\tag{4.30}
$$

Note that $\phi_{\text{effective}} = \phi_{\text{readout}}$ in Equation 4.17. While the effective loss angle is sometimes used, particularly in relation to direct thermal noise measurements (see Chapter 5), care needs to be taken as the effective loss depends on the beam radius used in a particular experiment.

For fused silica substrates coated with alternating layers of SiO_2 and Ta_2O_5, the Poisson's ratio of the coating is likely to be small enough that Equation 4.28 can be approximated to within $\approx 30\%$ by setting $\sigma' = \sigma = 0$ (Harry *et al.*, 2002):

$$S_x(f) = \frac{2k_B T}{\sqrt{\pi^3} f} \frac{1}{w_m Y} \left\{ \phi_{\text{substrate}} + \frac{1}{\sqrt{\pi}} \frac{d}{w_m} \left(\frac{Y'}{Y}\phi_\parallel + \frac{Y}{Y'}\phi_\perp \right) \right\}. \tag{4.31}$$

This expression is a significant simplification of Equation 4.28, and is useful for estimating the expected level of thermal noise. It should be noted that Equations 4.10 to 4.31 all represent the sum of the power spectral density of Brownian thermal noise from the coating and the substrate. The thermal noise power spectral density associated with the coating alone can be obtained explicitly by expanding the second term in Equation 4.31,

$$S_x(f)_{\text{coating}} = \frac{2k_B T}{\pi^2 f Y} \frac{d}{w_m^2} \left(\frac{Y'}{Y}\phi_\parallel + \frac{Y}{Y'}\phi_\perp \right). \tag{4.32}$$

4.1.4 Loss parallel and perpendicular to the coating layers

Coating loss measurements using ringdown techniques, as discussed in Section 4.2.1, can only measure $\phi_\parallel$ and not $\phi_\perp$. However, the following expressions for $\phi_\parallel$ and $\phi_\perp$ can be used for a coating made up of alternating layers of two materials and where the Poisson's ratios are taken to be zero:

$$\phi_\perp = Y_\perp \frac{(\phi_1 d_1/Y_1 + \phi_2 d_2/Y_2)}{d_1 + d_2}, \tag{4.33}$$

$$\phi_\| = \frac{Y_1 \phi_1 d_1 + Y_2 \phi_2 d_2}{Y_\|(d_1 + d_2)}, \tag{4.34}$$

$$Y_\perp = \frac{d_1 + d_2}{d_1/Y_1 + d_2/Y_2}, \tag{4.35}$$

$$Y_\| = \frac{Y_1 d_1 + Y_2 d_2}{d_1 + d_2}, \tag{4.36}$$

where Y, d and ϕ are the Young's modulus, thickness and loss respectively and the subscripts 1 and 2 refer to the two materials making up the coating. Equation 4.28 for the total thermal noise now becomes

$$
\begin{aligned}
S_x = \frac{2k_B T}{\sqrt{\pi^3} f} \frac{1 - \sigma^2}{w_m Y} \Bigg\{ & \phi_{\text{substrate}} + \frac{d}{\sqrt{\pi} w_m Y_\perp} \\
\times & \left[\left(\frac{Y}{1 - \sigma^2} - \frac{2\sigma_\perp^2 Y Y_\|}{(Y_\perp(1 - \sigma^2)(1 - \sigma_\|))} \right) \phi_\perp \right. \\
& + \frac{Y_\| \sigma_\perp (1 - 2\sigma)}{(1 - \sigma_\|)(1 - \sigma)} (\phi_\| - \phi_\perp) \\
& \left. + \frac{Y_\| Y_\perp (1 + \sigma)(1 - 2\sigma)^2}{Y(1 - \sigma_\|^2)(1 - \sigma)} \phi_\| \right] \Bigg\}.
\end{aligned}
\tag{4.37}
$$

An alternative formula for the thermal noise, which does not rely on any approximations for the Poisson's ratios, can be obtained using the following definitions of the effective mechanical loss parallel and perpendicular to the coating layers;

$$\phi_{\text{effective}} = \phi_{\|\text{effective}} + \phi_{\perp\text{effective}}, \tag{4.38}$$

$$\phi_{\|\text{effective}} = \frac{(1 + \sigma)(1 - 2\sigma)^2}{\sqrt{\pi} w_m Y(1 - \sigma)} \left[\frac{Y_1 d_1 \phi_1}{(1 - \sigma_1^2)} + \frac{Y_2 d_2 \phi_2}{(1 - \sigma_2^2)} \right], \tag{4.39}$$

$$
\begin{aligned}
\phi_{\perp\text{effective}} = \frac{Y}{\sqrt{\pi} w_m (1 - \sigma^2)} \Bigg[& \frac{(1 + \sigma_1)(1 - 2\sigma_1) d_1 \phi_1}{(1 - \sigma_1) Y_1} \\
& + \frac{(1 + \sigma_2)(1 - 2\sigma_2) d_2 \phi_2}{(1 - \sigma_2) Y_2} \Bigg],
\end{aligned}
\tag{4.40}
$$

where the subscripts 1 and 2 refer to the two materials making up the coating.

4.1.5 Refinements to coating thermal noise theory

The formulas given above for coating thermal noise assume that the mirror is half-infinite. A correction factor for the thermal noise of a finite mirror substrate was calculated by Bondu *et al.* (1998) (with some corrections by Liu and Thorne (2000)). More recently, Somiya and Yamamoto (2009) extended Bondu's method to calculate the Brownian thermal noise of a coating on a finite cylindrical mirror. For a mirror of thickness h and radius a, it was shown that when h is sufficiently larger than a, the thermal noise of a finite mirror agrees with the infinite mirror approximation. For thinner mirrors, where $h < a$, the coating thermal noise increases as h^{-2}. In this region, the thermal noise is significantly higher than predicted by the infinite mirror model. The effect of laser beam radius was also investigated, and for larger laser beams the finite mirror model predicts a slightly lower level of thermal noise than the infinite mirror model. As an example, for a mirror with radius $a = 17$ cm, thickness $h = 20$ cm and beam radius $w_m = 6.2$ cm (the Advanced LIGO mirror geometry, see Section 14.5.2) the correction for finite mirror size reduces the predicted coating thermal noise by about 2.6%. See Section 6.3 for more on finite mirror effects.

All of the theory discussed so far assumes that the laser beam is completely reflected at the front surface of the coating, and therefore senses the thermal motion of the entire coating. In practice, each interface in the coating reflects some fraction of the total laser power and some of the laser beam will penetrate to various depths in the coating. Recent calculations have shown that, because of this effect, Equation 4.28, when considering an HR coating with < 10 ppm transmission, overestimates the magnitude of the noise typically at the 5% level (Gurkovsky and Vyatchanin, 2010; Kondratiev *et al.*, 2011).

4.1.6 Reducing coating thermal noise

Examination of Equation 4.28 suggests several strategies for reducing coating thermal noise. Cooling the mirror is perhaps the most intuitive method of reducing thermal motion, and generally a direct reduction in thermal noise is expected on lowering the operating temperature of a material. However, in addition to the practical challenges and costs of operating instruments at cryogenic temperature, the temperature dependence of the mechanical loss and the other relevant mechanical properties of the substrate and coating materials must be considered. For example, fused silica has a large and broad mechanical loss peak centered on approximately 40–60 K (Braginsky *et al.*, 1985). Silica is therefore not a promising material for use as a mirror substrate in high precision measurement systems at low temperature. Studies of the temperature dependence of the dissipation of mirror coating materials have been initiated (Yamamoto *et al.*, 2004, 2006a; Martin *et al.*, 2008, 2009). In addition to allowing coating thermal noise to be estimated at low temperatures, the temperature dependence of the loss of coating materials can give insights into the underlying dissipation mechanisms (Nowick and Berry, 1972). Low temperature loss peaks have been observed in tantala, a common high-index coating material, and will be discussed more fully in Section 4.2. The role of cryogenics in coating thermal noise and other properties is considered in detail in Chapter 8.

Another approach to reduce coating thermal noise is to reduce the mechanical loss of the coatings, either by specific treatments (for example, doping or annealing) or by using

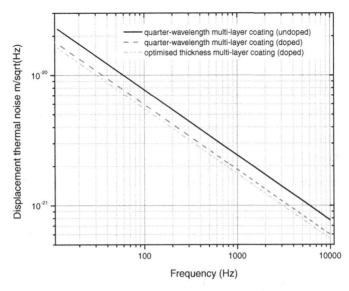

Figure 4.1 Brownian coating thermal displacement noise ($\sqrt{(S(x))}$) calculated for (a) a standard silica/tantala quarter-wavelength coating, (b) a standard quarter-wavelength silica/tantala coating in which the tantala layers are doped with 22.5% titania and (c) a thickness optimized (see Chapter 12) silica/doped-tantala coating (Villar *et al.*, 2010b). All coatings were modeled on a fused silica substrate using a laser beam radius $w_m = 0.06$ m.

alternative coating materials. The observed improvement in mechanical loss resulting from doping the tantalum pentoxide component of a standard silica/tantala coating with titanium dioxide is shown in Figure 4.1. Research into the specific mechanisms responsible for the observed coating mechanical loss and methods of reducing the loss will be discussed in Section 4.2.

Where the mechanical loss of the two component materials is different, it is possible to reduce the thermal noise by tailoring the coating design to minimize the volume of higher-loss material in the coating. This can be achieved by decreasing the thickness of the high loss component, increasing the thickness of the low loss component, and varying the number of coating layers to maintain the same reflectivity. These coating optimization procedures are discussed in detail in Chapter 5 and Chapter 12, and the achievable improvement in displacement thermal noise is shown in Figure 4.1.

The magnitude of the substrate Young's modulus has an inverse effect on the overall level of coating thermal noise, as can be shown from Equation 4.32. The calculated coating thermal noise for a typical silica/tantala coating, deposited on three different substrate materials, is shown in Figure 4.2. Another feature of coating thermal noise that is worth highlighting from Equation 4.32 is the extent to which the thermal noise depends on the how well matched the Young's moduli of the substrate and coating are (Harry *et al.*, 2002). A plot of the variation in thermal noise for a standard silica/tantala coating, when the Young's

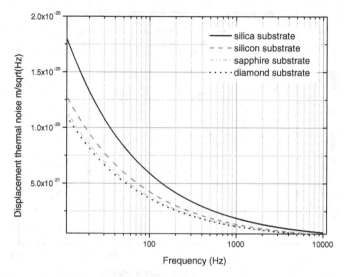

Figure 4.2 The effect of varying the substrate on the Brownian coating thermal displacement noise ($\sqrt{(S(x))}$) calculated for a standard quarter-wavelength silica/tantala coating in which the tantala layers are doped with 22.5% titania. This is intended to illustrate the effect on coating thermal noise only, and that thermal noise associated with the different substrate materials would also need to be taken into account to evaluate the total thermal noise.

modulus is artificially varied, is shown in Figure 4.3. Given these two factors, it is clear that different combinations of coatings materials may be optimal for different substrate materials, depending on their combinations of mechanical losses and elastic properties. See Chapter 7 for more on the interaction between substrates and coatings.

The magnitude of coating thermal noise is also related to the radius of the laser beam used to sense the position of the mirror, varying as w_m^{-2}, while the substrate thermal noise, arising from the homogeneous dissipation in the mirror itself, varies as w_m^{-1} (see Chapter 7). This suggests a possible method of reducing thermal noise in general, and coating thermal noise in particular, is to use laser beams of larger radius, as illustrated in Figure 4.4. In addition, the use of different laser beam geometries has been proposed, as a Gaussian beam profile averages the thermal displacement of the mirror over a relatively small region (Thorne *et al.*, 2000; D'Ambrosio, 2003; D'Ambrosio *et al.*, 2004a; O'Shaughnessy *et al.*, 2004; Bondarescu and Thorne, 2006; Agresti, 2008) and is discussed in detail in Chapter 13.

4.2 Coating mechanical loss

4.2.1 Background

The coatings used in optical precision measurement systems are generally required to be highly reflective (see Chapter 12) and to have a very low absorption (see Chapter 10) and scatter (see Chapter 11), requirements which are best met by dielectric multi-layer coatings

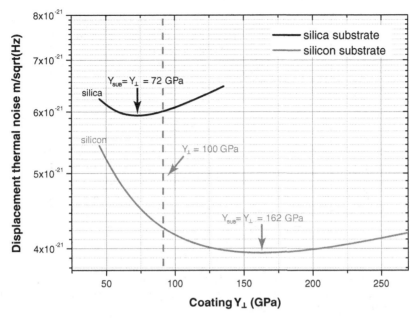

Figure 4.3 The effect of varying the coating Young's modulus on the coating thermal noise, where a reduction of thermal noise can be achieved in theory by matching the substrate and perpendicular coating Young's moduli. It should be noted that this is illustrative, where the coating parameters are taken to be those of a standard quarter-wavelength silica/tantala coating in which the tantala layers are doped with 22.5% titania, and where the Young's modulus of each coating material is artificially varied. The actual perpendicular Young's modulus for such a coating is 100 GPa, as marked on the plot.

(see Chapter 2). These coatings typically consist of alternating layers of two amorphous dielectric materials with different refractive indices. In a standard highly reflective coating the optical thickness of each layer is typically chosen to be $\lambda/4$ where λ is the wavelength to be reflected (although see Chapter 12 for exceptions). Optical thickness, δ, is given by

$$\delta = nt, \tag{4.41}$$

where t is the physical thickness and n is the refractive index of the layer. The reflectivity of the coating is related to the difference in refractive indices of the two materials and to the number of layers in the coating, see Equation 2.1. The reflectivity of a given coating can be increased either by increasing the number of pairs of layers, or by increasing the difference in refractive index between the two components of the coating. Since coating thermal noise is related to the total thickness of the coating, the refractive index can be an important parameter when comparing different coating materials.

Much of the early research into coating thermal noise was undertaken for interferometric gravitational wave detectors, for which the coating materials were required to have very low optical absorption at 1064 nm. The coatings used in these instruments were silica/tantala

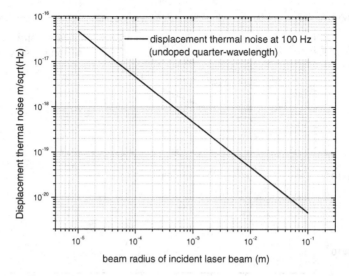

Figure 4.4 The effect of varying the beam radius on the Brownian coating thermal displacement noise ($\sqrt{(S(x))}$), calculated for a standard quarter-wavelength silica/tantala coating in which the tantala layers are doped with 22.5% titania.

multi-layer stacks deposited by ion-beam sputtering. Ion-beam sputtering is a relatively high-energy deposition process which tends to produce coatings with good adhesion, dense structures, fewer impurities and lower optical absorption than other techniques (Seshan, 2002). Post-deposition thermal annealing is used in most ion-beam sputtered coatings to reduce optical absorption losses (Netterfield *et al.*, 2005). Annealing temperatures vary from vendor to vendor, but silica/tantala coatings are typically annealed at 450–550 °C. Tantala films are known to crystallize when heated to between 600 ° and 700 °C (Netterfield *et al.*, 2005), to the detriment of their optical properties. See Chapter 2 for details on the coating process.

Following the work of Levin, it was clear that the mechanical dissipation of coating materials was likely to be a significant source of thermal noise in gravitational wave interferometers. A research program involving a number of groups around the world was established with the aim of quantifying and studying ways of reducing the mechanical dissipation of the coating materials. Early studies focussed on identifying the source of the loss in silica/tantala multi-layer coatings. Three potential sources of energy dissipation were considered; friction at the interfaces between the coating layers, dissipation at the interface between the coating and the substrate and internal dissipation in the coating materials.

A careful series of measurements was carried out to establish which of these mechanisms contributed to the measured coating loss (Penn *et al.*, 2003). By varying the number of pairs of silica and tantala layers, while keeping the total thickness of the coating constant, it was found that the number of coating layer interfaces had no effect on the coating loss. Similarly, by varying the coating thickness, it was found that the loss occurred in the coating, and not at the coating–substrate interface. To test the hypothesis that the loss occurs in the coating materials, coatings with varying relative proportions of silica and tantala were studied. A coating with a higher tantala content showed a significantly higher loss factor than a standard $\lambda/4$ coating, and a coating with a reduced tantala content (i.e. a higher silica content) showed a lower loss factor. This result demonstrated that the multi-layer coating loss was predominantly due to losses in the coating materials, and in particular that the loss associated with the tantala layers dominated the loss of the multi-layer coating. The loss of the silica and tantala layers was found to be $(0.5 \pm 0.3) \times 10^{-4}$ and $(4.4 \pm 0.2) \times 10^{-4}$ respectively (Penn *et al.*, 2003).

Measuring coating mechanical loss

Mechanical loss is often measured using a ring-down technique in which a resonant mode of a sample is excited and the exponential decay of the amplitude recorded. If all external sources of damping are suitably minimized, then the mechanical loss $\phi(f_0)$ is related to the amplitude decay $A(t)$ as

$$A(t) = A_0 e^{-2\pi \phi(f_0) f_0 t/2}, \tag{4.42}$$

where A_0 is the initial excitation amplitude and f_0 is the frequency of the resonant mode. Penn *et al.* (2003) give a detailed description of typical measurement techniques.

The loss of a coating is generally measured by comparing the mechanical loss of two identical substrates, one of which is coated while the other is kept uncoated as a control sample. The coating loss can be calculated from the difference in loss of the two samples, with a scaling factor to account for the fact that only a small fraction of the elastic energy is stored in the coating (Berry and Pritchet, 1975). In addition, the coating thermoelastic loss (Fejer *et al.*, 2004; Braginsky and Vyatchanin, 2003a), arising from the differing thermomechanical properties of the coating and the substrate, must be subtracted to obtain the intrinsic loss of the coating material. The coating thermoelastic loss can be calculated using the formulas derived by Fejer *et al.* (2004), although in practice the effect is usually very small. Thus, the intrinsic coating loss, $\phi(f_0)_{\text{coating}}$, can be found from (Crooks *et al.*, 2004)

$$\phi(f_0)_{\text{coating}} = \frac{E_s}{E_c} \left(\phi(f_0)_{\text{coated}} - \phi(f_0)_{\text{substrate}} \right) - \phi(f_0)_{\text{coating thermoelastic}}, \tag{4.43}$$

where $\phi(f_0)_{\text{coated}}$ is the measured loss of the coated sample, $\phi(f_0)_{\text{substrate}}$ is the measured loss of the un-coated substrate and E_s and E_c are the elastic energies stored in the substrate and in the coating respectively. For simple geometries and resonant mode shapes this energy ratio can be easily calculated analytically, while in more complex situations finite element modeling can be used to find the distribution of elastic energy in a sample. Thermoelastic effects are discussed in detail in Chapter 9.

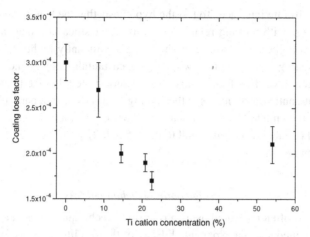

Figure 4.5 Mechanical loss for TiO_2-doped Ta_2O_5/SiO_2 as a function of the measured Ti cation concentration in the Ta_2O_5 layers (Harry et al., 2007).

4.2.2 Coating loss research

Different strategies have been pursued to attempt to reduce the coating mechanical loss without adversely affecting the optical properties of the coating. Some work has focussed on finding methods of reducing the loss of the more lossy tantala layers, while other research has concentrated on finding suitable alternative high refractive index materials.

Doping the tantala layers with titania has been shown to significantly reduce the coating loss. Titania was suggested as a dopant as the atomic size of the Ti atom potentially allows it to be densely packed within the tantalum–oxygen structure, it has a high Young's modulus, and the tantala-titania alloy has a high melting point which suggests that it has a stable amorphous structure. Coatings deposited by Laboratoire des Matériaux Avancés (LMA) in Lyon, France, where the tantala layers were doped with various concentrations of titania doping have been studied (Harry et al., 2007). The presence of any titania doping was found to reduce the coating loss and the magnitude of the dissipation was found to decrease as the titania concentration was increased from zero to 22.5% and appeared to increase at higher titania concentrations, as shown in Figure 4.5. The largest reduction in coating loss, of approximately 40%, occurred at a titania concentration of 22.5%.

Similar coatings produced by another vendor, Australia's Commonwealth Scientific and Industrial Research Organisation (CSIRO) in Sydney, where the tantala layers were doped with 15% titania also showed a reduction in loss of 40% compared to an un-doped coating. However, the loss of both un-doped and doped coatings from CSIRO were found to be higher than the equivalent coatings deposited by LMA, $(3.7 \pm 0.2) \times 10^{-4}$ for a tantala/silica multi-layer coating and $(2.4 \pm 0.1) \times 10^{-4}$ for a titania doped tantala/silica multi-layer coating (Murray, 2008). Similar variations in the mechanical loss of coatings produced by different vendors have often been observed, suggesting that the precise sputtering parameters (see Chapter 2) may have a significant effect on the coating loss. Some research has been carried out into the effect of coating process parameters, however because of the proprietary nature of much of this information it is difficult to draw any definitive conclusions. A coating in which the tantala layers were doped with silica to a level of 35% also showed a reduction in loss of ~40% (Murray, 2008).

A study of the effect of various dopant materials carried out by LMA, using 500 nm thick single layers of doped tantala, showed that, in addition to reducing mechanical loss, doping decreases optical absorption (see Chapter 10) and tends to increase the refractive index, reducing the thickness of tantala required to produce a coating of a particular reflectivity (Flaminio *et al.*, 2010). However, doping with tungsten, a tungsten–titanium alloy, and cobalt were all found to increase the optical absorption and the mechanical loss of the tantala. Cobalt was found to have a particularly detrimental effect, increasing the optical absorption by a factor of ~ 4000 and the mechanical loss by a factor of ~ 4. Of the dopant materials studied, only titanium was found to reduce optical absorption and the mechanical loss of tantala. An initial experiment on doping tantala with lutetium has been carried out, showing no significant improvement in loss over un-doped tantala (Murray, 2008).

4.2.3 Alternative high index materials

In addition to attempts to reduce the mechanical loss of tantala, some research has been directed towards alternative high-index materials with studies of hafnia/silica and niobia/silica coatings. Niobia was found to have a mechanical loss of $\sim 7 \times 10^{-4}$, which is approximately 50% higher than the mechanical loss of un-doped tantala (Harry, 2004). Hafnia/silica coatings were found to have poor adhesion and poor absorption, possibly due to crystallization.

As shown in Equation 4.28, the coating thermal noise can be reduced by matching the Young's modulus of the coating to that of the substrate. In an attempt to do this, CSIRO made a coating in which the high-index layers were composed of 65% silica and 35% titania, and the low-index layers were composed of silica. There is evidence that this coating may have a particularly low loss, approximately 60% lower than the loss of a standard silica/tantala coating from CSIRO (Murray, 2008).

More recent studies of single layers of un-doped zirconia, and of zirconia doped with titanium and with tungsten showed an increased loss in the doped coatings, although it should be noted that un-doped zirconia had a similar loss to titania-doped tantala (Flaminio *et al.*, 2010). However, the optical absorption of all of the zirconia coatings was found to be at least ten times higher than that of tantala, making them less suitable for use with high laser powers in precision measurements. Niobia, while having lower absorption than zirconia, had a higher mechanical loss of $(4.6 \pm 0.9) \times 10^{-4}$. Interestingly, this is lower than previous measurements of the loss of niobia supplied by MLD Technologies of Mountain View, California, again indicating that coating materials from different vendors can vary significantly. While the loss is higher than tantala, the high refractive index of niobia would allow the same reflectivity to be obtained with less thickness of high-index material. Furthermore, the Young's modulus of niobia is closer to the Young's modulus of silica, so that for a silica mirror substrate, the coating thermal noise would be further reduced.

Alumina has been found to both have a relatively low mechanical loss, $(3.1 \pm 0.2) \times 10^{-4}$ for alumina from MLD and $(1.6 \pm 0.4) \times 10^{-4}$ for alumina from Wave Precision, Inc. of Moorpark, California (Crooks *et al.*, 2006). A reflective coating consisting of alumina with silica has been studied. While alumina and silica have similar refractive indices ($n = 1.45$ and $n = 1.65$ respectively) it is still possible to obtain a high reflectivity by using significantly more coating layers. However, mechanical loss measurements and thermal noise studies showed that the potential benefits of the lower loss were outweighed by the increased coating thickness required for high reflectivity (Crooks *et al.*, 2006). In the case of a silica substrate, the Young's modulus of a silica/alumina coating is also less well matched to the substrate modulus, again increasing the level of thermal noise.

While silica/tantala coatings appear to be one of the best combinations of materials for operation at room temperature, in terms of both the optical and mechanical requirements for high reflectivity, low absorption coatings, some other promising materials may be worthy of further investigation.

4.2.4 The effect of deposition parameters

The mechanical loss of ion-beam sputtered silica/tantala coatings deposited by different vendors has been found to vary considerably, suggesting that the precise details of the deposition process can significantly affect the magnitude of the loss. Significant research on the effect of deposition parameters on the optical properties of ion-beam sputtered coatings has been carried out (Netterfield *et al.*, 2005; Pinard and The Virgo Collaboration, 2004). However relatively little work on the effect of deposition parameters on the mechanical loss has been published. Experiments using xenon instead of argon as the sputtering ion have been carried out. There was, however, no significant effect on the mechanical loss of a silica/tantala coating (Murray, 2008). Oxygen deficient tantala coatings have been produced by reducing the oxygen present in the coating chamber. However, initial experiments indicated that the coating loss was higher under these conditions.

Ion-bombardment of a film during deposition is a commonly used technique for evaporated coatings, resulting in increased coating density and better stability. Although ion-beam sputtering is a higher energy process and naturally produces denser films, secondary bombardment with "assisting" ion-beams can result in changes to the coating stoichiometry and stress (Netterfield *et al.*, 2005). A comparison of standard silica/tantala coatings deposited using a collimated and an un-focussed secondary ion-beam of oxygen ions found that the latter gave a 20% lower coating loss (Harry, 2004). However, this loss was higher than the loss of a coating deposited without ion assistance. It is interesting to note that both oxygen deficient coatings and oxygen rich coatings (due to the high rate of implantation by a collimated oxygen ion-beam) gave higher mechanical loss factors. This may suggest that stoichiometry is an important factor in determining the mechanical loss. See Chapter 2 for more on coating deposition.

4.2.5 Mechanical loss of coatings suitable for other wavelengths

Much of the research into the mechanical loss of optical coatings has focussed on materials suitable for coatings operating at a wavelength of 1064 nm. At different wavelengths alternative materials can also be suitable. Amorphous silicon can be used as a high refractive index material at wavelengths above ~ 1100 nm. Studies of amorphous silicon deposited using various techniques have shown that it can have a similar loss to tantala at room temperature (Liu and Pohl, 1998), and, for ion-beam sputtered silicon the loss at temperatures below 100 K can be particularly low, of order 10^{-5}. Furthermore, it has been shown that hydrogenation can reduce the loss of amorphous silicon by up to two orders of magnitude (Liu and Pohl, 1998). The precise mechanism by which hydrogen reduces the loss is not understood, but there is evidence that the presence of hydrogen reduces the number of tunneling states in the material. Many other possible coating materials exist, but the mechanical loss of thin films of these materials tends to be under-studied.

4.2.6 Dissipation mechanisms in coating materials

While it has been shown that titania doping can reduce the dissipation of the commonly used silica/tantala coatings, neither the mechanism responsible for the dissipation in tantala nor the process by which it is reduced by doping with TiO_2 is understood. Developing models of the dissipation mechanisms in coating materials is likely to be critical if further significant reductions in coating loss are to be made. One method of probing the dissipation mechanisms in a material is to measure the dissipation as a function of temperature. Many materials exhibit temperature dependent dissipation peaks, with the temperature and frequency dependence of the peak associated with the activation energy of the dissipation process (Nowick and Berry, 1972).

Initial measurements of the temperature dependence of the loss of a single layer of ion-beam sputtered tantala heat treated at 600 °C revealed a dissipation peak at approximately 20 K (Martin *et al.*, 2008). Further measurements suggested that titania doping reduced the height and increased the width of the dissipation peak (Martin *et al.*, 2009), in addition to reducing the loss at room temperature. A dissipation peak at a particular temperature usually arises from a specific loss mechanism within the material. Most dissipation mechanisms can be described by a Debye peak, for which the loss can be expressed as

$$\phi(f) = \Delta \frac{2\pi f \tau}{1 + (2\pi f \tau)^2}, \tag{4.44}$$

where Δ is a constant related to the magnitude of the dissipation (Nowick and Berry, 1972).

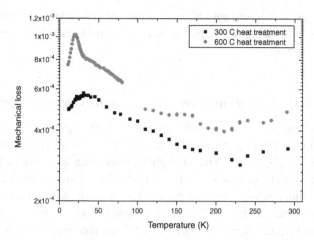

Figure 4.6 Comparison of the mechanical loss an undoped tantala coating heat treated at 300 °C and an otherwise identical coating heat treated at 600 °C (Martin *et al.*, 2010).

The temperature at which the loss peak occurs is found to increase at higher frequencies. This is characteristic of dissipation arising from a thermally activated relaxation process, in which the relaxation time, τ, is related to the activation energy, E_a, of the process by the Arrhenius law:

$$\tau = \tau_0 e^{E_a/k_B T}, \tag{4.45}$$

where τ_0 is the relaxation constant of the dissipation mechanism and k_B is Boltzmann's constant (Braginsky *et al.*, 1985). From Equation 4.44 it is clear that at the dissipation peak $2\pi f \tau = 1$ and thus Equation 4.45 can be written as

$$f = \frac{e^{-E_a/k_B T_{peak}}}{2\pi \tau_0}. \tag{4.46}$$

The activation energy and relaxation constant can then be found from the frequency dependence of T_{peak}. This analysis has been carried out for a pure tantala coating and for a tantala coating doped with 14.5% titania, giving activation energies of 28 ±1.2 meV and 38 ± 2.7 meV respectively (Martin *et al.*, 2009). Titania doping appears to increase the activation energy of the dissipation process responsible for the low temperature loss peak in tantala. Further studies have found that the dissipation peak at 20 K is not present in tantala coatings which are heat treated at lower temperatures (300 °C or 400 °C) after deposition. However, as shown in Figure 4.6, tantala films heat treated at lower temperatures were found to have a broader and lower dissipation peak at approximately 35 K with activation energy of 66 ± 10meV (Martin *et al.*, 2010).

The values of E_a and τ_0 can be substituted into Equations 4.45 and 4.44 to find the theoretical shape of a dissipation peak with these characteristics. For both of the peaks in tantala these theoretical peaks are significantly narrower than the observed dissipation peaks. This is a characteristic of disordered dissipative systems in which there is a range of values of the activation energy, of which the E_a found from the Arrhenius law is the average. It is likely therefore that the distribution of activation energies in tantala arises from the amorphous structure.

In many amorphous solids, the dissipation at temperatures above 10 K is thought to arise from thermally activated reorientations of atoms or molecules over potential barriers. One of the most extensively studied amorphous materials is fused silica, which has a large dissipation peak centered on 40–60 K with an average activation energy 44 meV (Anderson and Bömmel, 1955). Studies by several authors have associated the dissipation peak with energy dissipation arising from thermally activated transitions of oxygen atoms between two energy states in the amorphous SiO_2 network (Braginsky *et al.*, 1985; Bömmel *et al.*, 1956; Strakna, 1961), with the width of the dissipation peak associated with a distribution of activation energies resulting from the disordered structure of the silica (Fine *et al.*, 1954). Various models for the specific dissipation mechanism have been proposed. Anderson and Bömmel (1955) suggested that the dissipation occurs when certain oxygen atoms (the "bridge" oxygen atoms, linking together ring-like formations in the silica) move between two possible bond angles. Strakna (1961) proposed a similar model in which certain Si–O–Si bonds are naturally elongated. The loss then arises from vibrations of the oxygen atom between two positions of minimum potential (McLachlan and Chamberlain, 1964) in the elongated bond. Vukcevich (1972) proposed a third model in which the dissipation occurs when neighboring SiO_4 tetrahedral structures rotate between two stable states. The activation energy of this process depends on the microscopic environment of individual tetrahedra, again resulting in a spectrum of activation energies and a correspondingly wide dissipation peak. More recent work based on first principle atomic modeling has suggested that co-ordinated motion of larger groups of atoms may be responsible for the loss (Reinisch and Heuer, 2005).

In all of the above models, motion of small structural units in the silica network are postulated to be responsible for the low temperature dissipation peak. The models differ only in the type of motion which occurs and the nature of the associated potential barrier which must be overcome. In each model, the atom, or group of atoms, involved in the dissipation process can be considered to sit in a double-well potential in some generalized local configuration co-ordinate: that is, there are two possible positions which can be occupied with almost equal probability. Energy dissipation can occur if an atom gains enough energy (the activation energy) to surmount the potential barrier which separates the two equilibrium positions.

The activation energies for the two dissipation peaks observed in amorphous tantala are of the same order of magnitude as the activation energy found for the dissipation peak in silica. It has been postulated that similar dissipation mechanisms may be responsible for the loss in tantala, with the peaks at 20 K and 35 K corresponding to two different types of motion as discussed above for fused silica. Changes in the local structure of the tantala brought on by heat treatment at temperatures above 400 °C may allow a second mode of dissipation to occur, resulting in the dissipation peak at 20 K.

4.2.7 Barrier height distributions

As mentioned in Section 4.2.6, dissipation peaks in amorphous materials are often broader than a simple Debye peak characterized by a unique activation energy and a relaxation constant. The width of these peaks implies that there is a distribution of activation energies, or barrier heights, arising from the disordered structure of these materials. Topp and Cahill (1996) have developed a method of extracting the barrier height distribution from measurements of loss as a function of temperature, based on the asymmetric double-well potential model of Gilroy and Phillips (1981). While previous analysis had assumed the potential wells were symmetric, this model introduces an asymmetry in the depths of the wells, and had already been used to explain various types of anomalous low temperature behaviour in amorphous solids, including silica (Phillips, 1972; Anderson *et al.*, 1972). The asymmetric double-well potential model has since been corroborated by low frequency light scattering measurements (Wiedersich *et al.*, 2000).

Consider an asymmetric double-well potential with a barrier height V and an asymmetry Δ in the energies of the wells. In an amorphous solid there will be a distribution of potential barrier heights, $g(V)$, and also a distribution of the asymmetries, $f(\Delta)$, in the energies of adjacent potential wells. Gilroy and Phillips (1981) showed that the dissipation predicted by this model is given by

$$\phi = \frac{\gamma^2}{k_B T C_{ii}} \int_0^\infty \int_0^\infty \frac{2\pi f \tau}{1 + (2\pi f \tau)^2} \text{sech}^2 \left(\frac{\Delta}{2k_B T} \right) f(\Delta) g(V) d\Delta dV, \qquad (4.47)$$

where f is the frequency of measurement, C_{ii} is the appropriate elastic constant, γ is the elastic coupling constant which represents the coupling between the defect (e.g. the atom re-orienting within the asymmetric double-well potential) and the applied strain. The relaxation time, τ, associated with a barrier height, V, is given by the Arrhenius equation (Equation 4.45). Starting with this integral, and assuming that $g(V)$ is independent of temperature and that $f(\Delta) = f_0 = $ constant (Gilroy and Phillips, 1981), Topp and Cahill (1996) derive the relation between the mechanical loss and the function $g(V)$,

$$\phi = \frac{\pi \gamma^2 f_0}{C_{ii}} k_B T g(V), \qquad (4.48)$$

where

$$V = -k_B T \ln (2\pi f \tau_0). \qquad (4.49)$$

Thus the distribution of barrier heights, $g(V)$, can be calculated from the measured temperature dependence of the mechanical loss.

4.2.8 Atomic modeling of amorphous coatings

The results of many coating mechanical loss studies have suggested that level of loss can be strongly related to the coating deposition parameters, composition, and post-deposition

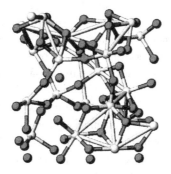

Figure 4.7 A structure for tantala heat treated at 400 °C, obtained from atomic modeling using the measured RDF data (Bassiri *et al.*, 2010). The lighter balls represent tantalum atoms and the darker balls represent oxygen atoms.

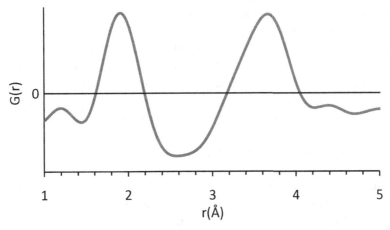

Figure 4.8 An example of an RDF from an electron diffraction measurement of a tantala coating heat treated at 400 °C (Bassiri *et al.*, 2011). The first peak at ∼1.9 Å corresponds to Ta–O nearest neighbor distances and the peak at ∼3.6 Å corresponds to Ta–Ta nearest neighbor distances.

heat treatment. In an effort to understand the relationship between deposition parameters, coating structure, and mechanical loss, studies of the atomic structure (see Figure 4.7) of ion-beam sputtered coating materials have been initiated using Reduced Density Function (RDF) analysis of electron diffraction data and sophisticated computer modeling techniques for amorphous materials (Cockayne, 2009). These modeling techniques have proven useful in recent studies of the atomic structure of $Ge_2Sb_2Te_5$ and N-doped $Ge_2Sb_2Te_5$ (Borisenko *et al.*, 2009a,b).

The RDF gives a statistical representation of the location of the nearest neighbour atoms with respect to a central atom. An example of an RDF is given in Figure 4.8. This in itself can yield interesting information about the local structure of tantala, and a preliminary study of heat treated tantala has been carried out (Bassiri *et al.*, 2010, 2011). However, using molecular dynamics and reverse Monte Carlo modeling it is possible to construct a detailed atomic model of the tantala structure which is consistent with the experimentally measured RDF. This allows direct access to information such as bond types and bond angle distributions, and further modeling can potentially allow prediction of material properties such as the Young's modulus and the index of refraction. It may be also possible to compare bond angles and barrier distributions obtained from models of coating materials with the barrier distributions obtained from mechanical loss measurements. The ultimate aim of this modeling is to relate the local structure to the mechanical loss, and to understand the effect of different deposition parameters, doping atoms, and heat treatment regimes on the structure and the mechanical loss of the material.

4.2.9 Conclusion and future outlook

Significant effort has been invested into quantifying the Brownian thermal noise associated with multi-layer dielectric coatings for use in precision measurement. The level of thermal noise observed is directly proportional to the mechanical loss (sometimes referred to as internal friction) of the coating material. These coatings typically exhibit a level of mechanical loss that is around a factor of 10^5 greater than that often observed in high-purity bulk materials, such as fused silica. Coating improvements are continuing to be pursued to reduce mechanical loss and thus thermal noise. These include doping, heat treatment, and thickness optimization (see Chapter 12). New materials are also being pursued, see Chapter 16 for a discussion of $Al_x Ga_{1-x} As$, for example.

Despite the success of many of these techniques, future experiments that demand large improvements in coating thermal noise set a formidable challenge to material scientists (Punturo and The Einstein Telescope Collaboration, 2007). For this reason, there have been many proposals for experimental configurations that can significantly reduce the thickness of coating material required, or eliminate deposited coatings altogether. In certain applications, traditional high-reflectivity coatings may be replaced by diffractively coupled mirrors (Sun and Byer, 1998; Burmeister *et al.*, 2010), or by using corner-reflectors (Braginsky and Vyatchanin, 2004) as detailed further in Chapter 6. Micro-machined surfaces have also been proposed and are being tested for fabricating monolithic waveguide "coatings" on silicon substrates (Brückner *et al.*, 2008, 2009, 2010), where specialized etching into the silicon surface can be used to fabricate regions of monolithic material with effective high and low indexes of refraction. These alternative strategies still require significant developments in order to achieve the level of optical efficiency (reflectivity, scattering, absorption, etc.) typically required for precision measurement. A wide range of R&D activities is therefore anticipated, developing various coating techniques and configurations, in order to pursue precision measurement beyond the current technology limits.

5

Direct measurements of coating thermal noise

KENJI NUMATA

5.1 Introduction

Mirror thermal noise has been directly measured in optical cavities with high displacement sensitivity. Direct comparisons between theory and measurement were done in dedicated setups which directly measure mirror thermal noise. In addition, frequency stabilities achieved by fixed-spacer cavities are limited by mirror thermal noise, ultimately by coating thermal noise. In this chapter, we will give an overview of such verification experiments and the experimental techniques which are used.

5.2 General considerations

There are two ways to form Fabry–Perot cavities to measure mirror thermal noise (see Figure 5.1).

- **Free-mirror configuration.** Two mirrors are mechanically isolated from each other, like in a gravitational-wave interferometer (see Chapter 14). This configuration more easily allows isolation of the mirror thermal noise from other thermal noises, but is more challenging for suppression of seismic noise. As a result, its measurement frequency range is limited to $\gtrsim 10$ Hz. The displacement measurement is obtained from a control (error) signal which keeps the free-mirror cavity on resonance (Black *et al.*, 2004a). Two identical test cavities allow for common-mode noise rejection. This has typically been done electronically. An optical recombination is also possible, as a Fabry–Perot Michelson interferometer, for the common-mode noise rejection. However, it adds another degree of freedom to be controlled.
- **Fixed-spacer configuration.** Two mirrors are mechanically connected by a fixed spacer, such as a rigid reference cavity used for frequency stabilization. Usually, information on mirror thermal noise is embedded in the laser frequency, and the signal is taken from beating against a second independent laser that has a stability comparable with or better

Optical Coatings and Thermal Noise in Precision Measurement, eds. Gregory M. Harry, Timothy Bodiya and Riccardo DeSalvo. Published by Cambridge University Press. © Cambridge University Press 2012.

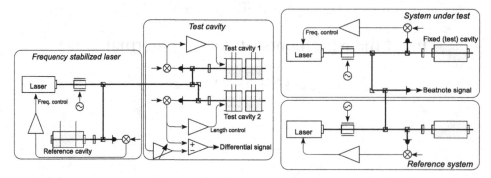

Figure 5.1 Free-mirror configuration (*left*) and fixed-spacer configuration (*right*).

than the one under test.[1] This configuration can be used from as low a frequency as milliHertz to as high a frequency as tens of kiloHertz. Although large common-mode rejection is provided by the spacer for seismic noise, vibration insensitive design is commonly adapted to cope with the seismic noise. See Chapter 15 for more on laser frequency stabilization.

Dedicated thermal noise measurement setups have mostly been based on the free-mirror configuration. The key features of such a setup are small beam radii ($\sim$100 µm) to amplify the effect of the surface fluctuation (see Chapter 4), short test cavities ($\sim$1 cm) to reduce the laser frequency stabilization requirement, frequency stabilization by a longer cavity which suppresses free-running laser frequency noise in the detection band, a relatively high optical power ($\geq$100 mW per cavity) to reduce shot noise, use of a single laser and two identical test cavities to permit common-mode noise rejection for external disturbances and laser noise, and vacuum and seismic isolation systems to reduce vibrational noise. The length variations of the two equivalent test cavities are measured with a single laser. Their differential signal carries the information of the thermal fluctuations in the mirrors.

5.3 Noise sources of interferometer

In order to enhance the displacement (frequency) sensitivity, the Fabry–Perot cavity and the Pound–Drever–Hall (PDH) locking scheme (Drever *et al.*, 1983) (see Section 15.2.1) are commonly adopted in sensitive optical measurements. There are many noise sources that limit the sensitivity of such measurements.

- **Thermal noise**: mirror thermal noise and suspension/spacer thermal noise.
- **Optical readout noise**: shot noise and radiation pressure noise.
- **Laser noises**: laser frequency noise and laser intensity noise.
- **Non-fundamental noise**: seismic noise, electric circuit noise, residual gas noise, and others.

---

[1] When one laser is used, instead of two, the signal can be taken from a demodulated signal of the reference system, using the reference cavity as an "analyzer". In that case, a controlled frequency shift is required to make the laser resonant on the two fixed-spacer cavities at the same time.

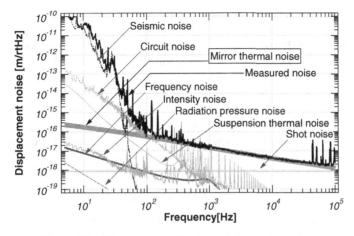

Figure 5.2 Noise sources in the free-mirror configuration.

Figure 5.2 shows typical noise sources in the free-mirror configuration. Chapter 15 has a summary of noise sources in the fixed-spacer configuration and an overview of the PDH locking.

5.3.1 Thermal noise

According to the Fluctuation–Dissipation Theorem (see Chapter 1), a body in a heat bath with a finite temperature is buffeted by a fluctuating force whose amplitude is proportional to the square root of mechanical losses in the body itself. The surface thermal motion of the mirror is called mirror thermal noise, and the fluctuating motion of the entire mirror body as a pendulum is called suspension thermal noise. In the case of the fixed-spacer configuration, there also exists thermal noise in the spacer. The coating thermal noises are discussed in Chapters 3, 4, and 9 and the substrate thermal noise in Chapter 7. These noises become the "signal" in direct measurement experiments.

Suspension thermal noise

When the mirror is suspended as a pendulum, it is buffeted by fluctuating thermal forces. In a simplified harmonic-oscillator model, the suspension thermal noise is expressed as

$$S_{\text{sus}}(f) = \frac{4k_B T \left(2\pi f_p\right)^2 \phi_p}{m} \frac{1}{\left(2\pi f\right)^5}, \tag{5.1}$$

at $f \gg f_p$, by considering only the high-frequency range of the force-to-displacement transfer function.[2] Here, ϕ_p is the loss angle of the suspension (pendulum), m is the mass of the mirror, and f_p is the resonant frequency. This noise is usually much smaller than seismic noise in a table-top experiment.

---

[2] $H(f) = [-m \left(2\pi f\right)^2 + m \left(2\pi f_p\right)^2 (1 + i\phi_p)]^{-1}$.

Spacer thermal noise

By considering the low-frequency portion of its transfer function,[3] the longitudinal thermal fluctuation of one end of a spacer is written, using the spacer's loss, ϕ_{spacer}, as

$$S_{\text{spacer}}(f) = \frac{4k_B T}{2\pi f} \frac{L}{3\pi R^2 Y} \phi_{\text{spacer}}, \tag{5.2}$$

assuming the system to be a one-dimensional cylindrical elastic bar. Here, R is the radius, L is the length, and Y is the Young's modulus of the spacer. In many fixed-spacer configurations, this noise is less important than the mirror thermal noise (Numata *et al.*, 2004).

5.3.2 Optical readout noise

There are two unavoidable noise sources in all optical measurements, shot noise and radiation pressure noise. Together, these are called optical readout noise.[4]

Shot noise

Shot noise originates in the photon counting statistics at the signal detection port of an interferometer. When a photocurrent, i_{DC}, flows in a photodetector, the output spectrum of the photocurrent fluctuations is $i_{\text{shot}} = \sqrt{2e i_{\text{DC}}}$, where e is the charge on the electron, because of Poisson counting statistics of the incoming photons. The signal-to-noise ratio is proportional to the square root of the incident light power (to the cavity), P_0, since the signal is proportional to P_0. In the case of a Fabry–Perot cavity, the equivalent mirror displacement of the shot noise is given by

$$S_{\text{shot}}(f) \approx \frac{h\lambda c}{32 \mathcal{F}^2 P_0} \left[1 + (2\pi f \tau_s)^2 \right], \tag{5.3}$$

with approximations.[5] Here, $\mathcal{F}$ is the finesse of the cavity, h is Planck's constant, and $\tau_s (= 2L\mathcal{F}/(c\pi))$ is the storage time. This shot noise is typically a sensitivity limiting factor for the free-mirror configuration (generally above 10 kHz). In the fixed-spacer configuration, other technical noises are usually more important (see Chapter 15).

Radiation pressure noise

The mirror position is buffeted by the back-action of the reflected photons, whose number fluctuates by the photon counting statistics. In the case of a suspended Fabry–Perot cavity, the radiation pressure induced displacement noise is

$$S_{\text{radi}}(f) \approx \frac{64 \mathcal{F}^2 P_0 h}{\left(\pi m (2\pi f)^2 \right)^2 \lambda c} \left[1 + (2\pi f \tau_s)^2 \right]^{-1}, \tag{5.4}$$

---

[3] $H(f) = -\cot(2\pi f L \sqrt{\rho/(Y(1 + i\phi))})/(2\pi^2 f R^2 \sqrt{\rho Y(1 + i\phi)})$.

[4] The sum of the two optical readout noises is always larger than the standard quantum limit, $S_{\text{SQL}}(f) = 4\hbar/(m (2\pi f)^2)$, which cannot be overcome by conventional techniques. See Section 1.4 for more on the standard quantum limit and Section 11.3 for non-conventional ways of beating this limit.

[5] Unity conversion efficiency of the photodetector, critical coupling cavity, high reflectivity mirror with no loss, small modulation, no demodulator loss, perfect mode matching, and $2\pi f L/c \ll 1$.

with the same approximations as above. In the table-top experiment, this radiation pressure noise is generally relevant only when very small masses ($m \lesssim 1$ mg) are used.

5.3.3 Noises in the laser source

A laser has fluctuations in its frequency and its intensity. They appear as equivalent displacement noise in the interferometer.

Frequency noise

In the Fabry–Perot cavity, the frequency noise of the laser, S_ν, directly couples to the displacement noise, S_L, by

$$\sqrt{S_L(f)} = \frac{L}{\nu}\sqrt{S_\nu(f)}, \tag{5.5}$$

since $\nu = Nc/(2L)$ is satisfied on resonance with N an integer. In the free-mirror configuration, this effect can be minimized by making the cavity length, L, shorter for a given frequency noise, S_ν. In frequency stabilization experiments using the fixed-spacer configuration, longer cavities should result in higher frequency stability for a given displacement noise, S_L.

Intensity noise

The displacement signal in the cavity is obtained as the intensity change at the photodetector. Thus, the fluctuation in intensity of the laser source can couple to the read-out displacement signal as

$$\sqrt{S_{\mathrm{int}}(f)} = \frac{\sqrt{S_P(f)}}{P}\langle \Delta x \rangle_{\mathrm{rms}}, \tag{5.6}$$

where $\sqrt{S_P(f)}/P$ is the relative intensity noise of the laser and $\langle \Delta x \rangle_{\mathrm{rms}}$ is the root mean square of the residual motion fluctuation of the mirror around the ideal lock point (i.e. resonance center). By suppressing the residual motion fluctuations, the effect of the intensity noise is effectively reduced. Experimentally, this effect can be evaluated by determining the (frequency dependent) "coupling factor" $\langle \Delta x \rangle_{\mathrm{rms}}$, including any other effects coupled to the laser intensity noise. This can be done by measuring the transfer function from an applied intensity modulation to the output (displacement) signal. Electronic subtraction of this laser noise is possible if two equivalent Fabry–Perot cavities are locked to one laser simultaneously. Intensity noise at an RF modulation frequency is down-converted by the demodulation process, imitating the error signal. Therefore, the modulation should be applied at a frequency where intensity noise is limited by shot noise.

Another type of intensity noise, called residual amplitude modulation (RAM), contaminates the error signal and affects stability, especially in a low-frequency measurement in the fixed-spacer configuration. The RAM occurs at the phase (frequency) modulation frequency due to misalignment between a crystal axis and polarization, parasitic etalons, and so on (see Section 15.2.2). Along with careful alignment and stray light control, temperature stabilization of a phase modulator effectively suppresses the RAM-induced instability.

5.3.4 Non-fundamental noise

There are other noise sources which are not inherent in interferometers but inevitable during its operation.

Seismic noise

Seismic motion is one such inevitable noise source for every ground-based experiment. Generally speaking, the ground motion has a spectrum of about

$$\sqrt{S_{gnd}(f)} \sim 10^{-7} \left(\frac{1\,\text{Hz}}{f} \right)^2 [\text{m}/\sqrt{\text{Hz}}], \tag{5.7}$$

above ~ 1 Hz in a typical lab environment.[6] This, of course, varies with location, time of year, time of day, weather, etc.

An isolation system, whose displacement-to-displacement transfer function is given by $H_{iso}(f)$, suppresses the motion to $\sqrt{S_{gnd}} \times |H_{iso}|$. A simple harmonic oscillator (pendulum) has $H_{iso}(f) \approx -f_p^2/f^2$ at $f \gg f_p$. In the free-mirror configuration, the test cavity is highly isolated from seismic motion by using a cascade of harmonic oscillators (e.g. a pendulum and an isolation stack) above a few Hz. This configuration typically provides an isolation ratio of $>10^8$ above 100 Hz.

In the fixed-cavity configuration operated at low frequency (<1 Hz), where a passive isolation system is not realistic, use of a mounting geometry that reduces acceleration sensitivity is required for higher sensitivity. The geometry of the cavity and its supporting structures can be designed by numerical analysis (see Chapter 15).

Electric circuit noise

In the free-mirror configuration, the cavity length has to be controlled and kept on resonance. The noise in a servo and actuator circuit can disturb the mirror position. Usually, the servo circuit is designed to have a sufficient gain for stable operation and for suppression of other noise sources (e.g. laser intensity noise). On the other hand, a gain set too high can introduce excess noise. Thus, the balance between high gain and low noise determines the servo-loop design. Similarly, the actuator should have enough dynamic range (sufficiently strong coupling) to suppress the mirror's residual motion. Once the coupling is determined, the actuator-induced noise is also practically determined, because the actuator circuit noise cannot be lower than a certain value, typically of order nV/$\sqrt{\text{Hz}}$ (input equivalent). Therefore, in order to make the actuator noise smaller than the target sensitivity, the uncontrolled

---

[6] Below 1 Hz, the frequency dependence of seismic noise is reduced by $\sim 1/f$, having a micro-seismic peak at $\sim$0.15 Hz and Earth tides at $\sim 10^{-5}$ Hz.

motion of the mirror has to be sufficiently damped, and the actuator circuit has to have low noise. Similar consideration applies to the fixed-spacer configurations. Temperature drift of a demodulator in the PDH method may become important for long time-scale measurements.

In general, output voltage noise from the demodulator represents the sum of the shot noise and of the detector noise (Shoemaker *et al.*, 1988). To make detector noise negligible, the DC photo-current (i.e. input optical power) must be large compared to the noise equivalent current. Use of a detector followed by an RF resonant circuit and/or a low-noise amplifier can help in improving the signal-to-noise ratio at the detector.

Residual gas noise

The gas molecules along the optical path affect the phase of light through their index of refraction fluctuations. This effect can be expressed as

$$S_{\text{gas}} = \sqrt{8\pi L} \frac{(n_0 - 1)^2}{(A_0/V_0)u_0\sqrt{\lambda}} \left(\frac{p}{p_0}\right) \left(\frac{T_0}{T}\right)^{\frac{3}{2}}, \tag{5.8}$$

where n_0 is the refractive index of the gas, V_0 is the volume of one mole of gas[7] at standard temperature T_0, A_0 is Avogadro's number,[8] and u_0 is the mean velocity of the gas molecule at the standard temperature (Shoemaker *et al.*, 1988). For short cavity lengths ($L \sim 1$ cm) the required vacuum level to reach 10^{-18} m$/\sqrt{\text{Hz}}$ is $\lesssim 10^{-3}$ Torr. This level of vacuum can generally be obtained by a turbo-molecular pump and/or an ion pump. Such vacuum level is required, in addition, for the operation of a high-finesse cavity, which is often used in the fixed-spacer configuration.

Other noise sources

Another type of mirror "thermal" noise, called photothermoelastic noise has been pointed out. It is not related to the Fluctuation–Dissipation Theorem, but caused by absorbed power on the coating and/or substrate, which fluctuates statistically. It becomes a displacement noise through a thermal expansion of the substrates and the coatings (Braginsky *et al.*, 1999; Cerdonio *et al.*, 2001; De Rosa *et al.*, 2002). See Chapter 3 for more on this and other thermal noises. In practice, similar heating effects caused by (classical) laser intensity noise cause observable drifts at low frequency. Therefore, it is important to use materials with low absorption (see Chapter 10 for more on absorption and thermal effects). Intensity stabilization of the laser can also be effective.

Classical coupling between temperature variations and thermal expansion of the cavity is always an issue at low frequency, and thus active temperature control is implemented in many cases. Other noise sources may include beam pointing (jitter) noise, acoustic noise (air turbulence), polarization drifts, phase noise in optical fibers (if used), noise in oscillators and counters, etc.

---

[7] $V_0 = 2.24 \times 10^{-2}$ m^3.   [8] $A_0 = 6.02 \times 10^{23}$.

5.4 Direct thermal noise measurements by free-mirror cavities

5.4.1 Background

Until the early 2000s, because of the small size of the fluctuations, direct measurement of mechanical thermal noise had been limited to around resonances in cantilevers (Kajima *et al.*, 1999; Yamamoto *et al.*, 2001), torsion pendulums (Gonzalez and Saulson, 1995; Smith *et al.*, 1999), mirrors (Cohadon *et al.*, 1999; Hadjar *et al.*, 1999) and other systems (Hirakawa and Narihara, 1975; Tittonen *et al.*, 1999). At the California Institute of Technology (Caltech) there was a 40 m interferometer that had the world record for displacement sensitivity, ($\sim 3 \times 10^{-19}$ m/$\sqrt{\text{Hz}}$) around 400 Hz, in the 1990s (Abramovici *et al.*, 1992, 1996). Because of the narrow bandwidth that could be analyzed, it is still unclear whether it was limited by mirror thermal noise (probably from the coatings) or not. In order to characterize thermal noise effects in gravitational wave detectors, two independent direct measurement setups were developed at the University of Tokyo and at Caltech. They successfully validated coating thermal noise theory (see Chapter 4) around 2003. In this section, these two free-mirror experiments are described.

5.4.2 Experiment at University of Tokyo

In the University of Tokyo setup, the effect of thermal noise was highly enhanced by using a small beam radius. The details of the experiment can be found in Numata *et al.* (2003); and Numata (2003).

Experimental setup

Figure 5.1 (*left*) and Figure 5.3 (*left*) show the experimental setup. The test cavity mirrors were monolithic cylinders, with a height of 6 cm and a diameter of 7 cm. Each cavity was formed by a flat mirror and a concave mirror with a curvature of 15 mm. The beam radii on the mirrors were 49 μm and 85 μm, respectively, with a cavity length of 1 cm. The cavity finesse, about 500–2000, was selected to make shot noise comparable with the thermal fluctuations of 10^{-18} m/$\sqrt{\text{Hz}}$ at 100 kHz in each measurement. The four test cavity mirrors were suspended as double pendulums from a common platform. The PDH control signal was fed back through the coils to the neodymium magnets, which were glued onto the mirrors. The unity gain frequency of this servo loop was about 1 kHz. To eliminate common mode noise, two error signals from the two test cavities were subtracted after multiplying one of them by an adjustable gain. The common mode rejection ratio for the frequency noise was higher than 100. The differential voltage signal was converted to displacement based on the coil–magnet actuation efficiency, which was separately measured with a Michelson interferometer formed by the two front mirrors. A commercial Nd:YAG laser with a wavelength of 1064 nm and a power of 500 mW was used as a light source. A small fraction of the beam was introduced into a high-finesse reference fixed-spacer cavity, and the remaining light was divided and sent to the two test cavities. As well as the

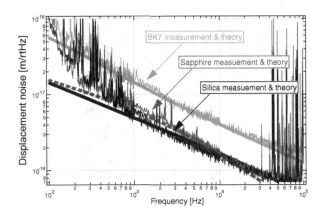

Figure 5.3 Setup photo (*left*) and measured thermal noise in different mirror substrate materials at the University of Tokyo (*right*). At low frequency (below ~300 Hz), the result is limited by seismic and circuit noise. Peaks above 40 kHz are the thermal noise of mirror resonances.

differential measurement, the longer (110 mm) reference cavity was used to suppress the laser frequency noise. Noise sources of this interferometer are shown in Figure 5.2.

Experimental results

Figure 5.3 (*right*) shows the results obtained by this experiment. Three kinds of thermal noise were validated in the setup by swapping the mirror substrate on the suspension. First, substrate Brownian thermal noise was measured by using optical glass, BK7, as a mirror substrate. The substrate had a large frequency-independent loss of $\phi = 1/3600$, which enhanced substrate Brownian noise. Second, substrate thermoelastic noise was measured with crystalline substrates, calcium fluoride and sapphire. They have large thermal linear expansion coefficients ($\alpha \sim 1.8 \times 10^{-5}/\mathrm{K}$) and/or large thermal conductivities ($\kappa \sim 40$ W/m/K), respectively, and these enhance the substrate thermoelastic noise (see Chapter 7 for more on substrate thermal noise). Third, coating Brownian noise was measured with fused silica substrates, which have small intrinsic loss and small thermal linear expansion. This combined made substrate thermal noises negligible. A coating thickness d of 3.5 μm and coating mechanical loss ϕ_{coat} of 4×10^{-4} were adopted to compare the experimental result with the theory, see Chapter 4. As shown in Figure 5.3 (*right*), all of the measured noise agrees with the theoretical predictions between ~300 Hz and ~100 kHz.

5.4.3 Experiment at Caltech

More intensive and complete coating thermal noise studies have been performed at Caltech using a dedicated setup referred to as the Thermal Noise Interferometer (TNI). The TNI is currently used for coating development by comparing the measured noise level and the theory in detail.

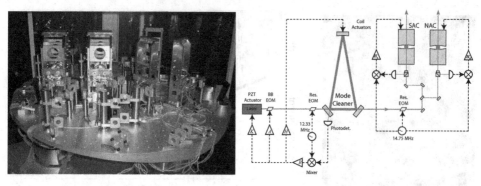

Figure 5.4 Photo (*left*) and optical configuration (*right*) of the Caltech TNI. From Villar *et al.* (2010a), courtesy of A. Villar.

Experimental setup

The experimental setup of the TNI (Figure 5.4) is similar to the University of Tokyo setup, using two short test cavities, one 500 mW Nd:YAG laser source, and a control system actuated with magnets. In order to reduce the thermal noise to a level comparable to the expected value in gravitational wave detectors, the radii of curvature of these mirrors (1 m) were chosen to be much larger than the cavity length (1 cm), giving a relatively large beam size (160 μm). The test cavity mirrors are 10 cm in both diameter and thickness. The typical finesse of the test cavity is 10 000. The mirrors are suspended as single pendulums in a vacuum chamber with magnet-and-coil damping and actuation systems. A triangle cavity with 100 cm round trip length serves as a frequency-reference and spatial-filtering mode cleaner. The three mirrors (3 inches in diameter, 1 inch in thickness) are made of fused silica and suspended similarly to the test cavity mirrors. Above 12 Hz, where suspended optics typically exhibit very low noise, the laser frequency follows the mode cleaner cavity. This is done with actuation on the laser's internal piezoelectric actuator stabilizing the frequency noise up to 30 kHz, and actuation on an external broadband Pockel's cell acting at higher frequencies. At frequencies below 12 Hz, the mode cleaner is locked to the laser to suppress seismic noise. All of the main optics are put on a stack, composed of steel and elastomer, to provide additional seismic isolation.

In addition to the measurement of substrate thermoelastic noise in sapphire (Black *et al.*, 2004c), the TNI tested Brownian noise in various coatings on fused silica substrates. The measured noise floor is fitted to the Brownian noise formula, Equation 4.29,

$$S(f) = \frac{2k_\mathrm{B}T}{\sqrt{\pi^3}f}\frac{1-\sigma^2}{w_m Y}\phi_\mathrm{eff}, \tag{5.9}$$

where ϕ_eff is dominated by the loss angle of the coating, approximated as

$$\phi_\mathrm{eff} \approx \frac{2}{\sqrt{\pi}}\frac{(1-2\sigma)}{(1-\sigma)}\frac{d}{w_m}\phi_\mathrm{coat}. \tag{5.10}$$

Table 5.1 *Measured coating φ's by the Caltech TNI.* t_L *and* t_H *represent total layer thickness of the low index material* (SiO$_2$) *and that of high index material* (Ta$_2$O$_5$), *respectively, and* $d = t_L + t_H$.

Coating type	t_L [μm]	t_H [μm]	ϕ_{eff} [×10⁻⁶]	ϕ_{coat} [×10⁻⁴]
(1) λ/4 SiO$_2$/Ta$_2$O$_5$	2.72	1.83	8.3 ± 0.3	∼3.3
(2) Optimized SiO$_2$/Ta$_2$O$_5$	4.05	1.36	6.9 ± 0.3	∼2.3
(3) λ/4 SiO$_2$/TiO$_2$–Ta$_2$O$_5$	2.54	1.67	6.0 ± 0.3	∼2.6
(4) Optimized SiO$_2$/TiO$_2$–Ta$_2$O$_5$	2.36	1.45	5.5 ± 0.5	∼2.6

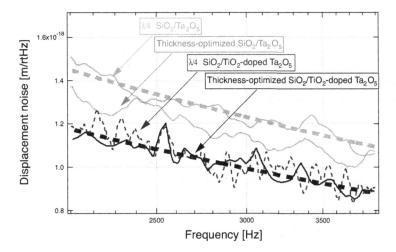

Figure 5.5 Measured thermal noise in different coatings on fused silica substrates at the Caltech TNI. Results for the four different coatings in Table 5.1 are shown. Thick dotted curves show a fit to the thermal noise model for the highest and lowest loss coatings. From Villar *et al.* (2010a).

Experimental results

Figure 5.5 and Table 5.1 summarize results obtained by the TNI. Four types of coatings have been tested.

(1) The λ/4 layers of SiO$_2$ and Ta$_2$O$_5$ (silica/tantala) coatings (Black *et al.*, 2004a), manufactured by Research-Electro Optics (REO) of Boulder CO, USA. The coating was composed of alternating λ/4 layers of two materials, which realizes the minimum number of layers to achieve a prescribed reflectance. ϕ_{eff} was found to be $(8.3 ± 0.3) × 10^{-6}$ in a fit to the data, corresponding to $\phi_{coat} \sim 3.3 × 10^{-4}$. This loss value was close to (but slightly smaller than) the ones measured by the ringdown measurements (see Chapter 4 for details).

(2) Thickness-optimized SiO$_2$/Ta$_2$O$_5$ coatings (Villar *et al.*, 2010b), manufactured by *Laboratoire Matériaux Avancés* (LMA) of Lyon, France. The geometry of the standard

λ/4 coating was modified to reduce the total amount of tantala while preserving the reflectivity of the coating (see Chapter 12 for details of this technique). The primary source of mechanical dissipation in these coatings was known to be in the tantala layers rather than in the silica layers or at the interfaces between layers (Penn et al., 2003). A systematic procedure for designing minimal-noise coatings featuring a prescribed reflectivity was established. The optimal configuration was found to be a periodic stack of identical high/low index layer pairs or doublets, with the exception of the terminal (top/bottom) layers. With this coating with reduced tantala layer, ~9% reduction of coating thermal noise (in linear spectral density) was obtained, as expected from theory.

(3) λ/4 layers of SiO_2 and TiO_2-doped Ta_2O_5 (Harry et al., 2007), manufactured by LMA. Adding TiO_2 as a dopant to Ta_2O_5 was tried because it has a high Young's modulus, its atomic size allows for dense packing in the Ta and O matrix, and the melting point of the TiO_2/Ta_2O_5 alloy is relatively high, indicating a stable amorphous structure. Thermal noise became smaller than undoped coatings by ~15%.

(4) Thickness-optimized[9] SiO_2 and TiO_2-doped Ta_2O_5 (Villar et al., 2010a), manufactured by LMA. This coating was based on a combination of the above two techniques, and showed the smallest thermal noise directly measured to date.

In a simplified model, ϕ_{eff} can be expressed as $\phi_{coat} = b_L t_L \phi_L + b_H t_H \phi_H$, where b is the weighting factor and t is the physical thickness (see Chapter 12). Here, L and H denote low and high index materials (e.g., silica and tantala, respectively). By inversely solving the equation using the experimental results, values of $(0.5 \pm 0.3) \times 10^{-4}$, $(4.7 \pm 0.5) \times 10^{-4}$, and $(3.6 \pm 0.5) \times 10^{-4}$ for the ϕ of SiO_2, Ta_2O_5, and TiO_2-doped Ta_2O_5 layers, were obtained, respectively. They agree with mechanical loss measurements performed on single-layer and multi-layer coatings (see Chapter 4).

5.5 Direct thermal noise measurements by fixed-spacer cavities

5.5.1 Background

Until 2004, there was no realization that fixed-spacer cavities were limited by mechanical thermal noise (Numata et al., 2004). In a rigid cavity, thermal noise contributions mainly come from the Brownian thermal noise of the substrates, coatings, and rigid spacer (see Equations 15.20, 15.21, and 15.22), neglecting other thermal noise sources.[10] The contribution from the spacer can be negligible for frequencies off mechanical resonance, and thus frequency stabilization systems based on fixed-spacer cavities can be ideal systems to test mirror thermal noise models. Assuming standard parameters, the thermal-noise-limited frequency stability (noise), $\sqrt{S_\nu}$, is at a level of $0.1/\sqrt{f}$ Hz/$\sqrt{Hz}$, using Equation 5.5 and

---

[9] The term "dichroic" is sometimes used to refer to this coating, since, in addition to optimizing for low thermal noise, it was designed for both 1064 nm and 532 nm for use in Advanced LIGO. These two optimizations end up requiring very similar coatings.

[10] Other thermal noise sources can include thermoelastic loss, loss due to any mechanical support, surface loss (other than coating), loss due to mirror bonding, etc.

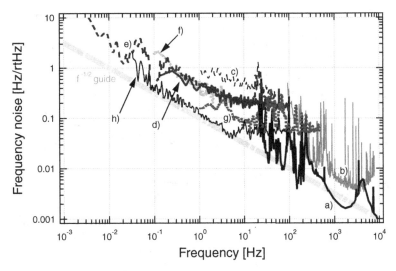

Figure 5.6 Frequency noise spectra achieved by fixed-spacer cavities; (a) 24 cm cavity at 1064 nm (Bondu *et al.*, 1996), (b) 20 cm cavity at 1064 nm (Conti *et al.*, 2003), (c) 3.5 cm cavity at 1064 nm (Notcutt *et al.*, 2006), (d) 7 cm cavity at 698 nm (Ludlow *et al.*, 2007), (e) 7.75 cm cavity at 972 nm (Alnis *et al.*, 2008b), (f) 10 cm cavity at 1064 nm (Webster *et al.*, 2008), (g) 10 cm cavity at 1064 nm (Millo *et al.*, 2009), (h) 29 cm cavity at 578 nm (Jiang *et al.*, 2011).

$\sqrt{S_L} \sim 10^{-15}/\sqrt{f}$ m/$\sqrt{\mathrm{Hz}}$. It can be converted to a constant[11] fractional Allan deviation (variance), σ_A, of $\sim 10^{-15}$, using a relationship of $\sqrt{S_\nu} = \sigma_A \nu / \sqrt{2(\ln 2) f}$. According to the model, the thermal-noise-limited frequency stability approximately scales with the cavity length, L (i.e. smaller frequency noise with longer cavity).

In a frequency stabilization setup used for precision spectroscopy and optical atomic clocks, long-term stability is more important than thermal noise. Thus, low thermal expansion glasses (such as ULE and Zerodur) have been preferred as substrate materials in spite of their large substrate Brownian thermal noise.[12] The input optical power is chosen to be as low as possible (~ 10 µW) in order to avoid the classical thermal expansion-to-length coupling from optical absorption. For high frequency applications (up to ~ 10 kHz), higher input power (>5 mW) is often used to reduce shot noise. Vibration isolation can be provided by suspending the cavity as a pendulum for the high-frequency measurements. Figures 5.6 and 5.7 summarize results obtained with fixed-spacer cavities at different frequencies and time scales. Even before Numata *et al.* (2004), some results had shown the characteristic frequency (or averaging gate time, τ) dependence of mirror thermal noise (i.e. $1/\sqrt{f}$ and τ^0) (Young *et al.*, 1999; Bondu *et al.*, 1996; Conti *et al.*, 2003). More recently, additional experiments have been reported to be at (Notcutt *et al.*, 2006; Ludlow *et al.*, 2007; Webster *et al.*, 2008; Jiang *et al.*, 2010, 2011) or close to (Alnis *et al.*, 2008a,b; Eisele

---

[11] In general, Allan deviation (variance) (Allan, 1966) and a frequency noise spectrum cannot be interconverted unless integration from zero to infinite frequency (averaging time) can be performed (Rutman and Walls, 1991; Greenhall, 1997).

[12] Mechanical losses, ϕ, of ULE and Zerodur are 1.6×10^{-5} and 3.2×10^{-4}, respectively (Numata *et al.*, 2004).

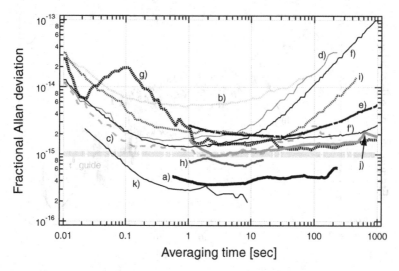

Figure 5.7 Fractional Allan deviation (variance) achieved by fixed-spacer cavities; (a) 24 cm cavity at 563 nm (Young *et al.*, 1999), (b) 3.5 cm cavity at 1064 nm (Notcutt *et al.*, 2006), (c) 7 cm cavity at 698 nm (Ludlow *et al.*, 2007), (d) 7.75 cm cavity at 972 nm (Alnis *et al.*, 2008b), (e) 8.4 cm cavity at 1064 nm (Eisele *et al.*, 2008), (f) 10 cm cavity at 1064 nm (second order drift removed for f') (Webster *et al.*, 2008), (g) 5.5 cm cavity at 1064 nm (Herrmann *et al.*, 2009), (h) 10 cm cavity at 1064 nm (Millo *et al.*, 2009), (i) 7.5 cm cavity at 934 nm (Sherstov *et al.*, 2010), (j) 7.75 cm cavity at 1064 nm (Jiang *et al.*, 2010), (k) 29 cm cavity at 578 nm (Jiang *et al.*, 2011).

et al., 2008; Ludlow *et al.*, 2009; Herrmann *et al.*, 2009; Millo *et al.*, 2009; Sherstov *et al.*, 2010) the thermal noise limit. As examples of thermal-noise-limited frequency stability, two fixed-spacer cavity experiments are discussed in the following sections.

5.5.2 Experiment at JILA/NIST

The most systematic investigation of the thermal noise model in a fixed-spacer cavity experiment was performed by the Hall/Ye group at JILA (National Institute of Standards and Technology/University of Colorado) (Notcutt *et al.*, 2006). They compared the frequency noise of lasers locked to a variety of rigid cavities of differing lengths and mirror substrate materials. The mirror thermal noise was enhanced by using relatively short cavities (10–30 mm).

Experimental setup

The frequency reference was a Nd:YAG laser at 1064 nm locked to a 50 mm ULE cavity that had sub-Hz linewidth (Notcutt *et al.*, 2005). The optical system of the test cavity was very similar to the reference laser system. Frequency noise spectral densities were measured from the beat note between the two systems. Zerodur spacers of different geometries and different mirror substrate materials were swapped out from the mounting structure, with other parts

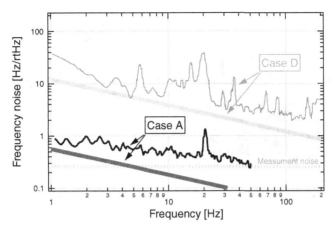

Figure 5.8 Frequency noise measured with different cavity configurations by NIST/JILA group (Notcutt *et al.*, 2005). Case A: 35 mm long Zerodur spacer, fused silica substrate cavity with 250 μm spot size. Case D: 10 mm long Zerodur spacer, Zerodur silica mirror with 130 μm spot size. Thick lines represent corresponding thermal noise estimates based on the simple thermal noise model.

remaining largely unchanged. With typically 100 μW of incident power on the cavity, measurement noise limited the sensitivity above ∼10 Hz at about the 0.3 Hz/$\sqrt{\text{Hz}}$ level for the cavity with highest stability. A third laser system was constructed at 698 nm (Ludlow *et al.*, 2006) and the frequency instability of this laser was measured (Hollberg *et al.*, 2005a) against the reference laser with a broadband, frequency stabilized, femtosecond comb.

Experimental result

The best stability (lowest frequency noise) was obtained with the fused silica substrate, the largest beam radius (250 μm), and the longest (35 mm) cavity (Case A of Figure 5.8, and Table 1 of Notcutt *et al.* (2006)). In this test cavity, ∼70% of thermal noise is expected to be from the coating, according to the simplified model. They experimentally showed that higher loss mirror (Zerodur), shorter cavity, and smaller beam size make the frequency noise higher. Case D in Figure 5.8 shows the result with a Zerodur substrate, 10 mm cavity, and 130 μm beam radius. In this case, the majority of the noise is expected to be from the substrate. Those results were comparable with the simplified theory within factor of ∼3. The difference can be attributed to model uncertainties and excess losses in surfaces and support systems that contribute more with the smaller cavities. The cavity at 698 nm with the same dimensions (but slightly larger beam size) showed similar stability to the 1064 nm cavity, giving independent confirmation.

A more carefully designed fixed-spacer cavity experiment showed better agreement with theory and ∼ 5 times higher frequency stability (Ludlow *et al.*, 2007). The cavity was made of all ULE, and had a longer length (70 mm). The 698 nm diode laser stabilized to this cavity enabled the observation of a narrow, 2 Hz, optical atomic resonance of ultracold strontium

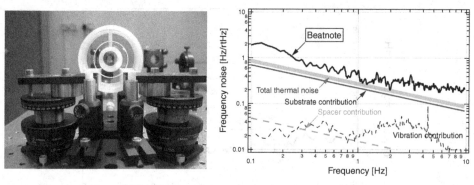

Figure 5.9 Photo of the cutout cavity (*left*) and frequency stabilization result (*right*) by NPL (Webster *et al.*, 2008).

(see Section 15.5 for details). NIST's oblong-shaped longer ULE cavities (24 cm long ULE mirror cavity (Young *et al.*, 1999) and 29 cm long, fused silica mirror cavity (Jiang *et al.*, 2011)) have the smallest fractional Allan deviation seen to date ($\sigma_A \sim 3 \times 10^{-16}$) in a cavity-stabilized laser (Curves (a) and (k) of Figure 5.7). The stabilities are believed to be limited by the substrate and coating Brownian noises, respectively.

5.5.3 Experiment at NPL

Mirror thermal noise has also been revealed[13] in a frequency-stabilization fixed cavity at the National Physical Laboratory (NPL), UK (Webster *et al.*, 2008). This experiment used a relatively simple optical setup, but carefully designed cavity geometry.

Experimental setup

Two Nd:YAG lasers were locked to two identical cavities independently and a beat note between the two lasers was detected, shown in Figure 5.1 (*right*). The cavities were made entirely from ULE and set horizontally on active vibration isolation platforms. The cavity had a cylindrical shape with a cutout (Figure 5.9 (*left*)), which was carefully designed to be vibration insensitive (Webster *et al.*, 2007). Square cutouts were made to the underside of the cavity's cylindrical spacer. For a certain separation of the support points, the response to vertical vibrations was nulled, and such points were identified experimentally. The temperature of the two cavities were also independently adjusted so that the thermal expansivity of ULE was nulled. The cavity was situated within an aluminum cylinder, cooled and temperature controlled by a Peltier element. By simultaneously monitoring the beat note and the actively controlled temperature, the zero crossing temperatures were experimentally identified.

---

[13] As most of the other fixed-spacer cavity setups, the primary goal of this experiment was to develop narrower linewidth lasers for atomic frequency standards, not to observe the mirror thermal noise.

Experimental result

Figure 5.9 (*right*) shows the result in the frequency domain. In the range 0.3–3 Hz, the frequency noise had a slope of $1/\sqrt{f}$, which quantitatively agrees with the thermal noise model. The deviation of the data from the model may be attributed to measurement noise at high frequency, drift at low frequency, and remaining uncertainties in the model. In the model, the majority of thermal noise was from the ULE substrate and a small correction factor was added by coating thermal noise from the 5.3 μm thick coating with a 220 μm beam radius. The frequency noise due to external vibrations was at least an order of magnitude lower than the measured level. The thermal noise limit in fractional Allan deviation was calculated to be $\sigma_A = 1.15 \times 10^{-15}$. For short time scales ($\tau < 200$ ms), the experimental Allan deviation exhibited $1/\sqrt{\tau}$ behavior (white frequency noise), but a limit of $\sigma_A \sim 1.5 \times 10^{-15}$ was met at an averaging time of 0.5 s. Having taken out second-order drift, the Allan deviation was proportional to $\tau^{0.07}$ between 1 s and 1000 s averaging time, τ (curve (f′) in Figure 5.7).

5.6 Summary

Through the direct measurements described above, thermal noise theory has been validated for coating Brownian noise, substrate Brownian noise, and substrate thermoelastic noise. Among them, coating Brownian noise is enough to cause limiting noise in precision measurements, such as next generation gravitational-wave interferometers and optical standards based on optical transitions with tens of mHz linewidth. According to current understanding obtained through the direct measurements, reduction of mirror thermal noise can be achieved by the following techniques.

(1) Low mechanical loss substrates with large Young's modulus (compromised by larger thermal expansion) (see Chapter 7).
(2) Large beam radius (compromised by coupling to higher order axis modes and complex mode matching) (see Chapter 4).
(3) Longer cavity length (compromised by acceleration sensitivity and classical thermal-mechanical noise couplings).
(4) Optimized layer coatings instead of standard $\lambda/4$ coatings (see Chapter 12).
(5) Thinner coatings (i.e. lower finesse, compromised by shallower error signal and instability due to RAM) (see Chapter 6).
(6) Low mechanical loss coating materials (e.g. TiO_2 doping of Ta_2O_5 and annealing) (see Chapter 4).
(7) Cryogenics (see Chapter 8).
(8) Other techniques like special coating layers (Kimble *et al.*, 2008), beam shaping and higher order modes (see Chapter 13), coating-free mirrors (see Chapter 6), and other possible ideas.

For direct measurements of coating thermal noise, possible future research directions include the following.

(1) Understanding and reducing mechanical losses in coatings materials and validating this in direct thermal noise measurements.

(2) Direct measurement of coating thermo-optic noise and its minimization.[14]

(3) Development of a new dedicated, simpler, direct measurement setup to investigate coating thermal noise (e.g. suspended short fixed-spacer cavity operated at high frequency).

For these unfinished works, mutual interactions between fields will become more important.

---

[14] An experiment designed to measure the possible coherent interaction between the coating thermoelastic noise and thermorefractive noise is ongoing at Caltech, see Section 9.5.2.

6

Methods of improving thermal noise

STEFAN BALLMER AND KENTARO SOMIYA

6.1 Introduction

The literature contains a number of ideas for minimizing thermal noise or its impact on an experiment's sensitivity. They can be split into two general categories: reducing thermal noise itself or reducing its impact on the measurement. Direct reduction of coating thermal noise is addressed in Chapter 4, through material losses, Chapter 8, through thermal dependence, and Chapter 12, through minimization of lossy materials. In this chapter, five other thermal noise reduction techniques are discussed; using shorter wavelength readout lasers (Section 6.2), optimizing the aspect ratio of the mirrors (Section 6.3), using Khalili cavities and etalons (Section 6.4), minimizing or eliminating the need for coatings on mirrors (Section 6.5), and correlating different coating thermal noise sources so there is cancelation (Section 6.6). One method of reducing the impact of coating thermal noise, changing the shape of the readout beam, is addressed in Chapter 13. Another coating thermal noise reduction method, displacement-noise-free interferometry, is discussed below in Section 6.7.

6.2 Shorter wavelengths

A simple way to reduce coating thermal noise, all other things being equal, is to use light with a shorter wavelength. The noise level is proportional to the square-root of the thickness of the coating (see Equation 1.4). Each layer is typically $\lambda/(4n_j)$ thick (although see Chapter 12 for exceptions), where λ is the light wavelength and n_j is the index of refraction (j indicates the material). Reducing λ is thus a way to decrease thermal noise, but there are a number of trade-offs in using shorter wavelength light.

The first trade-off is that the beam radius decreases and thermal noise then increases with the shorter wavelength, provided that the geometry of the mirrors is unchanged. When the waist is located at the middle point of the cavity, the beam radius on the mirrors is given by

$$w_m = \sqrt{\frac{\lambda L}{\pi} \frac{1}{\sqrt{1 - g^2}}} . \tag{6.1}$$

Optical Coatings and Thermal Noise in Precision Measurement, eds. Gregory M. Harry, Timothy Bodiya and Riccardo DeSalvo. Published by Cambridge University Press. © Cambridge University Press 2012.

Here L is the cavity length, R the radius of curvature of the mirror, and g is given by $1 - L/R$ (see Chapter 13 for a more detailed discussion of the g-factor). As g^2 gets closer to unity, higher-order spacial modes start to increase in the cavity (i.e. the cavity becomes more susceptible to tilts or mode mismatches). Since coating thermal noise is inversely proportional to w_m and the beam radius is proportional to $\sqrt{\lambda}$, there is no improvement in thermal noise at all if the beam radius is limited by this cavity instability issue. There is an improvement only if the instability issue is tolerable and the beam radius is limited by other conditions such as a diffraction loss due to the finite mirror size.

The index of refraction typically decreases with wavelength so that the thickness of each layer does not decrease linearly with λ, although the deviation is small for wavelengths around a few hundred nanometers. The empirical relationship between the index of refraction and the wavelength is given by the Sellmeier equation,

$$n^2(\lambda) = 1 + \sum_i \frac{B_i \lambda^2}{\lambda^2 - C_i} . \tag{6.2}$$

The coefficients B_i and C_i are experimentally determined for the case of silica to be $B_1 = 0.696$, $B_2 = 0.408$, $B_3 = 0.897$, $C_1 = 4.679 \times 10^{-3}$ µm, $C_2 = 1.351 \times 10^{-2}$ µm, and $C_3 = 97.93$ µm. The index of refraction for 1064 nm infrared light is 1.45 and that for 472 nm blue light is 1.46. The deviation is less in the case of tantalum pentoxide.

The second trade-off is that Rayleigh scattering increases when shorter wavelength light is used (Born and Wolf, 1999). Scattering is often the dominant source of optical losses in the cavity and also a possible noise source through the vibration of the inner wall of the vacuum pipe if any light is reflected back into the main beam. See Chapter 11 for a more detailed discussion of scatter. In addition, absorption typically increases with shorter wavelength and may cause various kinds of thermal problems, as is considered in Chapter 10.

One benefit of using light with a shorter wavelength is that cavity length sensing noise decreases for the same power. The energy of each photon is inversely proportional to wavelength, so if the incident laser power is the same both the number of photons and their fluctuations decrease.

There is the possibility of using a mirror with less coating layers, an approach which trades shot noise for improved thermal noise. The power reflectivity of a mirror with alternating layers of silica and tantala, for example, is 99% with 8 doublets (see Chapter 12 for a discussion of coating design). This would typically be enough if the measurement device is a simple Michelson interferometer, but shot noise can end up being the limitation in sensitivity. If, to reduce shot noise, an optical cavity is configured to store more power, the reflectivity of the mirror will need to be higher. The optical power gain of a cavity with 99% reflectivity in both mirrors is 100, assuming no optical losses. The gain will be 1000 with 99.9% mirrors, but each mirror now needs 12 doublets and the corresponding coating thermal noise increases.

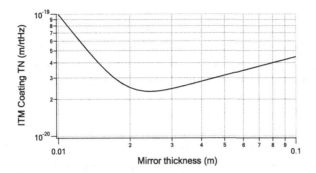

Figure 6.1 Coating thermal noise of a 100 g mirror with different mirror thickness (h). The beam radius is set $1/2.5$ of the mirror radius (a). The minimum occurs at the values $h = 2.42$ cm and $a = 2.45$ cm.

6.3 Optimal mirror aspect ratio

The thermal noise level of mirror coatings is inversely proportional to the radius of the beam spot on the mirror. Thus, increasing the beam radius is a natural way to reduce thermal noise. However, to avoid diffraction losses, the beam radius should not be too large compared with the radius of the mirror. This would suggest that, for a fixed mirror mass, wide and thin mirrors may be preferable. Unfortunately, coating thermal noise starts rapidly increasing when the mirror thickness becomes comparable to the beam size (Somiya and Yamamoto, 2009), as discussed in Section 4.1.5.

The coating thermal noise of a finite-size mirror calculated from the Fluctuation–Dissipation Theorem (see Chapter 1) shows that coating expansion along the beam path is not a major contributor to the total thermal fluctuations. The dominant part is due to the shear motion that deforms the boundary of the substrate and coating. If the mirror is too thin compared with the beam size, this deformation becomes dominant, leading to an increase in coating thermal noise.

Figure 6.1 shows the results of an analysis of a prototype gravitational wave interferometer at the Albert Einstein Institute Hannover in Germany. The mirror mass was constrained to be 100 g and the beam radius was set to be $1/2.5$ of the mirror radius. For a given mirror thickness the mirror and beam radii were derived, and the corresponding thermal noise calculated. The noise level is minimized for a Gaussian beam when the aspect ratio ($\equiv 2a/h$) is about 2, while for a Mesa beam (see Section 13.3.2) when the aspect ratio is 2.4 (Agresti, 2008).

6.4 Khalili cavity and Khalili etalon

A reflective, dielectric mirror coating is a series of very short anti-resonant cavities. The thermal noise seen by a laser beam is the spatially averaged phase of the beam reflected off this stack of cavities. Normally the motion of all the coating interface layers is primarily coherent, leading to a reflected phase jitter equal to twice the coating motion times the wave vector k. An elegant way to circumvent this coherent motion of all coating interface layers was proposed by Khalili (2005). He suggested mechanically separating the first few layers, which reflect most of the light power, from the remaining ones. This can be achieved by replacing a mirror with an anti-resonant cavity consisting of an initial mirror (M1) with a few layers of coatings and a second mirror (M2) with as many layers as a conventional mirror. This allows for the same reflectivity as a conventional mirror, but the coating interfaces of mirror M1 now move independently of those on M2. Furthermore, the interface of M1 is driven only by the thermal motion of a few lossy coating layers. The interface of M2 is still driven by many layers but, the motion of M2 is not sensed as much as M1 by the laser beam due to the anti-resonant cavity. The contribution of the M2 motion to the reflected phase, φ_{M2}, is roughly given by

$$\varphi_{M2} \simeq \frac{1 - r_1^2}{4} \times 2kx_2 \tag{6.3}$$

where r_1 is the amplitude reflectivity of M1, k is the wave number, and x_2 is the motion of M2. This scheme will introduce some thermorefractive noise (thermal fluctuations in the index of refraction, see Chapter 9) due to the beam traveling through the M1 substrate. Decreasing the number of coating layers on M1 reduces the coating thermal noise of M1 but increases the thermorefractive noise. The optimal number in terms of the total noise level will be around 2 doublets, while practical constraints due to the increase of the laser power in the substrate may give a different number. A Khalili cavity is planned to be tested at the prototype experiment at the Albert Einstein Institute Hannover in the near future (Goßler et al., 2010).

The obvious disadvantage of Khalili cavities is that the number of required mirrors is doubled. One can avoid this by using the mirror itself as an etalon. The front surface of such a Khalili etalon is coated with a few layers, and the remaining layers of the coating are applied to the back surface. The thickness of the etalon then needs to be thermally controlled to maintain the anti-resonance condition. Such a setup, however, does not provide a perfect mechanical separation for the coatings. The light reflected by the front surface probes some fraction of the thermal fluctuations caused by the mechanical loss of the back-surface coatings, and vice versa. In order to calculate the coupling coefficient, the Fluctuation–Dissipation Theorem is used with a finite-size mirror, adding gedanken forces on both ends of the mirror, and summing up the dissipated power over the coating layers (see Chapter 1 for details on this technique). Figure 6.2 (right panel) shows such an optimization result with a 100 g silica mirror. The optimization leads to a reduced mirror radius for an increased mirror thickness. The beam radius is set to $1/2.5$ of the mirror radius. This reduced beam radius results in a noise increase, see Section 4.1.1. Owing to the strong coupling of the back-surface thermal energy to the motion of the front surface, the optimal thickness of a Khalili etalon is larger than that of Khalili-cavity mirrors. The minimum noise level that can be reached by the Khalili etalon is slightly worse than that by the Khalili cavity.

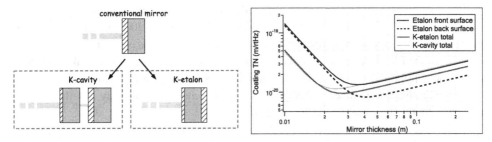

Figure 6.2 Schematic view (*left panel*) and noise spectrum (*right panel*) of a Khalili cavity and a Khalili etalon. Coating thermal noise of the Khalili etalon consists of fluctuations originating from the dissipation in the front-surface coatings and from the dissipation in the back-surface coatings.

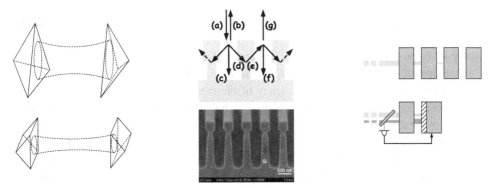

Figure 6.3 *Left panel*: Proposed 3-facet (top) and 2-facet (bottom) corner reflectors (Braginsky and Vyatchanin, 2004). *Center panel*: Schematic (top) and the photo (bottom) of monolithic waveguide gratings (Brückner *et al.*, 2010). *Right panel*: Series of Khalili cavities with bald substrate bulks (top) and rigidly controlled Khalili cavity (Somiya, 2009b) (bottom).

6.5 Coatingfree mirrors

The largest contribution to mirror thermal noise is coating thermal noise, which is due to the high mechanical loss material in the coatings. There are several possible ways to avoid a coating altogether.

6.5.1 Corner reflectors

Braginsky and Vyatchanin (2004) introduced a reflective device based on total internal reflection. Snell's law tells us that the light with an incident angle larger than the critical angle ($\arcsin(1/n_s)$) on the boundary of the substrate (e.g. silica, index of refraction $n_s = 1.45$) to the air will be totally reflected. The left panel of Figure 6.3 shows two types of corner reflectors; a 3-facet type (top) and a 2-facet type (bottom). The incident angle of the light on the boundary in the 3-facet type is $\arcsin(\sqrt{2/3})$ and that in the 2-facet is $\arcsin(\sqrt{1/2})$. In either case, the incident angle is larger than the critical angle, assuming a silica substrate, allowing for total internal reflection. One issue for this reflector is scatter losses at the edges of the corner cube. Braginsky and Vyatchanin (2004) roughly estimated the optical losses to be of the order 10 ppm, but polishing near edges is generally worse than away from them and imperfections will likely make it even worse. Additional losses are due to the substrate absorption and the required anti-reflective coating. This anti-reflective coating will also reintroduce some high-loss material, albeit significantly less than a full reflective coating because it need not be as thick.

The corner reflector also suffers from thermorefractive noise. Temperature fluctuation of the bulk causes phase noise on the light in the reflector through the temperature dependence of the index of refraction $\beta = \partial n_s / \partial T$ (see Chapter 9 for a discussion of thermorefractive noise). The power spectrum of thermorefractive noise, in the case of a simple transmissive bulk with no edge, is given by (Benthem and Levin, 2009)

$$\mathrm{TR}_x(f) = \frac{4k_B T^2 \beta^2 \kappa \ell}{\pi^3 C^2 \rho^2 w_m^4 f^2} \left[1 + \frac{k^2 w_m^2}{1 + 4k^4 \kappa^2 / (\rho^2 C^2 \pi^2 f^2)} \right], \qquad (6.4)$$

where k_B is the Boltzmann constant, T is the temperature, κ is the thermal conductivity, ℓ is the optical path inside the bulk, C is the heat capacity per mass, ρ is the density, w_m is the beam radius, f is the frequency of the measurement, and $k = 2\pi n_s / \lambda$ is the wave vector (λ is the wavelength of the light). The optical path will be slightly different because of the difference in shape of the corner reflector compared to the simple bulk. As an example, using values for the Advanced LIGO gravitational wave detector (Harry and The LIGO Scientific Collaboration, 2010) (see Chapter 14) and comparing the noise level, thermorefractive noise is smaller than coating Brownian noise by a factor of ~ 2 at around 100 Hz and the gain decreases at high frequencies.

6.5.2 Waveguide grating mirrors

Another way to realize a coatingfree mirror is to apply a nanostructured surface on a substrate forming a waveguide grating mirror. By adjusting the grating parameters one can realize resonant excitation in the structured surface leading to a near 100% reflectivity under normal incidence. The mechanical loss of the microstructure can be very low (Nawrodt et al., 2007b), and gratings do not need any high index material that tends to dominate the mechanical loss of traditional dielectric coatings (see Chapter 4). Thus waveguide grating mirrors are an attractive candidate for replacing conventionally coated mirrors. The manufacturing challenge is to reach the required high reflectivity and sufficiently low scattering losses. See also Section 11.4 for more on diffractive gratings as replacements for transmissive optics.

The center panel of Figure 6.3 shows a schematic of a monolithic reflector based on a nanostructured surface (top) and a photo taken by a scanning electron microscope (bottom). The reflector consists of the upper grating, which supports the first-order modes, the lower grating, which is designed not to support the first-order modes, and the substrate at the bottom. The effective index of refraction in the middle part of the structure is low due to its "T-structure", and it works as a waveguide without accompanying lossy materials. The input light field (a) under normal incidence produces the zero-order reflecting field (b), the zero-order transmitting field (c), and the first-order transmitting fields (d); the higher order modes are omitted. The field (d) is totally reflected to make (e) on the boundary to the waveguide as the incident angle is larger than the critical angle. The field (e) makes the first-order reflecting field (f) and the first-order transmitting field (g). The parameters are tuned in such a way that the fields (b) and (g) are in the same phase and the fields (c) and (f) are in the opposite phase to realize the perfect reflectivity.

A power reflectivity of 99.8% for 1550 nm laser light has been achieved with monocrystalline silicon (Brückner *et al.*, 2010). Even higher reflectivity is expected with improved lithography and etching technologies.

6.5.3 Rigidly controlled Khalili cavity

A more complex variant of the Khalili cavity discussed in Section 6.4 is possible using uncoated substrates. The interface between vacuum and an uncoated substrate has a non-zero reflectivity of $(n - 1) / (n + 1)$, where n is the substrate index. A silica substrate, for example, has an index of 1.45, so the amplitude reflectivity is $(1.45 - 1)/(1.45 + 1) = 0.184$. The substrate can be thought of as an etalon with this reflectivity on each end. If the thickness of the etalon is controlled to keep it anti-resonant (a Khalili etalon), the reflectivity of the etalon is $2 \times 0.184/(1 + 0.184^2) = 0.356$. If we use four etalons in series, as is shown in the right panel of Figure 6.3 (top), and keep the thickness of each etalon to make an anti-resonant Khalili cavity, the power reflectivity of this 4-mirror system is about 81%. We would need 8 etalons to achieve a reflectivity of 99%.

A better proposal to realize a coating-free system is made in Somiya (2009b). Only the first mirror of the Khalili cavity is an uncoated substrate. The end mirror is a conventional mirror with many coating layers. This is not truly a coating-free system but just an extreme case of a Khalili cavity. We now add one additional beam that is resonant in the Khalili cavity which is used to probe the cavity fluctuations, including the thermorefractive noise of the first mirror and the coating thermal noise of the end mirror. Feeding back the information obtained with the probe beam, the Khalili cavity can be rigidly locked, suppressing this thermal noise (Figure 6.3, right panel, bottom). The thermal noise of the front surface of the first mirror cannot be removed as it cannot be separated from the motion of the mirror. But since that surface is free of a coating there is no coating thermal noise from that surface.

However, this rigid control imposes sensing noise, namely shot noise of the probe beam. The power of the probe beam needs to be high enough so its shot noise is below the thermal noise level of a conventional mirror. The required power depends on the transmittance of the first mirror. In order to realize this coating-free system, the power of the probe beam should be higher than that of the main beam, which may be impractical. Nevertheless, this rigidly controlled Khalili cavity is interesting because it illustrates the possibility of trading thermal noise for quantum noise (shot and radiation pressure noise). In principle, methods developed for improving quantum noise (e.g. squeezed light injection, see Section 11.3) could also be used to address thermal noise.

6.6 Systematic approach for minimizing total thermal noise

The thermal noise literature, including other chapters in this book, has many excellent papers describing one particular type of thermal noise or another (see, in particular, chapter 3 for an overview of many types of thermal noise). But it is hard to get an overview of all relevant phenomena and their relation to thermal noise. This section provides such an overview, and in the process shows a possible path to improving thermal noise. In the process, this Section will revisit many noise sources, but with a focus on their relation rather than their detailed calculation.

6.6.1 Review of the Fluctuation–Dissipation Theorem

The Fluctuation–Dissipation Theorem was introduced in Chapter 1 for both mechanical and optical (Brownian and thermo-optic) thermal noise. Here we sketch another derivation which provides insight into their relationship.

One dimension

The Fluctuation–Dissipation Theorem can be introduced in a simple one-dimensional system, driven by an initially unknown thermal force, F_{noise}. The equation of motion for a point mass with mass m, spring constant k and (viscous) damping constant γ is

$$m\ddot{x} + \gamma\dot{x} + kx = F_{\text{noise}}. \tag{6.5}$$

The complex impedance Z is defined as the ratio between the driving force and the resulting velocity, i.e. $Z\dot{x} = F_{\text{noise}}$. Switching to the Fourier domain ($\dot{x} \rightarrow i\omega x, \omega = 2\pi f$), Z is given by

$$Z = i\omega m + \gamma + \frac{k}{i\omega}. \tag{6.6}$$

For this impedance it can be shown that

$$\frac{1}{m} = 4 \int_0^\infty \Re\left(Z^{-1}\right) df, \tag{6.7}$$

by simply performing the integration. Next, we know from the Equipartition Theorem that the average kinetic energy is given by

$$\frac{1}{2}m\langle\dot{x}^2\rangle = \frac{k_{\text{B}}T}{2}. \tag{6.8}$$

For this to be correct in the presence of a non-zero damping constant γ, the driving force F_{noise} also has to be non-zero, because the thermal energy would otherwise damp out to zero, which is clearly unphysical. More formally we can use the definition of the velocity power spectrum $S_{\dot{x}\dot{x}}(f)$,

$$\int_0^\infty S_{\dot{x}\dot{x}}(f)df = \langle \dot{x}^2 \rangle \tag{6.9}$$

and find

$$\int_0^\infty S_{\dot{x}\dot{x}}(f)df = 4k_B T \int_0^\infty \Re\left(Z^{-1}\right)df. \tag{6.10}$$

Although we have not shown it here, the integrand equation

$$S_{\dot{x}\dot{x}}(f) = 4k_B T \Re\left(Z^{-1}\right) \tag{6.11}$$

also holds, and is true for more general impedances Z than that defined by Equations 6.6 and 6.7.

Generalization to multiple dimensions

We first define the impedance matrix $\mathbf{Z}$ for more than one degree of freedom as

$$\sum_j \mathbf{Z}_{ij} \dot{x}_j = F_i, \tag{6.12}$$

where F_i is the noise force driving the ith degree of freedom, and $\dot{x}_j$ is its velocity. Equation 6.11 then generalizes to

$$S_{\dot{x}_i \dot{x}_j}(f) = 2k_B T \left(\mathbf{Z}^{-1} + \mathbf{Z}^{\dagger^{-1}}\right)_{i,j}, \tag{6.13}$$

where $\mathbf{Z}^{-1}$ is the matrix inverse of the impedance matrix $\mathbf{Z}$, and $\mathbf{Z}^{\dagger}$ is the Hermitian conjugate of $\mathbf{Z}$. This can be shown rigorously by diagonalizing the equation of motion and applying the one-dimensional Equation 6.11. While the diagonal terms $S_{\dot{x}_i \dot{x}_i}$ are the velocity power spectra of each degree of freedom, the off-diagonal terms $S_{\dot{x}_i \dot{x}_j}$ are the cross-power spectra. Equation 6.13 is impractical for a real physical system but, fortunately, we are usually only interested in one observed degree of freedom which is a linear combination of the internal degrees of freedom. Therefore, we are looking for a weighted quantity

$$e = f_i x_i = \vec{f}^T \vec{x}. \tag{6.14}$$

The power spectrum for the quantity e is then given by projection onto this degree of freedom, i.e.

$$S_{ee} = \frac{S_{\dot{e}\dot{e}}}{\omega^2} = \frac{2k_B T}{\omega^2} \vec{f}^T \left(\mathbf{Z}^{-1} + \mathbf{Z}^{\dagger^{-1}}\right) \vec{f}. \tag{6.15}$$

Dissipation

The matrix element $\vec{f}^T \left(\mathbf{Z}^{-1} + \mathbf{Z}^{\dagger-1} \right) \vec{f}$ is connected to dissipation in an elegant way. In a general multi-degree-of-freedom system the average dissipation can be calculated as

$$P_{\text{diss}} = \frac{1}{T} \int_{-T/2}^{T/2} \vec{F}^{\dagger} \vec{\dot{x}} dt. \tag{6.16}$$

Using Parseval's Theorem to switch to the frequency domain, the definition of $\mathbf{Z}$ (Equation 6.12), and the identity $\mathbf{Z}(-f) = \mathbf{Z}^*(f)$, we find

$$P_{\text{diss}} = \frac{1}{T} \int_{0}^{\infty} \vec{F}^{\dagger} \left(\mathbf{Z}^{-1} + \mathbf{Z}^{\dagger-1} \right) \vec{F} df. \tag{6.17}$$

From this we get the following very useful formulation of the Fluctuation–Dissipation Theorem, which was introduced in Chapter 1.

Fluctuation–Dissipation Theorem for mechanical fluctuations

Assume that we drive the system with a force $\vec{F} = F_0 \vec{f} \cos \omega t$, where $\vec{f}$ is dictated by the degree of freedom for which we want to calculate the thermal noise (Equation 6.14). We then calculate the dissipation $P_{\text{diss}}(\omega)$ for this *driven* system. The thermal noise of e in the *non-driven* system is given by

$$S_{ee}(f) = \frac{8 k_B T}{(2\pi f)^2} \frac{P_{\text{diss}}}{F_0^2}. \tag{6.18}$$

In this form the Fluctuation–Dissipation Theorem was first used by Gonzalez and Saulson (1994) and Levin (1998).

6.6.2 Fluctuation–Dissipation Theorem applications (1)

Brownian noise

The above form of the Fluctuation–Dissipation Theorem can be used to calculate the Brownian thermal noise of a mirror surface (Levin, 1998) (see also Chapter 4). First, we need to know how the surface position fluctuations $\delta x(r_\perp)$ are averaged by the circulating beam. Typically, the mirror is interrogated interferometrically by a laser beam. The reflection translates position fluctuations $\delta x(r_\perp)$ into phase fluctuations $\delta \varphi(r_\perp) = 2k \delta x(r_\perp)$. The reflected field amplitude ψ_{out} is thus

$$\psi_{\text{out}}(r_\perp) = \psi_{\text{in}}(r_\perp) e^{i2kx(r_\perp)}. \tag{6.19}$$

The interferometric beat signal against a reference beam ψ_{ref} is then proportional to

$$\int d^2 r_\perp \psi_{\text{ref}}^*(r_\perp) \psi_{\text{in}}(r_\perp) e^{i2kx(r_\perp)}. \tag{6.20}$$

In most applications, a Gaussian beam is used (although see Chapter 13 for alternatives). A Gaussian beam has an intensity profile $I(r_\perp) = |\psi(r_\perp)|^2$ given by

$$I(r_\perp) = \frac{2P}{\pi w_m^2} e^{-2r_\perp^2/w_m^2}, \qquad (6.21)$$

with P the total laser power and w_m the Gaussian beam spot size. If both input beam $\psi_{\rm in}$ and reference beam $\psi_{\rm ref}$ are Gaussian, then our readout variable $\vec{f}$ from Equation 6.14 is given by this Gaussian intensity profile (Equation 6.21).

The Fluctuation–Dissipation Theorem, as formulated in Section 6.6.1, now tells us to drive the surface of the optic with a periodic Gaussian force profile,

$$F(r_\perp) = \frac{2F_0}{\pi w_m^2} e^{-2r_\perp^2/w_m^2} \cos \omega t, \qquad (6.22)$$

solve the resulting elasticity problem, and calculate the mechanical loss for this motion. If the size of the mirror d is much smaller than the sound wavelength, $d \ll c_s/\omega$ (with c_s the speed of sound in the mirror substrate), the elasticity problem can be solved statically. This was done in Levin (1998) for an uncoated mirror and in Harry *et al.* (2002) for coated mirrors, and is discussed in Chapter 4.

The shape of the reference beam used for the interferometric readout makes a difference. We could, for instance, use a plane wave as the reference beam $\psi_{\rm ref}$. Then the amplitude profile (and not the intensity profile) of the interrogation beam $\psi_{\rm in}$ defines our readout variable $\vec{f}$ from Equation 6.14 (for simplicity we stay in the near field). This leads to an effective averaging area increase of a factor of two, or a thermal noise reduction of $\sqrt{2}$. The problem with this approach is the bad overlap integral between interrogation and reference beam, which leads to poor shot noise performance. See Chapter 13 for a detailed discussion of beam shapes and coating thermal noise.

Thermoelastic noise

The formulation of Section 6.6.1 for the Fluctuation–Dissipation Theorem has also been applied to calculate thermoelastic noise (Fejer *et al.*, 2004), using the following steps.

- Apply a Gaussian force profile (Equation 6.22) and calculate the elastic response.
- Calculate the local heat (or entropy) increase due to compression.
- Calculate the additional loss due to this heat (or entropy) field.

We will come back to this result in Section 6.6.4. See also Section 1.3 and Chapter 9 for more on thermoelastic noise.

6.6.3 Temperature fluctuations and the Fluctuation–Dissipation Theorem

So far we described our system with positions $\vec{x}$ and forces $\vec{F}$, or equivalently by a strain field ϵ and its thermodynamic conjugate, the stress field $\vec{\sigma}$. However, we know that we need a second pair of conjugate variables for a full thermodynamic description of a system: temperature δT and entropy $\delta S = \delta Q / T_0$.

The Fluctuation–Dissipation Theorem can also be extended to this case, as was shown in Levin (2008). In analogy to Equation 6.12, we can define a thermal impedance Z given by the heat equation

$$Z\delta\dot{T} = dV \left(\frac{\rho C}{i\omega T_0} + \frac{\kappa \, \Delta}{\omega^2 T_0} \right) \delta\dot{T} = \delta S = \frac{\delta Q}{T_0}, \tag{6.23}$$

where dV is the volume element under consideration, ρ is the density, C is the heat capacity per mass, κ is the thermal conductivity, T_0 is the average temperature, δQ is the injected heat, and Δ is the Laplace operator. In analogy to the discussion in Section 6.6.1, we can define a readout variable which is some linear combination of values of the temperature field $\delta\vec{T}$. For simplicity, and to keep the analogy, we use a discrete index notation,

$$e = \vec{f}^T \delta\vec{T}. \tag{6.24}$$

Now the Fluctuation–Dissipation Theorem for fluctuations in e can be stated.

Fluctuation–Dissipation Theorem for temperature fluctuations

Assume that we drive the system with an entropy (heat) field $\delta\vec{S} = \delta\vec{Q}/T_0 = S_0 \vec{f} \cos \omega t$, where $\vec{f}$ is dictated by the degree of freedom for which we want to calculate the thermal noise (Equation 6.24). Then calculate the dissipation $P_{\text{diss}}(\omega)$ for this *driven* system. The thermal noise of e in the *non-driven* system is given by

$$S_{ee}(f) = \frac{8k_{\text{B}}T}{(2\pi f)^2} \frac{P_{\text{diss}}}{S_0^2} \tag{6.25}$$

This form of the Fluctuation–Dissipation Theorem was derived by Levin (2008).

6.6.4 Fluctuation–Dissipation Theorem applications (2)

Coating thermo-optic noise

As an example application, Levin (2008) rederived the surface temperature fluctuations as seen by a Gaussian laser beam. If the readout variable (Equation 6.24) is given by

$$\delta\hat{T} = \int d^2 r_\perp \, dz \; \delta T(r_\perp, z) \, \frac{2}{\pi w_m^2} e^{-2r_\perp^2/w_m^2} \delta(z), \tag{6.26}$$

the thermal noise can be calculated by driving the surface with a Gaussian heat (entropy) profile

$$\frac{dQ(r_\perp)}{\delta A} = T_0 S_0 \frac{2}{\pi w_m^2} e^{-2r_\perp^2/w_m^2} \cos \omega t. \tag{6.27}$$

If the beam spot w_m is bigger than the diffusion length $r_T = \sqrt{\frac{\kappa}{\rho C \omega}}$, and the reflected phase only depends on temperature fluctuations in a thin layer $d \ll r_T$, then the heat dissipation can be calculated by solving the one-dimensional heat diffusion equation. The resulting thermal noise for $\delta\hat{T}$ is

$$S_{\delta\hat{T}\delta\hat{T}} = \frac{2\sqrt{2}}{\pi w_m^2} \frac{k_B T^2}{\sqrt{\kappa\rho C\omega}} \tag{6.28}$$

which agrees with Braginsky *et al.* (2000).

Evans *et al.* (2008) then used this formalism to describe thermo-optic noise (i.e. both thermoelastic and thermorefractive noise, see Chapter 9) coherently. Here we just mention how the reflected phase fluctuations, $\delta\varphi(r_\perp)$, depend on the temperature field. Since the total coating thickness is typically small compared to the beam spot, $d \ll w_m$, this dependence can be written as

$$\delta\varphi(r_\perp) = \frac{\partial\varphi}{\partial T_i}\delta T_i(r_\perp), \tag{6.29}$$

where $\delta T_i(r_\perp)$ is the temperature in the *i*th coating layer at the transverse position $r_\perp$. Since a coating is essentially a series of thin, anti-resonant cavities, the partial derivative $\frac{\partial\varphi}{\partial T_i}$ contains two parts: (1) the change in the reflective phase due to a change in the round-trip phase $\delta\varphi_i$ in the *i*th coating layer, and (2) the reflective phase change due to the overall expansion of the coating. These two terms appear with a negative sign, resulting in destructive noise interference. The equivalent surface position fluctuation $\delta x(r_\perp)$ due to thermal fluctuations in the coating can thus be written as

$$\delta x(r_\perp) = \sum_{i=1}^{N}\left(-\frac{\partial\varphi}{\partial\varphi_i}\frac{\partial(n_i d_i)}{\partial T_i} + \frac{\partial(d_i)}{\partial T_i}\right)\delta T_i(r_\perp), \tag{6.30}$$

where φ_i is the round trip phase of the cavity formed by coating layer i, and n_i and d_i are the index of refraction and thickness of that layer. If we again assume that we are working with thin coatings compared to the thermal diffusion length ($d \ll r_T$), the temperature fluctuations across all layers are identical, and we immediately get the instructional form

$$S_{\delta x\delta x} = S_{\delta\hat{T}\delta\hat{T}}\left[\sum_{i=1}^{N}\left(-\frac{\partial\varphi}{\partial\varphi_i}\frac{\partial(n_i d_i)}{\partial T} + \frac{\partial(d_i)}{\partial T}\right)\right]^2, \tag{6.31}$$

with $S_{\delta\hat{T}\delta\hat{T}}$ given by Equation 6.28. The term $\frac{\partial\varphi}{\partial\varphi_i}\frac{\partial(n_i d_i)}{\partial T}$ describes the thermorefractive noise coupling, while the term $\frac{\partial(d_i)}{\partial T}$ is responsible for the thermoelastic noise. See Chapter 9 for more details.

Thermoelastic noise, again?

Thermoelastic noise has here been derived twice. Once by using the Fluctuation–Dissipation Theorem for mechanical fluctuations, in Sections 6.6.1 and 6.6.2, and once through the Fluctuation–Dissipation Theorem for thermal fluctuations in Section 6.6.3. The results are indeed identical. This is curious, since the two formulations are typically used to calculate the uncorrelated Brownian noise and thermo-optic noise, respectively. To calculate

Table 6.1 *Possible choices of thermodynamic variables for thermal noise calculations. Sets A and B differ in their definition of the strain. In set A, strain is defined to include thermal expansions. Thus, Hooke's law gets an explicit temperature dependence, but the expansion is only strain dependent. In set B, the situation is reversed. Since the thermoelastic noise calculation by Evans* et al. *(2008) explicitly assumes a strain dependent expansion, they implicitly work with variable set B. The derivation by Fejer* et al. *(2004), on the other hand, uses set A.*

	Set A	Set B
Thermodynamic variables	$\delta T ; \vec{\epsilon}$	$\delta T ; \vec{\epsilon}_\sigma = (\vec{\epsilon} - \vec{\alpha}\delta T)$
Hooke's law	$\vec{\sigma} = \mathbf{H} \cdot (\vec{\epsilon} - \vec{\alpha}\delta T)$	$\vec{\sigma} = \mathbf{H} \cdot \vec{\epsilon}_\sigma$
Lin. expansion (e.g. coating thickness)	$\delta d = \epsilon_1 d$	$\delta d = (\vec{\epsilon}_\sigma + \vec{\alpha}\delta T)_1 d$
Implicitly used by	Fejer *et al.* (2004)	Evans *et al.* (2008)

the total noise we should calculate both mechanical and thermal fluctuations, which suggests that we have to be careful not to double-count thermoelastic noise. To understand this apparent paradox we need to consistently choose a full set of thermodynamic variables (see Section 6.6.7 and Table 6.1).

6.6.5 Generalization of the Fluctuation–Dissipation Theorem

Based on the discussion in Sections 6.6.1 and 6.6.3, formulating a version of the Fluctuation–Dissipation Theorem that includes both mechanical and thermal fluctuations is straightforward. First, we note that our internal degrees of freedom are now given by the combined set of positions and temperatures $\left[\vec{x}, \vec{T}\right]$. Next we specify our readout variable e, which can depend on all internal degrees of freedom:

$$e = \vec{f}_x^{\,T}\vec{x} + \vec{f}_T^{\,T}\delta\vec{T}. \tag{6.32}$$

The generalized Fluctuation–Dissipation Theorem for fluctuations in e can now be stated.

Generalized Fluctuation–Dissipation Theorem

Assume that we drive the system with a force and an entropy (heat) field

$$\begin{bmatrix} \vec{F} \\ \delta\vec{S} \end{bmatrix} = G_0 \begin{bmatrix} \vec{f}_x \\ \vec{f}_T \end{bmatrix} \cos\omega t. \tag{6.33}$$

Note that this drive is *coherent*, i.e. force and heat are driven in phase. The quantities $\vec{f}_x$ and $\vec{f}_T$ are given by the choice of the readout variable e (Equation 6.32). We then calculate all forms of dissipation, $P_{\text{diss}}(f)$, for this *driven* system. The thermal noise in e of the *non-driven* system is given by

$$S_{ee}(f) = \frac{8k_B T}{(2\pi f)^2} \frac{P_{\text{diss}}}{G_0^2}. \tag{6.34}$$

If we work in the framework of continuum mechanics, then three of the four thermodynamic variables will change: force $F \to$ stress $\vec{\sigma}$; position $x \to$ strain $\vec{\epsilon}$; entropy $\delta S \to$ entropy per volume δs. The generalized Fluctuation–Dissipation Theorem will, however, still be valid.

6.6.6 Hooke's law

Notation for Hooke's law

Hooke's law states that there is a linear relation between the strain field $\vec{\epsilon}$ and stress field $\vec{\sigma}$, both of which have 6 independent components. Here we choose to write both $\vec{\epsilon}$ and $\vec{\sigma}$ as a 1 by 6 vector, i.e. $\vec{\epsilon} = [\epsilon_{11}, \epsilon_{22}, \epsilon_{33}, \epsilon_{12}, \epsilon_{23}, \epsilon_{31}]$ and $\vec{\sigma} = [\sigma_{11}, \sigma_{22}, \sigma_{33}, \sigma_{12}, \sigma_{23}, \sigma_{31}]$. Therefore, the Hooke tensor $\mathbf{H}$ becomes a 6 by 6 matrix. In a homogeneous and isotropic material only two of the 36 elements are independent. Unfortunately, there are six different elastic moduli (proportionality constants) that are commonly used in the literature – bulk modulus (K), Young's modulus (Y), Lamé's first parameter (λ), shear modulus (G), Poisson's ratio (σ, not to be confused with stress) and P-wave modulus (M). Any two of them can be used to fully describe the Hooke tensor, but none of the combinations leads to particularly simple or readable equations. Thus, for the purpose of this section, we will simply write Hooke's law in matrix notation,

$$\vec{\sigma} = \mathbf{H} \cdot \vec{\epsilon}. \tag{6.35}$$

Thermal expansion and Hooke's law

A homogeneous and isotropic system with a linear expansion coefficient α that undergoes a small temperature change without the influence of external stresses will expand according to

$$\vec{\epsilon}_{\text{thermal}} = \vec{\alpha}\delta T, \tag{6.36}$$

where we define $\vec{\alpha} = (\alpha, \alpha, \alpha, 0, 0, 0)^T$. If we define the strain $\vec{\epsilon}$ as the total strain due to both thermal expansion and elastic deformation, Hooke's law can be generalized to

$$\vec{\sigma} = \mathbf{H} \cdot (\vec{\epsilon} - \vec{\alpha}\delta T). \tag{6.37}$$

6.6.7 Choice of thermodynamic variables

We can pick the strain $\vec{\epsilon}$ as defined above for Equation 6.37 as well as the temperature fluctuations δT as our set of thermodynamic variables. Since this definition of $\vec{\epsilon}$ already includes thermal expansion, there will be no explicit δT dependence in the expression for expansion (see Table 6.1).

An alternative choice is to use only the stress-induced strain $\vec{\epsilon}_\sigma$

$$\vec{\epsilon}_\sigma = \vec{\epsilon} - \vec{\alpha}\delta T \tag{6.38}$$

and the temperature fluctuations δT as the fundamental set of variables. This leads to a temperature independent Hooke's law,

$$\vec{\sigma} = \mathbf{H} \cdot \vec{\epsilon}_\sigma, \tag{6.39}$$

and an explicit δT dependence in the expression for expansion.

Table 6.1 shows the two self-consistent ways to define stress. Fejer *et al.* (2004) and Evans *et al.* (2008) implicitly picked different sets of thermodynamic variables, which explains why they got the same result for thermo-elastic noise using seemingly different approaches.

6.6.8 Example

As an example on how to apply the generalized Fluctuation–Dissipation Theorem, we can again look at the noise contribution due to a coating. The total effective position noise due to the coating is given by

$$\delta x(r_\perp) = \sum_{i=1}^{N} \left(-\frac{\partial \varphi}{\partial \varphi_i} \delta(n_i d_i) + \delta(d_i) \right). \tag{6.40}$$

Thus, we need to express the variations in the index of refraction δn as a function of our choice of thermodynamic variables. For this we introduce the thermo-optic and stress-optic constants as they are typically measured,

$$\beta = \frac{\partial n}{\partial T}|_{\vec{\sigma}}, \tag{6.41}$$

$$\vec{C}^T = -\frac{2}{n^3} \frac{\partial n}{\partial \vec{\sigma}}|_T. \tag{6.42}$$

If the material is birefringent, the index of refraction in Equation 6.42 is the one seen by the polarization and propagation direction of the laser beam. Therefore the index of refraction effectively remains a scalar. We will also need the identities

$$\frac{\partial \vec{\sigma}}{\partial \vec{\epsilon}}|_T = \mathbf{H} \tag{6.43}$$

$$\frac{\partial \vec{\sigma}}{\partial T}|_{\vec{\epsilon}} = -\mathbf{H}\vec{\alpha}, \tag{6.44}$$

which derive directly from Hooke's law (Equation 6.37). If we now choose the variable set A of Table 6.1, we get

$$\delta n = \frac{\partial n}{\partial T}|_{\vec{\epsilon}}\ \delta T + \frac{\partial n}{\partial \vec{\epsilon}}|_T\ \delta \vec{\epsilon} \tag{6.45}$$

$$= \left(\frac{\partial n}{\partial T}|_{\vec{\sigma}} + \frac{\partial n}{\partial \vec{\sigma}}|_T \frac{\partial \vec{\sigma}}{\partial T}|_{\vec{\epsilon}}\right)\ \delta T + \left(\frac{\partial n}{\partial \vec{\sigma}}|_T \frac{\partial \vec{\sigma}}{\partial \vec{\epsilon}}|_T\right)\ \delta \vec{\epsilon} \tag{6.46}$$

$$= n \left(\frac{\beta}{n} + \frac{n^2}{2}\vec{C}^T \mathbf{H}\vec{\alpha}\right)\ \delta T - n \left(\frac{n^2}{2}\vec{C}^T \mathbf{H}\right)\ \delta \vec{\epsilon}. \tag{6.47}$$

Similarly, for the variable set B we find

$$\delta n = \frac{\partial n}{\partial T}|_{\vec{\epsilon}_\sigma}\ \delta T + \frac{\partial n}{\partial \vec{\epsilon}_\sigma}|_T\ \delta \vec{\epsilon}_\sigma \tag{6.48}$$

$$= \beta\ \delta T - n \left(\frac{n^2}{2}\vec{C}^T \mathbf{H}\right)\ \delta \vec{\epsilon}_\sigma. \tag{6.49}$$

Next, we express the variations in optical and physical thickness, δd and $\delta(nd)$, required for Equation 6.40 as a function of our choice of thermodynamic variables. For variable set A we get

$$\frac{\delta d}{d} = \delta \epsilon_1 \tag{6.50}$$

$$\frac{\delta(nd)}{nd} = \left(\vec{\delta}_1^T - \frac{n^2}{2}\vec{C}^T \mathbf{H}\right)\delta \vec{\epsilon} + \left(\frac{\beta}{n} + \frac{n^2}{2}\vec{C}^T \mathbf{H}\vec{\alpha}\right)\delta T, \tag{6.51}$$

with $\vec{\delta}_1^T = (1, 0, 0, 0, 0, 0)$. Similarly, for variable set B we get

$$\frac{\delta d}{d} = \delta \epsilon_{\sigma 1} + \alpha \delta T \tag{6.52}$$

$$\frac{\delta(nd)}{nd} = \left(\vec{\delta}_1^T - \frac{n^2}{2}\vec{C}^T \mathbf{H}\right)\delta \vec{\epsilon}_\sigma + \left(\frac{\beta}{n} + \alpha\right)\delta T. \tag{6.53}$$

Equations 6.40, 6.50 and 6.51 (or alternatively Equations 6.40, 6.52 and 6.53) describe the stress-optic effect in mirror coatings.

6.6.9 The stress-optic effect

The stress-optic effect introduced in Equation 6.42 is traditionally neglected for thermal noise calculations. To understand the circumstances under which it can become important, we first look at the size of the stress-optic tensor. For silica it was measured to be (Stone, 1988)

$$C_1 = 4.22 \times 10^{-12} \text{m}^2/\text{N} \tag{6.54}$$

$$C_2 = C_3 = 0.65 \times 10^{-12} \text{m}^2/\text{N}, \tag{6.55}$$

where we chose the laser propagation direction to be 1. Using the Young's modulus and Poisson's ratio for silica from Table 9.1, we find for the relevant stress-optic terms

$$\frac{n^2}{2}\vec{C}^T\mathbf{H} = (0.36, 0.13, 0.13, 0, 0, 0) \tag{6.56}$$

$$\frac{n^2}{2}\vec{C}^T\mathbf{H}\vec{\alpha} = 0.63\,\alpha. \tag{6.57}$$

This illustrates that, for silica, the stress-optic effect leads to about a 30% reduction in the longitudinal strain sensitivity of the coating layer optical thickness, while introducing some transversal strain sensitivity.

The stress-optic effect would have to be considered in situations where optical thickness variations can contribute significantly to the total thermal noise. An example might be Khalili cavities and etalons, see Section 6.4. In addition, the size of the stress-optic effect in dielectric coatings can be changed by moving away from a solely anti-resonant cavity coating design.

However, the influence on the total thermal noise for regular dielectric mirrors from this stress-optic effect is relatively small. First, for the Brownian noise, the appropriate driving stress field will see about a 30% change inside the coating. However, this will hardly affect the resulting deformation pattern for the whole optic. This includes the resulting deformation inside the coating, which is mostly responsible for the mechanical losses. Second, the thermo-optic noise is not changed at all. This can be seen in Equation 6.53 from the absence of the stress-optic coefficient in the temperature fluctuation term. (Recall that variable set B is best suited for calculating the thermo-optic noise alone, since neglecting driving stresses is equivalent to setting $\delta\vec{\epsilon}_\sigma$ to zero.)

6.6.10 Recipe for a systematic approach to calculating thermal noise

The example in Section 6.6.8 illustrates, for the effective one-dimensional case of a thin coating, how the total thermal noise of a specific mirror geometry can be calculated systematically. The first step is to express the phase of the probe laser beam as a function of the strain and temperature fields across the optic. This provides the readout functions $\vec{f}_x$ and $\vec{f}_T$. Next, the general form of the Fluctuation–Dissipation Theorem in Section 6.6.5 requires a calculation of the total dissipation in the system when driven by a force and heat given by Equation 6.33. In general, this involves solving a three-dimensional elastic and thermal diffusion problem. This task, however, can be done numerically using a finite-element software package.

The numerical approach has the advantage that the geometry of both the mirror substrate and coating can be varied to minimize the total thermal noise. In the coating Brownian noise dominated case this will imply finding a geometry that minimizes the strain in the coating layer. Doing this accurately however requires properly modeling details such as the strain-optic effect in the coating, the size of which can also be tuned through coating variations.

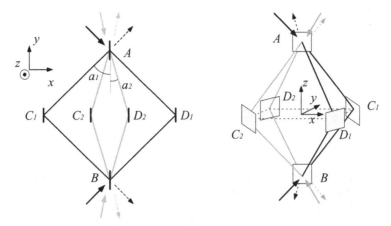

Figure 6.4 Configuration of the two-dimensional (2-D) (*left*) and the three-dimensional (3-D) (*right*) DFI. A proper combination of the output fields of the four beams contains no thermal noise but does contain gravitational-wave signals.

6.7 Displacement-noise-free interferometers

A displacement-noise-free interferometer (DFI) proposed by Kawamura and Chen (Kawamura and Chen, 2004; Chen and Kawamura, 2006; Chen *et al.*, 2006b) is an elegant solution to reduce thermal noise in a gravitational-wave detector. Gravitational waves can be approximated as a force on the test mass if the wavelength is much shorter than the baseline length of the interferometer. In fact, the gravitational waves actually affect the light not only at the test mass but on the entire way between the test masses, thus the gravitational-wave signal and displacement noise on the test mass are different and can be separated. Although this is a special case which can reduce the effects of thermal noise in gravitational-wave detectors, and not other precision measurements, let us here introduce the elegance of the noise separation. See Chapter 14 for more on gravitational waves and gravitational-wave detectors.

Figure 6.4 shows two-dimensional (2-D) and three-dimensional (3-D) configurations of a DFI that achieves the cancelation of displacement noise from all the optical components while still being sensitive to gravitational waves, albeit AC coupled. In both configurations, there are four beams entering from the beamsplitters, i.e. a total of eight beams inside the interferometer. Combining the beams in a proper way, displacement noise is canceled. For example, subtracting the signals in the beam $A \rightarrow C_1 \rightarrow B$ from the signals in the beam $B \rightarrow C_1 \rightarrow A$, we cancel out the motion of the optical component C_1. When a gravitational wave comes from the z-direction, the phase shift imposed on the beam $A \rightarrow C_1$ at time $[0, \tau]$ and that on $B \rightarrow C_1$ at time $[0, \tau]$ are the same but with opposite signs. Thus, they add up at the subtraction of the signals ($\tau \equiv L_{AC_1}/c = L_{BC_1}/c$). The phase shift on the beam $C_1 \rightarrow B$ at time $[\tau, 2\tau]$ and that on $C_1 \rightarrow A$ at time $[\tau, 2\tau]$ are again the same but with opposite signs. However, the phase shift on the beam $A \rightarrow C_1$ at time $[0, \tau]$ and that

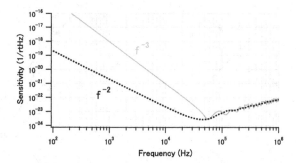

Figure 6.5 Shot-noise-limited sensitivity of 2-D and 3-D DFIs. The 2-D version has a f^{-3} low frequency dependence, while the 3-D DFI is somewhat better with a f^{-2} low frequency dependence. The (longer) baseline is 3 km and the effective input power is 100 MW. The sensitivity decreases with decreasing frequency due to the reduction of the response to gravitational waves at low frequencies.

on the beam $C_1 \to A$ at time $[\tau, 2\tau]$ can be different as the phase of the gravitational waves has changed in the time τ. Therefore, the differential signal of the beam $A \to C_1 \to B$ and the beam $B \to C_1 \to A$ contains a phase shift caused by the gravitational wave, but is free from displacement noise of the component C_1. Doing the same for all the optical components, we can realize a DFI.

Comparing the 2-D and 3-D configurations, one can see that both of them are free from displacement noise but the responses to gravitational waves are different. With the 3-D configuration, displacement noise can be canceled more efficiently, i.e. the gravitational-wave signal remains stronger after the cancelation of displacement noise. Figure 6.5 shows the calculated sensitivity curves of the 2-D and 3-D DFIs, which is given by the ratio of white shot noise to the frequency response of the gravitational waves. The response of the 2-D DFI decreases by f^{-3}, while the response of the 3-D DFI decreases by f^{-2}. Here the baseline (distance between A and C_1) is 3 km, effective input power is 100 MW, and the angles a_1 and a_2 in the 2-D DFI are $45°$ and $30°$, respectively. Even with this high input power, the sensitivity of the DFI is not better than that of conventional gravitational-wave detectors as the response decreases significantly at low frequencies where displacement-noise-free interferometry would be useful.

The DFI requires the beams on the mirror surfaces to be overlapping for the thermal noise cancelation to occur. In addition, the motion of the beamsplitters (A and B) is measured by the beams approaching from different angles, so that thermal fluctuation inside the substrate cannot be canceled. This includes thermorefractive noise (phase noise caused by temperature driven index of refraction fluctuations, see Chapter 9). Therefore, true displacement-noise-free interferometry is quite challenging, but big reductions in thermal noise would be possible.

7

Substrate thermal noise

SHEILA ROWAN AND IAIN W. MARTIN

7.1 Introduction

Other chapters in this book have discussed the limits to precision measurement capabilities set by the thermal noise of the optical coatings which are used to form the highly reflective mirrors commonly used in the relevant experimental arrangements. The substrates of such mirrors can also be a very important source of thermal noise and other limitations in practical situations, and thus merit further discussion here.

As discussed in Chapters 1, 4, and 9 the direct application of the Fluctuation–Dissipation Theorem to the optically sensed position of the front face of a substrate allows the power spectral density of the thermal noise, $S_x(f)$, to be calculated. Using the technique described in Levin (1998) and Chapter 1 of applying a notional pressure of the same spatial profile as the intensity of the sensing laser beam to the front face of the substrate, we find that $S_x(f)$ can then be described by the relation

$$S_x(f) = \frac{2k_B T}{\pi^2 f^2} \frac{W_{\text{diss}}}{F_0^2},$$ (7.1)

where F_0 is the peak amplitude of the notional oscillatory force and W_{diss} is the power dissipated in the mirror. This is the same as Equation 1.2, reproduced here for convenience.

The thermal displacement noise sensed at the surface of optical substrates arises from dissipation from a combination of sources, which are most commonly classified into three types. Brownian noise, as described in Chapter 4, is the descriptor given to the fluctuations resulting from localized sources of mechanical dissipation associated with, for example, defects distributed inside the mirror substrate. Thermoelastic noise is associated with statistical fluctuations in temperature which cause heat flow (and thus energy loss) in the substrate. These temperature fluctuations then couple to the motion of the surface of a substrate via the thermal expansion coefficients of the material, as described by Braginsky *et al.* (1999) (see Chapter 9). Other categories of thermal noise are discussed in Chapter 3. The formalism of Levin can be used in the first two cases above to develop quantitative

Optical Coatings and Thermal Noise in Precision Measurement, eds. Gregory M. Harry, Timothy Bodiya and Riccardo DeSalvo.
Published by Cambridge University Press. © Cambridge University Press 2012.

expressions for the resulting power spectral density of thermal noise. Initially this is most easily considered for the case where the surface motions of a substrate are sensed by a beam whose radius is very much less than the diameter of the substrate, i.e. the substrate is assumed to occupy a "half-infinite" volume of space.

7.1.1 Thermal noise in "half-infinite" substrates

Brownian thermal noise

Let us first consider the case of Brownian thermal noise. Using the formalism above, Bondu *et al.* (1998) developed an expression for the power spectral density of Brownian thermal noise of a half-infinite test mass (ITM), $S_x^{\text{ITM}}(f)$, which is given by

$$S_x^{\text{ITM}}(f) = \frac{2k_{\text{B}}T}{\sqrt{\pi^3}f} \frac{1-\sigma^2}{Yw_m} \phi_{\text{substrate}}(f, T), \tag{7.2}$$

where $\phi_{\text{substrate}}(f)$ is the mechanical loss of the substrate material (here assumed to be spatially homogeneous), Y and σ are the Young's modulus and Poisson's ratio of the material, respectively, and w_m is the radius of the laser beam on the mirror where the electric field amplitude has fallen to $1/e$ of the maximum value.

From this expression it is clear that the level of thermal noise is a function of the temperature at which the substrate operates, the mechanical loss factor and the Young's modulus of the substrate material, along with the radius of the sensing laser beam. Thus in evaluating the expected thermal noise level of a substrate and in seeking to minimize it, the values of these parameters are of particular interest.

Thermoelastic thermal noise

Historically (see Nowick and Berry, 1972, for example), thermoelastic damping was identified with heat flow across thin flexing beams or fibers. However, Braginsky *et al.* (1999) recognized that the thermomechanical properties of many crystals are such that when used as mirror substrate materials, statistical fluctuations in temperature cause displacements of the front face of the mirror through the substrate coefficient of linear thermal expansion which can, in some cases, dominate over the thermal noise from Brownian dissipation. It can be shown that, for a test mass approximated as half-infinite compared with the beam size, the power spectral density of thermoelastic thermal noise, $S_{\text{TE}}(f)$, can be written as

$$S_{\text{TE}}^{\text{ITM}}(f) = \frac{4(1+\sigma)^2\kappa\alpha^2 k_{\text{B}}T^2}{\sqrt{\pi^5}C^2\rho^2 w_m^3 f^2}, \tag{7.3}$$

if

$$\frac{w_m^2\pi f\rho C}{\kappa} \gg 1, \tag{7.4}$$

where C is the specific heat capacity at constant mass, α the coefficient of linear expansion, κ the thermal conductivity and ρ the density of the substrate.

It can be seen from inspection of Equation 7.3 that the level of thermoelastic dissipation depends strongly on both temperature and a combination of the substrate thermomechanical properties; in particular the coefficient of linear thermal expansion. This expression for thermoelastic noise is valid only if Equation 7.4 is satisfied, i.e. in the adiabatic limit where the thermal diffusion length at measurement frequencies is very much less than the beam radius. It was pointed out by Cerdonio *et al.* (2001), following Rowan (2000), that this condition may not be satisfied in experimental situations that utilize very small beam spots, or instances where the substrate is cooled to very low temperatures, and an expression appropriate for use in these cases was developed. This is discussed in detail in Section 8.2.4.

An additional source of thermal noise can occur due to the temperature dependence of the refractive index of a material (see Braginsky *et al.*, 2000; Braginsky and Vyatchanin, 2003a). This is known as thermorefractive noise. The sum of the thermorefractive noise and the thermoelastic noise is referred to as thermo-optic noise, and is discussed in detail in Chapter 9 for coatings. A recent analysis in which the thermoelastic and thermorefractive effects were treated coherently has shown that these mechanisms can occur with a relative negative sign, and thus the total thermo-optic noise can be substantially lower than had been estimated previously (Evans *et al.*, 2008).

7.1.2 Thermal noise in finite-sized substrates

Equations 7.2 and 7.3 show that the power spectral densities of the thermal noises scale as $1/w_m$ and $1/w_m^3$, respectively, suggesting that it is desirable to make the beam-spot radius of the sensing beam large. However, when w_m stops being small relative to the dimensions of the mirror substrate, the analyses leading to the above expressions require modification. These modifications must take into account the effects of the finite size and boundaries of real substrates on the stored elastic energy.

Bondu *et al.* (1998) take the example of a finite sized substrate formed by a right circular cylinder of radius a and length H whose optical axis coincides with the axis of the cylinder. They develop an expression for the total strain energy, $U = (U_0 + \Delta U)$, stored in this geometry of substrate under the application of a notional Gaussian pressure. By application of the method in Levin (1998) and Chapter 1, they show that the Brownian substrate noise can be numerically evaluated for specific test mass geometries. Their resulting formulae were subsequently corrected by Liu and Thorne (2000), who also develop an expression for the thermoelastic noise in finite sized test masses.

From Liu and Thorne (2000), following Levin (1998), an expression for the Brownian thermal noise in a finite sized test mass (FTM), $S_x^{\text{FTM}}(f)$ is developed such that

$$S_x^{\text{FTM}}(f) = \frac{4k_B T}{\pi f} \phi_{\text{substrate}}(f, T)(U_0 + \Delta U), \qquad (7.5)$$

where U_0 can be expressed as

$$U_0 = \frac{(1 - \sigma^2)\pi a^3}{Y} \sum_{m=1}^{\infty} U_m \frac{p_m^2 J_0^2(\zeta_m)}{\zeta_m}, \tag{7.6}$$

with

$$U_m = \frac{1 - Q_m^2 + 4k_m H Q_m}{(1 - Q_m)^2 - 4k_m^2 H Q_m} \tag{7.7}$$

and $Q_m = \exp(-2k_m H)$. Following Liu and Thorne (2000), J_0 is the Bessel function of order zero, $k_m = \zeta_m/a$ where ζ_m is the mth zero of the first order Bessel function $J_1(x)$ and p_m are coefficients, given by

$$p_m = \frac{4}{a^2 J_0^2(\zeta_m)} \int_0^a \frac{e^{-2r^2/w_m^2}}{\pi w_m^2} \, dr, \tag{7.8}$$

ΔU is given by

$$\Delta U = \frac{a^2}{6\pi H^3 Y} \left[\pi^2 H^4 p_0^2 + 12\pi H^2 \sigma p_0 s + 72(1 - \sigma)s^2 \right], \tag{7.9}$$

and

$$s = \pi a^2 \sum_{m=1}^{\infty} \frac{p_m J_0(\zeta_m)}{(\zeta_m)^2}. \tag{7.10}$$

Assessment of the effect of the finite size, and in particular the aspect ratio, of a substrate on the level of thermal noise sensed requires numerical evaluation using the expressions above applied to a specific case. Figure 7.1 shows, as an example, the effect of varying the radius and the thickness of a substrate on the magnitude of the thermal noise sensed by a laser beam of radius $w_m = 5.9$ cm. Also shown is the level of thermal noise predicted by the half-infinite mirror approximation. The problem is further complicated by the aspect ratio also influencing coating thermal noise (see Section 6.3), so an overall optimization is needed.

Liu and Thorne (2000) define a numerical correction factor, C_{FTM}^2, such that

$$C_{FTM(x)}^2 = \frac{S_x^{FTM}}{S_x^{ITM}}. \tag{7.11}$$

For reference, in an advanced gravitational wave detector (see Chapter 14), the dimensions of a substrate mirror would be of order $H = 20$ cm and $a = 17$ cm (Advanced LIGO Team, 2007) and $C_{FTM(x)}$ is found to be about 0.82. Laser stabilization cavity experiments often use small and relatively thin mirrors (see Numata et al., 2004, for example, in which mirrors of $a = 12.7$ mm and $H = 5$ mm were studied) and specific examples of the thermal noise of such systems are discussed in Chapters 5 and 15.

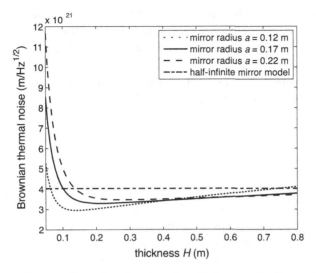

Figure 7.1 Plot showing the effect of varying the mirror thickness and radius on the thermal noise for a finite-sized silica substrate of mechanical loss $\phi = 5 \times 10^{-9}$ evaluated at 100 Hz. Also shown for comparison is the thermal noise calculated using the half-infinite mirror approximation.

For thermoelastic noise, a similar technique allows the correction factor $C^2_{\text{FTM(TE)}}$ for finite size to be evaluated

$$C^2_{\text{FTM(TE)}} = \frac{S^{\text{FTM}}_{TD}}{S^{\text{ITM}}_{TD}}, \qquad (7.12)$$

where

$$
\begin{aligned}
C^2_{\text{FTM(TE)}} = {} & \frac{\sqrt{\pi^3} w^3_m}{a^3} \frac{a^5 H \tilde{c}^2_1}{(1+\sigma)^2} \\
& + \sum_{m=1}^{\infty} \frac{a^5 k_m p^2_m (1 - Q_m) J^2_0(\zeta_m)}{[(1 - Q_m)^2 - 4H^2 k^2_m Q_m]^2} \\
& \times [(1 - Q_m)^2 (1 + Q_m) + 8 H k_m Q_m (1 - Q_m) \\
& + 4H^2 k^2_m Q_m (1 + Q_m)].
\end{aligned}
\qquad (7.13)
$$

Liu and Thorne (2000) investigate the dependence of the correction factor $C_{\text{FTM(TE)}}$ for thermoelastic noise on H, a and w_m and note that for a given beam radius, substrates shaped like thin discs result in higher levels of thermoelastic noise than substrates whose aspect ratio is more like that of a long cylinder.

Application of the expressions above suggests that for many practical situations, corrections to the thermal noise found using the half-infinite substrate approximation are typically at the level of $\approx 10\% - 20\%$, as seen in Figure 7.1. For finite-sized substrates, increasing the beam radius of the sensing laser beam results in lower levels of sensed thermal noise. However, this approach becomes limited by optical losses when the beam becomes "clipped" by the edges of the substrate. Using optical beams whose intensity profile is non-Gaussian is one way to reduce the sensed thermal noise. This works by filling the face of the substrate whilst keeping the optical beam profile within the substrate surface. This motivates much of the work discussed in detail in Chapter 13.

The above treatments have so far assumed that the mechanical dissipation of a substrate is spatially homogeneous. In practice, there are cases where this is not a good approximation, and this will be discussed further below in Section 7.2.2.

7.2 Sources of mechanical dissipation

As noted in Section 7.1.1 above, key parameters for determining the level of Brownian thermal noise from a substrate are its Young's modulus Y, the temperature at which it operates T and, of particular interest here, its mechanical loss $\phi(f, T)$.

Section 4.1 outlines how internal friction in a material arises from the process of anelasticity, (see, for example, Zener, 1937, 1938; Nowick and Berry, 1972). An oscillating stress applied to an anelastic material will result in a periodic strain, where the strain response is not instantaneous but develops with some phase lag ϕ with respect to the stress. This phase lag ϕ is known as the mechanical loss angle, or loss factor. Equivalently, when $\phi \ll 1$, this can be defined in terms of the dissipation of energy over one cycle of oscillation of a mechanical system. Anelasticity, and thus mechanical loss, can result from a variety of dissipation processes in a material, with some processes dependent on temperature and/or frequency. Hence mechanical loss may be expressed in a more general way as $\phi(f, T)$. There are a number of processes that can result in the dissipation of energy in a material and the overview given here cannot be comprehensive, thus some key mechanisms only are discussed.

Experimental studies in both crystalline and amorphous materials of interest as substrates indicate that mechanical dissipation can arise from contributions from both the bulk and the surface of samples (see, for example, Braginsky *et al.*, 1985; Penn *et al.*, 2006; Yasamura *et al.*, 2000, amongst others). Whilst loss mechanisms associated with the surfaces of substrates are not necessarily intrinsic to the material, neither are they truly external sources of dissipation such as gas damping or support losses, and so the relevance of surface loss will be discussed in Section 7.2.2 below.

7.2.1 Bulk loss

Phonon damping

Processes exist that result in the dissipation of mechanical energy in a body under some oscillatory deformation even in an ideal crystalline solid which is free from structural defects. Thermoelastic damping (Zener, 1937, 1938; Nowick and Berry, 1972; Braginsky *et al.*, 1999) as discussed in Section 7.1.1 above, is one such mechanism, and phonon damping (see Akhiezer, 1939) represents another. Very useful discussions of this mechanism can be found in Braginsky *et al.* (1985) and Ferreirinho (1991). Phonons represent lattice vibrational modes and have frequencies characterized by the Grüneisen parameter, $\hat{\gamma}$. An applied oscillatory strain having a wavelength larger than the average phonon mean-free-path modulates the phonon frequencies, shifting their distribution away from equilibrium. The relaxation of the phonon distribution back to a local equilibrium has associated with it mechanical dissipation. Braginsky *et al.* (1985) provide an approximate expression for the dissipation from phonon–phonon damping, ϕ_{phph},

$$\phi_{\text{phph}} = \frac{CT\hat{\gamma}^2}{v} 2\pi f \tau^* \tag{7.14}$$

valid for $2\pi f \tau^* \ll 1$ where v is the velocity of sound in the material, C is the heat capacity per unit volume, and τ^* is the phonon relaxation time, related to thermal conductivity by

$$\kappa = \frac{1}{3} C v_D \tau^*. \tag{7.15}$$

Here v_D is the mean Debye sound velocity, such that

$$\frac{3}{v_D^3} = \frac{1}{v_l^3} + \frac{2}{v_t^3}. \tag{7.16}$$

At room temperature, estimates for the level of phonon–phonon damping in materials are made in Braginsky *et al.* (1985) for crystalline quartz, $\phi_{\text{phph}} \simeq 2 \times 10^{-18}$ at 1 kHz and in Ferreirinho (1991) for sapphire, $\phi_{\text{phph}} \simeq 5 \times 10^{-12}$ at 1 kHz. As will be seen below, in many practical situations this level is well below that resulting from other sources of mechanical dissipation. At low temperature, however, phonon–phonon damping is thought to be responsible for a dissipation peak in single-crystal sapphire (Bagdasarov *et al.*, 1974; Braginsky *et al.*, 1985) of magnitude $\sim 3 \times 10^{-9}$. The peak occurs at about 30 K where the the phonon relaxation time, and hence the intrinsic thermal conductivity of bulk sapphire, is known to peak (Braginsky *et al.*, 1985).

Internal friction due to structural defects

When mechanical vibrations of a solid body containing defects or other kinds of structural imperfections occur, associated changes in the defect state or rearrangement of structural features result in dissipation of energy. As noted in Section 4.2.6, many dissipation mechanisms result in mechanical loss whose level and frequency dependence can be described by a Debye peak, where the loss can be expressed as

$$\phi(f) = \Delta \frac{2\pi f \tau}{1 + (2\pi f \tau)^2} , \tag{7.17}$$

where Δ is a constant related to the magnitude of the dissipation (Nowick and Berry, 1972) and τ is the characteristic time associated with the relaxation process responsible for the dissipation. Many relaxation processes are thermally activated, and thus the relaxation time is related to the activation energy for the process, E_a, by the Arrhenius equation,

$$\tau = \tau_0 e^{E_a / k_B T}, \tag{7.18}$$

where τ_0 is known as the relaxation constant for the dissipation mechanism.

Point defects in a crystal lattice such as substitutions, interstitial impurities, and vacancies can be identified as sources of loss (Nowick and Berry, 1972). Several dissipation peaks observed in silicon are believed to arise from relaxation mechanisms involving particular impurities (McGuigan et al., 1978; Lam and Douglass, 1981; Nawrodt et al., 2008). A range of specific candidate mechanisms associated with structural defects have been identified as possible sources of internal friction, and a thorough review of this area is beyond the scope of this chapter; a more comprehensive discussion of dissipation mechanisms can be found in Nowick and Berry (1972).

Dissipation due to inherent material microstructure

The inherent microstructure of a material can also result in mechanical dissipation. Fused silica is an interesting and rather thoroughly studied example of such a material, with losses as low as $\sim 5 \times 10^{-9}$ measured at room temperature (Ageev et al., 2004). However, at low temperature the loss increases to a peak occurring at approximately 30–50 K (Fine et al., 1954), with the peak dissipation of the order of 10^{-3}. This peak is significantly wider than a simple Debye peak, suggesting that the dissipation process responsible has a broad spectrum of activation energies (Anderson and Bömmel, 1955). As discussed in Section 4.2.6, several models for the dissipation process have been proposed in which the dissipation mechanism is linked to the structure of fused silica (see Anderson and Bömmel, 1955; Strakna, 1961; Vukcevich, 1972). In these models the dissipation occurs when oxygen atoms (or in one model, tetrahedral structures of silicon and oxygen atoms) undergo thermally activated transitions between stable positions separated by a potential barrier. The broad spectrum of activation energies is therefore directly related to the distribution of potential barrier heights in the amorphous Si–O network. In the simplest model, in which the loss occurs when certain Si–O bonds undergo transitions between two stable bond angles, the distribution of barrier heights arises from the distribution of bond angles in the silica network. Investigations of the distribution of potential barrier heights using precise low-frequency light scattering techniques have given good quantitative agreement with this model of dissipation in silica (see Wiedersich et al., 2000). The intrinsic loss of silica at room temperature is believed to be limited by the tail of the low temperature dissipation peak, and the frequency dependence of the loss can be explained by the model of dissipation discussed above (see Wiedersich et al., 2000). More extensive reviews of the literature on fused silica can be found in Braginsky et al. (1985) and Lunin (2005).

Dissipation due to electronic effects

In addition to the relaxation interactions described above, an applied stress can interact with the electrons in a material and a variety of resulting dissipation mechanisms have been identified. In silicon, for example, stress-induced changes to the band gap can result in changes of the density of charge carriers, with a relaxation time related to the time taken for charge carriers to recombine. This process produces a dissipation peak at approximately 600 °C (Nowick and Berry, 1972).

7.2.2 Surface loss

Losses associated with the surface of a substrate can often be significantly higher than the loss of the bulk material itself. The surface layers of a substrate can have very different properties to the bulk of the substrate, either as a result of damage during fabrication or through the absorption and adsorption of impurities. Sample preparation processes such as machining, grinding and polishing can damage the surface resulting in the presence of micro-cracks and a higher occurrence of dislocations and other lattice defects (see e.g. Braginsky *et al.*, 1985). Surface defects can also occur in amorphous materials, for example dangling oxygen bonds or surface microcracks in silica (Lunin, 2005). Impurities can be incorporated into surface layers during substrate fabrication, for example through the diffusion of atoms from polishing compounds into the surface. The use of diamond based polishing compounds results in carbon being a common surface contaminant (Braginsky *et al.*, 1985). Other common impurities are water, atmospheric gases, and organic compounds which can easily be absorbed into the surface layer (Lunin, 2005; Braginsky *et al.*, 1985).

When a damaged surface is mechanically deformed, relaxation processes can occur when the surface defects respond to the deformation, resulting in dissipation as described in Section 7.2.1. A general model allowing evaluation of the magnitude of surface loss in samples of specific geometries has been formulated by Gretarsson and Harry (1999). If thermoelastic loss and sources of extrinsic loss are neglected, then the total loss of a substrate, ϕ_{total}, can be considered as arising from the bulk material, ϕ_{bulk}, and from the surface, such that

$$\phi_{total} = \phi_{bulk} \left(1 + \mu \frac{d_s}{V/S} \right), \tag{7.19}$$

where V/S is the volume-to-surface ratio for a given sample and d_s is the "dissipation depth", given by

$$d_s = \frac{1}{\phi_{bulk} Y_{bulk}} \int_0^h \phi(n) Y(n) dn. \tag{7.20}$$

Here Y_{bulk} is the Young's modulus of the bulk material and h is the physical thickness of the lossy surface layer of the sample. The loss and Young's modulus of the surface layer, $\phi(n)$ and $Y(n)$, are allowed to vary with n, the depth from the surface. The parameter μ is a numerical factor dependent on both the mode-shape and the geometry of the sample such that

$$\mu = \frac{V}{S} \frac{\int_S \epsilon^2(\vec{r}) d^2 r}{\int_V \epsilon^2(\vec{r}) d^3 r}, \tag{7.21}$$

where ϵ is the strain amplitude. The dissipation depth quantifies the total dissipation arising from the surface layer, normalized to the dissipation of the bulk material. If the surface layer has a homogeneous loss, ϕ_{surf}, and the Young's modulus is approximately equal to the Young's modulus of the bulk material, then the dissipation depth is related to the thickness of the surface layer, h, the bulk dissipation, and the loss of the surface layer, ϕ_{surf}, as

$$d_s = h \frac{\phi_{\text{surf}}}{\phi_{\text{bulk}}}. \tag{7.22}$$

Evidence of surface loss in substrate materials

Evidence of excess loss associated with surfaces has been observed in several materials. The precise magnitude of the surface loss, ϕ_{surf}, of a particular material depends on a number of factors including the manufacturing process, roughness, and thermal history. Studies of sapphire show that the type of surface polish applied can significantly affect the total measured loss of a sample (Braginsky *et al.*, 1985). The loss of a sapphire sample polished using particles of grain size $\sim$50 μm can be reduced by more than an order of magnitude by re-polishing the surface using a finer grain size of $\sim$4 μm.

Evidence of surface loss has also been observed in fused silica. Studies of the mechanical loss of silica fibers show that the total measured loss is a function of the fiber diameter, with thinner fibers exhibiting higher total loss (Gretarsson and Harry, 1999; Penn *et al.*, 2001). This is consistent with the surface of the fiber having a higher loss than the bulk material. As the diameter decreases a larger fraction of the elastic energy is stored close to the surface, resulting in a higher measured loss factor. The magnitude of the surface loss of cylindrical fibers was found to be 1×10^{-5} (Gretarsson and Harry, 1999), while for rectangular cross-section fibers a value of 3×10^{-6} has been obtained (Heptonstall *et al.*, 2006). Both of these measurements assumed that the lossy surface layer was 1 μm thick (Doremus, 1979). The difference in the magnitude of the surface loss may be related to different levels of stress occurring in the cylindrical and rectangular geometries as the molten silica cooled (Heptonstall *et al.*, 2006). The lowest loss measured in bulk fused silica is 5×10^{-9} (Ageev *et al.*, 2004), which is approximately three orders of magnitude lower than the loss of the damaged surface layer.

Studies in which fused silica resonators were polished using progressively finer particles are consistent with a model in which the loss of the surface decreases as the size of the polishing particles, and thus the size of the surface defects, is reduced. Using this technique lossy surface layers can be reduced to ~ 0.1 μm in thickness. This residual lossy layer, related to the formation of hydroxide compounds when Si–O bonds are broken during polishing, has been shown to result in mechanical loss peaks at particular temperatures (Lunin, 2005). Removing this residual damaged layer, and thus reducing the magnitude of the loss peaks, can be pursued by chemical etching of the silica surface (Tellier, 1982; Lunin, 2005). The effects of chemical etching treatments are highly dependent on the prior mechanical treatment of the sample, with larger improvements observed when etching rougher surfaces. Thus both high-quality polishing and chemical etching are likely to be required to minimize the surface loss of a silica sample.

Surface loss mechanisms associated with absorbed water and atmospheric gases also exist. On exposure to air a hydrated layer rapidly forms on the surface of silica, and surface loss associated with this layer has been observed (Lunin, 2005). Heat-treatment under vacuum conditions has been shown to almost eliminate the losses associated with this hydrated layer (Lunin, 2005; Mitrofanov and Tokmakov, 2003). A thorough discussion of losses associated with absorbed water and other atmospheric gases is given in Lunin (2005).

Polishing silica with a flame can, in some circumstances, reduce the surface loss, particularly when the silica has not previously been machined (e.g. fibers pulled from a melt) (Ageev *et al.*, 2004; Penn *et al.*, 2001). However, flame-polishing a mechanically ground or polished surface can result in vitrification of the damaged layer, increasing the surface loss (Lunin, 2005).

Studies of silicon also find evidence for the presence of surface loss, with the loss measured in silicon generally observed to vary with the surface to volume ratio (Mohanty *et al.*, 2002; Yasamura *et al.*, 2000; Yang *et al.*, 2002; McGuigan *et al.*, 1978). The incorporation of oxygen into the surface layer of silicon has been postulated as a source of loss, and experiments with micro and nano-scale silicon cantilevers have shown that the loss can be reduced by approximately a factor of 10 by heating the cantilevers to 1000 °C in vacuum and thus removing oxygen from the surface (Yang *et al.*, 2000; Ono *et al.*, 2003; Liu *et al.*, 2005).

Inhomogeneous loss

In the discussions of the formalism for calculating thermal noise outlined earlier in this chapter it was assumed that the loss of the substrate was homogeneous. Thus, to include the effect of a known lossy surface layer into the calculation of thermal noise requires modification of the treatment given above. The method of Levin (see Levin (1998) and Chapter 1) is still applicable here, and we can note that, to first order, expressions given in Chapter 4 for the thermal noise of a lossy coating layer on the surface of the substrate provide an approximation to an analytical treatment for the surface loss.

Surface loss is only one example of inhomogeneous losses which can occur. Other examples include the use of lossy optical coatings (see Chapter 4) or the jointing of support structures to the mirror. In all of these situations finite element modeling is a powerful tool for evaluating the Brownian thermal noise. In general, the thermal noise is given by (see Chapter 1)

$$S_x(f) = \frac{2k_B T}{\pi^2 f^2} \frac{W_{diss}}{F_0^2}, \tag{7.23}$$

where W_{diss} is the power dissipated under the notional oscillatory Gaussian pressure of peak magnitude F_0. This dissipated power can be found from

$$W_{diss} = 2\pi f \int_V \epsilon(x, y, z) \phi(x, y, z, f) dV, \tag{7.24}$$

with $\epsilon(x, y, z)$ the energy density of the elastic deformation of the substrate when it is maximally deformed under the oscillating Gaussian pressure. In the case of inhomogeneously distributed loss, $\epsilon(x, y, z)$ can be obtained by using finite element analysis to calculate the energy density associated with the elastic deformation at each point in the mirror (Yamamoto, 2000; Yamamoto et al., 2002). This method has been used to calculate the thermal noise associated with localized lossy regions such as a magnet attached to the back face of a mirror (Yamamoto, 2000), or suspension fiber attachments bonded to the sides of a mirror (Cunningham et al., 2010). The finite element method also allows the thermal noise for alternative laser beam geometries to be calculated simply by varying the spatial profile of the pressure applied to the mirror surface in the finite element model.

7.3 Design choices for mirror substrates in precision experiments

Substrates for use in optical measurement systems are typically formed from one of a range of amorphous dielectrics (e.g. BK7 glass, fused silica, low expansion glass) or crystalline materials (e.g. sapphire, silicon, calcium fluoride), with the exact choice dependent on the wavelength of the light relevant for the system, and other experimental considerations (robustness, thermal loading, etc.).

As discussed in Section 7.2, the substrate Brownian thermal noise is a function of the Young's modulus, mechanical loss, and temperature. The substrate thermoelastic loss is a function of the thermal expansion coefficient, the thermal conductivity, the specific heat capacity, and the temperature. Of these material parameters, the one typically least well characterized is mechanical loss.

As design examples, we consider substrates for gravitational wave detectors in Section 7.3.1, which require larger, free-hanging optics, and for other experiments using optical cavities in Section 7.3.2, generally using smaller, fixed optics. See Chapter 14 for a more complete discussion of gravitational wave detectors and Chapters 15, 16, and 17 for more on other applications.

7.3.1 Gravitational wave detectors

In addition to having a low level of Brownian thermal noise and thermoelastic noise, there are a range of other requirements for mirror substrates for use in gravitational wave detectors. Where mirrors are partially transmissive for use as input couplers to arm cavities low optical absorption at the wavelength of interest is necessary to reduce the laser power absorbed by the mirror. High thermal conductivity and low thermal expansion are desirable to minimize the thermal deformation of the mirror from any absorbed laser power. The temperature dependence of the refractive index, dn/dT should also be low to minimize thermo-optic noise and thermal lensing effects; see Chapters 9 and 10, respectively.

Relatively few materials can meet all of the requirements for use as gravitational wave detector mirrors. Materials studied for use at a laser wavelength of 1064 nm include fused silica and sapphire. Silicon has been proposed as a suitable material for use in an all-reflective interferometer topology (see Section 11.4) or for operation at higher laser wavelengths e.g. 1550 nm (Winkler *et al.*, 1991; Rowan *et al.*, 2003; Schnabel *et al.*, 2010). See also Section 6.2. Mechanical loss factors of $\sim 3 \times 10^{-9}$ have been measured in sapphire at room temperature (Braginsky *et al.*, 1985) while the lowest losses measured at room temperature in silicon and silica are 0.9–1×10^{-8} (Numata *et al.*, 2001; Nawrodt *et al.*, 2008; Murray, 2008) and 5×10^{-9} (Ageev *et al.*, 2004), respectively. Thus, a low level of Brownian thermal noise would be expected from all three of these materials at room temperature. However, as pointed out by Braginsky *et al.* (1999), the thermal properties of silicon and sapphire (particularly the thermal expansion coefficient and the thermal conductivity, see Figures 8.13 and 8.15) would lead to levels of thermoelastic noise in a silicon or sapphire mirror at room temperature which can exceed that of fused silica. Fused silica, already adopted as the mirror substrate for the first generation of gravitational wave detectors, is also planned for use in the second generation detectors Advanced LIGO (Harry and The LIGO Scientific Collaboration, 2010) and Advanced Virgo. See Section 14.5.2 for more on these detectors.

To obtain increases in sensitivity beyond the level of the second generation detectors, further reductions in thermal noise are likely to be required. One method of achieving this is through cooling the mirror, as discussed in more detail in Chapter 8. Care needs to be taken as the substrate thermal noise is proportional to both $\sqrt{T}$ and $\sqrt{\phi}$, and the mechanical loss of many materials is strongly temperature dependent. As discussed above, the mechanical loss of fused silica increases at low temperature to $\sim 10^{-3}$, effectively precluding the use of silica substrates at cryogenic temperature. The loss of sapphire and silicon, however, can be as low as 2–5×10^{-10} at cryogenic temperature (McGuigan *et al.*, 1978; Braginsky *et al.*, 1985), making these materials attractive candidates for cooling. Furthermore, the thermoelastic noise in sapphire and silicon would be significantly lower at cryogenic temperature, due to the temperature dependence of the thermal expansion coefficient of these materials, as shown in Figures 8.13 and 8.15.

Silicon and sapphire have the potential to withstand significantly higher laser powers than fused silica for a given amount of thermally induced surface deformation, largely due to their relatively high thermal conductivity (see Winkler *et al.*, 1991). Thus silicon or sapphire substrates would potentially allow the use of higher circulating laser powers to reduce photon shot noise and enhance detector sensitivity at high frequencies. See Chapter 10 for more on high laser power and thermal effects in mirrors.

Sapphire mirrors are proposed for use in the Large-scale Cryogenic Gravitational wave Telescope (LCGT), a planned Japanese cryogenically cooled interferometer using transmissive optics (Kuroda and The LCGT Collaboration, 2010). See Section 14.5.2 for more on LCGT. While some experiments (see Section 7.2.1) have shown a dissipation peak of magnitude $\sim 3 \times 10^{-9}$ at 30 K in sapphire, this peak loss is of a similar magnitude to the loss at room temperature, and the general trend is for the loss to decrease below room temperature (Uchiyama *et al.*, 1999; Braginsky *et al.*, 1985).

Silicon mirrors are of particular interest for use in future gravitational wave detectors as there are two zeros in the thermal expansion coefficient at approximately 18 K and

125 K, suggesting that operation at either of these temperatures could in principle eliminate thermoelastic noise (Rowan *et al.*, 2003). Measurements by McGuigan *et al.* (1978) showed evidence of dissipation peaks at ∼13 K and ∼ 150 K. While these peaks are very close to the nulls in the thermoelastic loss, it has been speculated that the peaks arise from impurities in the silicon sample, and are not related to the zeros in the thermal expansion coefficient. Other measurements of silicon (Nawrodt *et al.*, 2008; Bignotto *et al.*, 2008; Reid *et al.*, 2006) have shown no evidence of these loss peaks, suggesting that the peaks are not intrinsic to all types of silicon. Thus 18 K or 125 K may be very attractive temperatures at which to operate a silicon-based detector.

7.3.2 Optical cavities

Ultra-stable Fabry–Perot cavities also form the critical components of frequency stabilized lasers for several precision experiments including optical atomic clocks (Ludlow *et al.*, 2006), high-resolution optical spectroscopy (Young *et al.*, 1999; Rafac *et al.*, 2000), fundamental quantum optics measurements, quantum information science (Schmidt-Kaler *et al.*, 2003), and cavity QED experiments (Miller *et al.*, 2005), see Chapters 15, 16, and 17. The frequency stability which can be achieved is dependent on the length stability of these cavities and thus thermal noise can form a significant limit to the precision of these experiments. The thermal noise limit has been calculated analytically (see Numata *et al.*, 2004) and thermal noise limited performance has been measured experimentally (Notcutt *et al.*, 2006). The cavities used in these measurements usually consist of two coated mirrors separated by a rigid spacer. One of the key criteria in material selection is a low coefficient of thermal expansion, to minimize the coupling of thermal fluctuations to the length of the cavity. Materials such as ULE, Zerodur, silica, silicon and sapphire have been used for this purpose. Initially, materials were primarily chosen for a low coefficient of thermal expansion, and silicon and sapphire are most often cooled to cryogenic temperature to achieve this goal (Richard *et al.*, 1990; Seel *et al.*, 1997; Müller *et al.*, 2003a).

Current experiments have reached, or are rapidly approaching, the limit set by thermal noise. Measurements of the effect of substrate thermal noise on cavity length, and thus frequency, stability have been carried out. Mechanical loss measurements indicate that Zerodur substrates have a relatively high mechanical loss of ∼3×10^{-4}, while the loss of ULE substrates was found to be ∼2×10^{-5} (Numata *et al.*, 2004). It has been demonstrated that for a cavity constructed from Zerodur spacers, the level of thermal displacement noise can be reduced in line with theoretical predictions when fused silica mirror substrates are used in place of Zerodur substrates (Notcutt *et al.*, 2006). When fused silica mirror substrates are used, the mirror coatings become the dominant source of thermal noise. While the use of fused silica mirrors can reduce thermal noise, it will also increase the thermal expansion of the cavity from ∼10^{-9} K^{-1} to ∼10^{-7} K^{-1} (Notcutt *et al.*, 2006), potentially making length changes associated with temperature fluctuations more significant. Thus the use of crystalline materials such as silicon and sapphire, which can provide both low mechanical loss and a low coefficient of thermal expansion at cryogenic temperatures, may be the most promising method of reducing cavity thermal noise, provided appropriate low noise mirror coatings can be identified. See Section 5.5 for more on thermal noise in fixed-mirror cavity experiments.

7.4 Conclusion and future directions

Thermal noise associated with mirror substrates is an important consideration in the design of many precision mechanical and optical experiments. In many precision measurement experiments, the thermal noise from the reflective coatings applied to the mirrors is larger than the thermal noise from the substrate materials. However, techniques for developing "coating free" mirrors based on diffractive optics, novel concepts such as nano-structured surfaces, and other techniques are currently being developed (see Chapter 6 and Brückner *et al.* (2008, 2009)). These techniques may remove the need for lossy dielectric coatings altogether, in which case the limiting source of thermal noise is likely to arise from the mirror materials. Thus, the thermal noise of optic materials, and in particular any losses associated with structuring the surface of these materials, may be of enhanced significance in future experiments.

8

Cryogenics

KENJI NUMATA AND KAZUHIRO YAMAMOTO

8.1 Introduction

The use of cryogenic mirrors can be an attractive way of minimizing thermal noise due to optical coatings and substrates. As long as special care is taken in choosing materials, thermally induced fluctuations should be "frozen" at low temperatures. In this chapter, we introduce the current understanding of temperature dependence of mirror thermal noise, some practical design issues, and examples of cryogenic experiments. We will mainly focus on off-resonant thermal noise and physical cooling issues of macroscopic optics. For more on cooling of resonant thermal noise, see Chapter 16.

8.2 Temperature dependence of mirror thermal noise

In sensitive optical experiments, thermal fluctuations of coatings and substrates are important sources of noise and both tend to be reduced at cryogenic temperatures. The gains from cryogenics could come from several directions: the direct reduction of temperature itself, the decrease in mechanical losses (occurring in some materials), and the precipitous drop in specific heat and the increase in the mean free path of phonons. Appropriate materials and beam parameters must be adopted so that every kind of thermal noise is minimized by cryogenics simultaneously. Typically, Brownian thermal noise in coatings improves very slowly from cryogenics. This is due to a weak temperature dependence of mechanical loss. Coating thermal noise is also usually dominant over the substrate Brownian noise, so improvements of coating thermal noise directly impact measurement sensitivity. Substrate thermoelastic noise (see Chapter 7), coating thermo-optic noise (see Chapter 9), and other thermal noises (see Chapter 3) generally have less importance at cryogenic temperatures and at low frequencies. In this section, we give an overview of the theoretical background of mirror thermal noise. Mechanical and thermal properties of materials are summarized in Section 8.5.

Optical Coatings and Thermal Noise in Precision Measurement, eds. Gregory M. Harry, Timothy Bodiya and Riccardo DeSalvo. Published by Cambridge University Press. © Cambridge University Press 2012.

8.2.1 Brownian noise due to structural damping

As described in Chapters 1, 4, and 7, thermal noise is caused by internal friction, which occurs within coating and substrate materials. Such noise is traditionally called mirror Brownian noise. A simplified form of mirror Brownian noise is given by Equation 4.10, reproduced here for reference,

$$S_x(f, T) \approx \frac{2k_B T}{\sqrt{\pi^3} f} \frac{1 - \sigma^2}{w_m Y} \left[\phi_{sub}(f, T) + \frac{2}{\sqrt{\pi}} \frac{(1 - 2\sigma)}{(1 - \sigma)} \frac{d}{w_m} \phi_{coat}(f, T) \right], \quad (8.1)$$

where $S_x(f, T)$ is the one-sided power spectrum of the surface displacement measured by a Gaussian beam with radius w_m, the substrate loss angle is ϕ_{sub}, and the coating mechanical loss is ϕ_{coat}. The first and the second terms correspond to contributions from the substrate and the coating, respectively. Temperature dependence of S_x and the ϕs are noted specifically, while the T dependence of Y, σ, and d are generally small[1] and less important. Frequency dependence of material loss, $\phi(f)$, has been studied experimentally and a constant ϕ model (called structural damping)[2] has been shown to be relatively common in low mechanical loss materials (Kovalik and Saulson, 1993; Saulson *et al.*, 1994) and is a good model to start with.[3] Therefore, the temperature dependence of loss, $\phi(T)$, becomes the main concern here.

8.2.2 Substrate contribution to Brownian noise

Different substrate materials show different temperature dependencies of mechanical loss, $\phi_{sub}(T)$. Fused silica has a large mechanical loss peak at around 50 K and has a tail down to 0 K due to energy dissipation by thermally activated transitions of oxygen atoms within the amorphous Si–O–Si network bonds (Bömmel *et al.*, 1956; Anderson *et al.*, 1972; Wiedersich *et al.*, 2000). Thus, fused silica and other optical glasses made up of silica are generally poor substrate materials at cryogenic temperatures. On the other hand, crystalline materials, such as sapphire (Braginsky *et al.*, 1985; Uchiyama *et al.*, 1999), silicon (Nawrodt *et al.*, 2008), and calcium fluoride (Nawrodt *et al.*, 2007a) are known to have low mechanical loss ($\phi < 10^{-8}$) at cryogenic temperatures. Figure 8.12 in Section 8.5 summarizes temperature dependence of mechanical loss in substrate materials.

These small mechanical losses and low temperatures cooperatively make the substrate's Brownian thermal noise negligible in many cases. Figure 8.1 shows the dependence of

---

[1] Among them, Young's modulus, Y, has the biggest relative change at a level of 3%, for example, in fused silica between room and cryogenic temperatures (McSkimin, 1953).

[2] When working at low frequency (below milli-Hz), one has to be aware of the fact that the loss, ϕ, should become zero at zero frequency (Saulson, 1990).

[3] Measurements indicate that some fused silica brands have lower loss at lower frequencies, all generally at the $\phi \sim 10^{-7}$ level, at room temperature (Penn *et al.*, 2006). This may hold for other low-loss materials and temperatures as well.

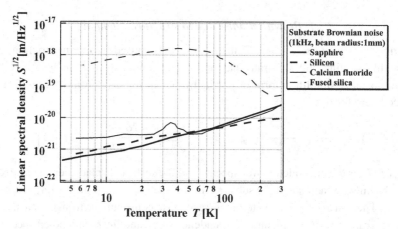

Figure 8.1 Temperature dependence of substrate Brownian noise. Frequency f is 1 kHz. Beam radius w_m is 1 mm. The linear spectral density $\sqrt{S}$ is proportional to $1/\sqrt{w_m f}$. Mechanical losses in Figure 8.12 were used for this calculation.

substrate Brownian noise on temperature with different substrate materials. With the exception of fused silica, they all become smaller by a factor $\sim$10 going from 300 K to 30 K.

8.2.3 Coating contribution to Brownian noise

The weak frequency dependence of coating mechanical loss, $\phi_{coat}(f)$, has been experimentally verified through mechanical loss measurements (Crooks et al., 2004; Yamamoto et al., 2006a; Martin et al., 2008, 2009) and direct measurements of coating thermal noise (Numata et al., 2003; Black et al., 2004a; Harry et al., 2007) between $\sim$100 Hz and $\sim$100 kHz. The temperature dependence of $\phi_{coat}(T)$ has been studied primarily in ion-beam sputtered SiO_2/Ta_2O_5 dielectric coatings, see Chapter 4. It has been experimentally shown that the loss is nearly constant at the $(4–6) \times 10^{-4}$ level between 4 K and 300 K (Yamamoto et al., 2006a). Other results indicate the existence of a small loss peak at the $\phi \sim 10^{-3}$ level around 20 K and peak suppression by annealing (Martin et al., 2010). These results support the model of very weak frequency and temperature dependence of coating loss at a level of $\phi_{coat} \sim 10^{-4}$, at least for this kind of coating.[4] Figure 8.12 shows the coating loss as a function of temperature.

Figure 8.2 shows the temperature dependence of coating Brownian noise. Comparing Figures 8.1 and 8.2, the linear spectral density of Brownian thermal noise, $\sqrt{S_{Brown}}$, is dominated by the coating contribution and thus proportional to $\sqrt{T\phi_{coat}}$. This holds true

---

[4] Recent studies indicate that the temperature dependence may be different for other coatings. A SiO_2/Ta_2O_5 coating formed by a magnetic sputtering process shows a temperature dependence of $\sim$10^{-3} at 300 K and 4×10^{-4} below 30 K (Nawrodt et al., 2007b). Other coating materials, such as zirconia, hafnia, and niobia, are being actively investigated, see Section 4.2.3.

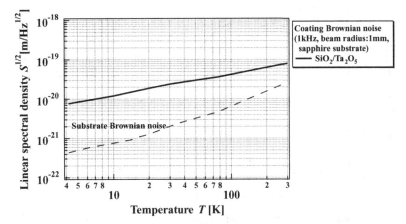

Figure 8.2 Temperature dependence of coating Brownian noise with a coating thickness $d = 4.7$ μm, a sapphire substrate, a beam radius $w_m = 1$ mm, and a frequency $f = 1$ kHz. The linear spectral density $\sqrt{S}$ is proportional to $1/\left(w_m\sqrt{f}\right)$.

as long as the beam radius w_m is smaller than $\sim 40\,000\,d$ (~ 20 cm for a typical coating), since $\phi_{sub} \sim 10^{-8}$, $\sigma \sim 0.2$, and $\phi_{coat} \sim 5 \times 10^{-4}$ in Equation 8.1.[5]

8.2.4 Substrate thermoelastic noise

There are other kinds of thermal noise in optics. Among them are substrate thermoelastic noise (see Chapter 7) and coating thermo-optic noise (see Chapter 9). See Chapter 3 for a detailed list of thermal noises. Thermoelastic and thermo-optic noise are associated with thermoelastic dissipation (Zener, 1937), which is caused by heat flow along temperature gradients. Unlike Brownian noise, this dissipation mechanism can be theoretically calculated from the equations of motion and heat conduction.

A formula for substrate thermoelastic noise can be found in Braginsky *et al.* (1999); Cerdonio *et al.* (2001) as well as in Chapter 7. The thermoelastic noise has a low-pass property whose characteristic frequency is given by a cutoff frequency,[6]

$$f_c = \frac{\kappa(T)}{\pi\rho C(T)w_m^2}. \tag{8.2}$$

This cutoff frequency corresponds to the relaxation time of thermal gradients of size w_m. When the observing frequency f is higher than this cutoff frequency, f_c, the spectrum is

---

[5] Coating thickness is determined by the desired reflectivity and laser wavelength. See Chapter 12 for more on dielectric coatings and coating design.

[6] There is another cutoff frequency, $f \sim \kappa/(\rho C a^2)$, which corresponds to the relaxation time in the scale of the mirror size, a. Below this cutoff frequency, the system can be treated as isothermal and the thermoelastic noise has a constant value over frequency. We neglect this region for simplicity.

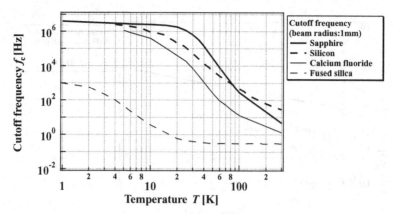

Figure 8.3 Temperature dependence of cutoff frequency f_c in substrate thermoelastic noise. The beam radius w_m is 1 mm. This cutoff frequency is proportional to $1/w_m^2$.

given by the adiabatic approximation,

$$S_{\text{adi}}(f, T) = \frac{4}{\sqrt{\pi^5}}(1 + \sigma)^2 \frac{\alpha(T)^2 k_{\text{B}} T^2}{(\rho C(T))^2} \kappa(T) \frac{1}{w_m^3 f^2}, \tag{8.3}$$

where $\alpha(T)$ is the coefficient of linear expansion (Braginsky et al., 1999). Below this cutoff frequency, the spectrum can be written by the semi-isothermal approximation,

$$S_{\text{iso}}(f, T) = \frac{\sqrt{2}}{\pi}(1 + \sigma)^2 \frac{\alpha^2(T) k_{\text{B}} T^2}{\sqrt{\rho C(T)}} \frac{1}{\sqrt{\kappa(T) f}}, \tag{8.4}$$

which does not depend on beam radius w_m (Cerdonio et al., 2001; Somiya et al., 2010). In both regions, the linear spectral density $\sqrt{S}$ is proportional to αT.

Figure 8.3 shows the dependence of the cutoff frequency f_c with temperature. Since the cutoff frequency becomes higher at lower temperature, the semi-isothermal condition, Equation 8.4, applies to most cryogenic situations where the observing frequency is <kHz.

Figure 8.4 shows substrate thermoelastic noise in different substrate materials. The noise level strongly depends on the temperature and the thermal properties of the material. In general, thermoelastic noise does not decrease drastically in the adiabatic region. Thus, mirrors typically have to be cooled substantially to reach the semi-isothermal region to minimize the noise. In this example, with beam radius $w_m = 1$ mm and frequency $f = 1$ kHz, the transition from the adiabatic to the semi-isothermal condition happens at $\sim$80 K, $\sim$80 K, $\sim$40 K, and $\sim$1 K in sapphire, silicon, calcium fluoride, and fused silica, respectively. Silicon has zero thermoelastic noise around 20 K and 120 K because its α is zero at these temperatures. A similar situation occurs in fused silica at $\sim$180 K. Because of thermoelastic noise, most crystalline substrates become advantageous over amorphous ones only at low temperatures. This is due to the crystalline material's relatively larger α and κ at room temperature.

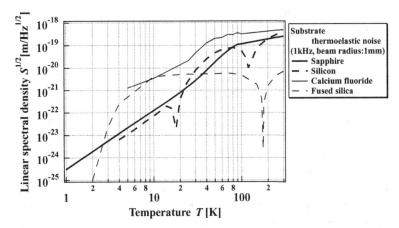

Figure 8.4 Temperature dependence of substrate thermoelastic noise. Frequency f is 1 kHz and beam radius w_m is 1 mm.

The temperature dependence of thermoelastic noise can be understood using a phonon model (Kittel, 1995). Because phonons carry heat, the specific heat C is roughly proportional to T^3 below the Debye temperature, which is higher than 300 K in most materials used for mirrors. In crystalline materials, α/C is independent of temperature (the Grüneisen relation),[7] and thus the thermal expansion parameter α is also proportional to T^3. Thermal conductivity is expressed as

$$\kappa(T) = \frac{1}{3}\rho C(T)l_{\text{ph}}(T)v, \tag{8.5}$$

where l_{ph} is the mean free path of phonons and v is the velocity of phonons. This velocity v is almost independent of temperature, T. Above $\sim$30 K, l_{ph} is limited by phonon–phonon Umklapp scattering. Below $\sim$30 K, where the number of thermally excited phonons decreases, the mean free path l_{ph} gets significantly longer. For example, l_{ph} of sapphire is about 10^5 times longer around 30 K than at 300 K. As a result, κ becomes about 10^2 times larger. Below this temperature, l_{ph} becomes nearly constant because it is limited by impurities, lattice defects, and/or the dimensions of the sample, not by any temperature-related process. As a result, κ is proportional only to C and thus T^3 in this region. Figures 8.13, 8.14, and 8.15 show the temperature dependence of α, C, and κ of various substrate materials, respectively.

These effects dynamically change the thermoelastic noise level. The temperature dependence of the cutoff frequency (Figure 8.3) is similar to that of l_{ph} due to Equations 8.2 and 8.5. The temperature dependence of substrate thermoelastic noise changes around 1–100 K, where the transition from the adiabatic condition to the semi-isothermal condition happens, and around 30 K, where the phonon mean free path becomes limited by impurities.

---

[7] The Grüneisen relation is not valid in silicon. Still, $|\alpha/C|$ of silicon at lower temperatures is comparable to, or smaller than, the room temperature value.

Above 30–100 K (the adiabatic limit), $\sqrt{S_{\mathrm{adia}}}$ is nearly constant, except for temperatures where α becomes zero, since α/C and $T\sqrt{\kappa}$ are nearly temperature independent. Below 30 K, $\sqrt{S_{\mathrm{iso}}}$ is proportional to $T^{2.5}$, both in the isothermal and adiabatic limits.

8.2.5 Thermo-optic noise

Thermo-optic noise is caused by the relaxation dissipation of temperature differences between the substrate and coating. This noise consists of coating thermoelastic noise (Braginsky and Vyatchanin, 2003a; Fejer et al., 2004) and thermorefractive noise (Braginsky et al., 2000). Coating thermoelastic noise is a fluctuation of the coating surface while thermorefractive noise originates in the fluctuations of the coating optical thickness. These fluctuations are detected as mirror displacement noise since some of the light is reflected by the inner layers of the coating. See Chapter 9 for a full description of coating thermoelastic and thermorefractive noise.

There is a correlation between thermoelastic and thermorefractive noise because of the common origin in the same temperature relaxation. Thus, they are treated coherently as thermo-optic noise (see Chapter 9). Above the cutoff frequency of Equation 8.2, Equation 9.21 in Chapter 9 can be simplified to

$$S_{\mathrm{to(adi)}} = \frac{2}{\sqrt{\pi^3}} [\alpha_{\mathrm{eff}}(T)d - \beta_{\mathrm{eff}}(T)\lambda]^2 \frac{k_B T^2}{\sqrt{\rho_0 C_0(T)}} \frac{1}{w_m^2 \sqrt{f \kappa_0(T)}}. \tag{8.6}$$

Below the cutoff,

$$S_{\mathrm{to(iso)}} = \frac{2}{\sqrt{\pi}} [\alpha_{\mathrm{eff}}(T)d - \beta_{\mathrm{eff}}(T)\lambda]^2 k_B T^2 \frac{1}{\kappa_0(T)} \frac{1}{w_m}. \tag{8.7}$$

Here, α_{eff} and β_{eff} can be approximated as

$$\alpha_{\mathrm{eff}} \approx \alpha_1 \frac{(1 + \sigma_0)}{(1 - \sigma_1)} \left[\frac{Y_1}{Y_0}(1 - 2\sigma_0) + \frac{1 + \sigma_1}{1 + \sigma_0} \right] - 2(1 + \sigma_0)\alpha_0 \frac{\rho_1 C_1}{\rho_0 C_0}, \tag{8.8}$$

and

$$\beta_{\mathrm{eff}} \approx \frac{1 + \sigma_1}{1 - \sigma_1}\alpha_1 n + \beta, \tag{8.9}$$

by using averaged coating values (instead of weighted averages in Equations 9.35, 9.39 and 9.40). The subscripts 0, 1 on the material properties denote substrate and coating, respectively. Terms with α_{eff} and β_{eff} represent the coating thermoelastic noise and thermorefractive noise, respectively. Like the substrate thermoelastic noise, the two approximations are valid at high and low temperature regions, respectively.[8]

Coating thermal parameters (α_1, β_1, and C_1) contribute to the noise only through α_{eff} and β_{eff}. The second term of Equation 8.8 includes material properties of the substrate as α_0/C_0.

---

[8] The thermo-optic noise also has a cutoff frequency that corresponds to the relaxation time of thermal gradients of thickness d. The cutoff frequency is usually high (>10 kHz) because of the thinness of the coating. Therefore, we focus on the region below this cutoff frequency.

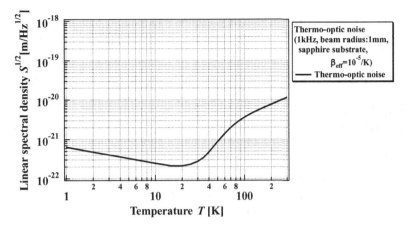

Figure 8.5 Temperature dependence of thermo-optic noise, where α_{eff} and β_{eff} are assumed to be 0 and 10^{-5} /K, respectively. The substrate is assumed to be sapphire. Frequency f is 1 kHz and beam radius w_m is 1 mm. The wavelength of light λ is 1064 mm.

This quantity has a constant value at low temperature (as discussed in Section 8.2.4). Since both α_1 and C_1 become smaller at lower temperature, in general, α_{eff} is expected to be small. The first term in β_{eff} is also expected to be low due to α_1. The temperature dependence of the second term, β, is not as well understood.[9] In order to simplify the discussion, we assume that β is temperature independent and α_{eff} is neglected. Then the thermo-optic noise $\sqrt{S_{\text{to(adi)}}}$ is approximately proportional to $\beta T C_0^{-1/2} l_{\text{ph}}^{-1/4}$, and $\beta T^{-1/2} l_{\text{ph}}^{-1/4}$. In the other limit, $\sqrt{S_{\text{to(iso)}}}$ is approximately proportional to $\beta T (C_0 l_{\text{ph}})^{-1/2}$, and $\beta (T l_{\text{ph}})^{-1/2}$.

Figure 8.5 shows the dependence of the thermo-optic noise on temperature, assuming constant $\beta_{\text{eff}} = 10^{-5}$ /K, a typical room temperature value (Evans *et al.*, 2008). When cooled from room temperature, the noise becomes smaller because l_{ph} gets longer.[10] Below ~ 30 K, the linear spectral density is proportional to $1/\sqrt{T}$, because of the constant l_{ph}. Even with this, thermo-optic noise is typically smaller than coating Brownian noise.

8.2.6 Summary of temperature dependence of thermal noise

Figures 8.6 and 8.7 show the frequency dependence of thermal noise in a sapphire mirror at 300 K and 4 K, respectively. The largest contribution at 300 K comes from the substrate thermoelastic noise (see Chapter 7). When the mirror is cooled down, the substrate Brownian noise and thermoelastic noise drop quickly. The coating Brownian noise depends on the temperature more weakly, $\sqrt{T}$, because the coating loss is nearly temperature independent. It has the largest contribution to the thermal noise at low temperatures, due to the large loss at the $\phi_{\text{coat}} \sim 10^{-4}$ level. Current estimates indicate that the thermo-optic noise should not

---

[9] There are some indications that β becomes smaller at lower temperature (Astrath *et al.*, 2005).
[10] In Figure 8.5, around 80 K, the cutoff frequency f_c in Equation 8.2 becomes 1 kHz. Above and below this temperature, the adiabatic and semi-isothermal approximations, Equations 8.6 and 8.7, become valid, respectively.

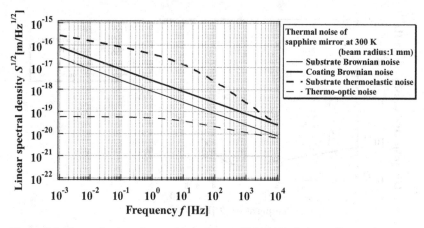

Figure 8.6 Thermal noise of a sapphire mirror at 300 K. The beam radius w_m is 1 mm.

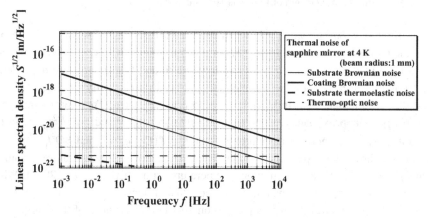

Figure 8.7 Thermal noise of a sapphire mirror at 4 K. The beam radius w_m is 1 mm.

have a large contribution as long as the temperature coefficient of the refractive index β is of the order of 10^{-5} /K.

8.3 Practical considerations in designing cryogenic experiments

Based on the theoretical predictions given in the previous section, a target temperature can be selected to minimize thermal noise taking into account the available materials. In order to reach and maintain the target temperature, heat in the system (mirror plus any supports) has to be extracted through a thermally conducting channel to a cooling medium. The final mirror temperature is set by the temperature of this medium, the heat conduction of the cooling channel, and the heat load on the mirror. There is an inherent trade off between high optical power (or high finesse), which increases heating, and cryogenics. In addition, the cooling process must not add any excess mechanical disturbances to the system. In this section, we discuss these and other general concepts for optical measurements, reviewing low-temperature experimental techniques (Ventura and Risegari, 2007).

8.3.1 Heat absorption in mirrors

There are two main mechanisms that can prevent mirrors from reaching a target temperature – absorption of laser power and thermal radiation from higher temperature surrounding structures. In order to minimize the effect of the absorption of laser power, one should use the lowest finesse (to minimize coating thickness), the lowest input power consistent with the sensitivity requirements, and materials with high thermal conductivity and low absorption. If high input optical power has to be used to minimize shot noise, heat generated in the mirror substrate and coating must be effectively extracted.[11] Typically, the absorption of dielectric coatings, a_c, is 1–10 ppm (see Chapter 10). As substrate materials, sapphire (Tomaru *et al.*, 2001; Blair *et al.*, 1997) and silicon (Green and Keevers, 1995) have 10–100 ppm/cm of absorption, δ, at their transmission bands.[12] In that case, a mirror in a simple Fabry–Perot cavity with finesse $\mathcal{F}$ and incident laser power P would have a heat load of

$$P_{\text{abs}} \sim 32 \ \mu\text{W} \frac{P}{1 \ \text{mW}} \frac{1}{2} \left(\frac{\delta}{50 \ \text{ppm/cm}} \frac{t}{1 \ \text{cm}} + \frac{2}{\pi} \frac{a_c}{5 \ \text{ppm}} \frac{\mathcal{F}}{10000} \right) \tag{8.10}$$

in the substrate and the coating.[13] Note that, t is the thickness of the mirror.

The other possible heating mechanism is radiation from the room temperature environment. When radiation shields have holes, e.g. for laser light transmission, the heat falls onto the mirror as

$$P_{300\text{K}} \sim \sigma_B A \epsilon \frac{N \pi r^2}{4 \pi L^2} (300 \ \text{K})^4, \tag{8.11}$$

$$\sim 57 \ \mu\text{W} \left(\frac{A}{10 \ \text{cm}^2} \right) \left(\frac{N}{2} \right) \left(\frac{2r}{1 \ \text{cm}} \right)^2 \left(\frac{10 \ \text{cm}}{L} \right)^2 \left(\frac{\epsilon}{0.1} \right), \tag{8.12}$$

where σ_B is the Stefan–Boltzmann constant, A and ϵ are the surface area and the emissivity of the mirror, N and r are the number and the radius of the holes, and L is the distance between the mirror and the holes. This 300 K radiation heat can be comparable to that from light absorption. One can apply appropriate optical filters (an etalon or short-pass filter) and/or shield pipes with buffers to minimize the radiation effects at far-IR wavelength (Tomaru *et al.*, 2008a,b). In some systems, optical fibers may also be used to introduce the laser into the cryostat to avoid such radiation effects.

---

[11] Thanks to generally higher thermal conductivity at lower temperatures, inhomogeneous heat deposition does not cause as large of a thermal gradient within the substrate as at room temperature (Tomaru *et al.*, 2002). Therefore, absorption affects noise levels through the elevated temperature (i.e. thermal noise) but not through thermal lensing and deformation effects, which are discussed in Chapter 10.

[12] Typically 50 ppm at 1064 nm for large samples of sapphire, occasionally lower in smaller samples. These values may also be dependent on temperature. For example, silicon shows lower absorption at lower temperature (Macfarlane *et al.*, 1958, 1959).

[13] The factors depend on coupling conditions of the cavity.

8.3.2 Heat extraction from mirror

Any heat generated must be extracted from the mirrors by heat conduction through heat links, since convection does not occur in vacuum and radiation is not efficient in cryogenic environments. The heat links should be as short and thick as possible and have large thermal conductivity. Pure copper and aluminum[14] are convenient materials for heat links, since they have a large thermal conductivity between 1 K and 300 K. At 300 K, the thermal conductivities of pure copper and aluminum are on the order of 100 W/m/K, which is about 10 times larger than that of sapphire. At lower temperatures, their thermal conductivity becomes even larger, taking maximum values of $\sim 10^4$ W/m/K around 10 K. Below ~ 10 K, the thermal conductivity falls approximately proportionally to T (not T^3 as in crystals[15]). Below 10 K, κ can be estimated by the Wiedemann–Franz law (Ventura and Risegari, 2007), $\kappa \sim T \times$ RRR (in SI unit), where RRR is the residual resistivity ratio (ratio of electrical resistances at 300 K and 4.2 K). The RRR in metals depends strongly on impurities and it can be measured by putting the sample in a ^{4}He pot and measuring its resistance. Through such measurements, appropriate heat link materials can be identified. Figures 8.16 and 8.17 show the κ of aluminum and copper with different RRRs, respectively.

Unfortunately, those metals with favorable thermal properties tend to have large mechanical losses. When thermal noise from the heat link (and/or its support structure) becomes an issue, crystalline materials must be used in spite of their smaller thermal conductivity and lower cooling efficiency. Keeping the connections to mechanically lossy heat links well away from the readout beam is another way of reducing their thermal noise effects.

8.3.3 Cooling media and vibration

Traditionally, liquid nitrogen (77 K boiling temperature at atmospheric pressure) and liquid helium (4.2 K for ^{4}He) have been used as cooling fluids, since they are cost effective. With such cooling fluids, cryogenic system design becomes relatively easy, because their temperatures are highly stable and they produce little mechanical disturbance from boiling. When required, the boiling vibrations can be mechanically filtered out, for example, by using a mechanical suspension. Most cryogenic rigid cavity experiments have been done with these two fluids, having multiple liquid pods and radiation shields (layers at 77 K, 4.2 K, and other temperatures) under vacuum. On the other hand, these cooling fluids require frequent refills, which may inhibit continuous operations of the experiment, and can cause periodic deformation of mechanical structures. They also have inconvenient logistics. Special attention must be paid to not produce any operational hazards.

A newer and more efficient "dry cooling" technology is replacing such liquid coolants. This new technology is represented by closed-cycle mechanical cryocoolers. They have

---

[14] Aluminum cannot be used as a heat link below 1.2 K because it becomes a superconductor.

[15] In pure metals, electrons play an important role in thermal conductivity, rather than phonons. Below 10 K, the mean free path of electrons is limited by impurities. Thus, from Equation 8.5, κ is proportional to C (electronic specific heat), therefore to T.

simplified logistics, low maintenance, and are cost effective when used continuously. They do not have the traditional temperature limitations, such as 77 K and 4.2 K. These chillers include the Joule–Thomson, Brayton, Stirling, Gifford–McMahon, and pulse tube cryocoolers (Radebaugh, 2009). More recently, adsorption cryocoolers are finding practical uses, especially for small cooling capacity applications. In spite of their easy turnkey operation, mechanical cryocoolers tend to cause larger mechanical vibrations because they have movable parts in cyclic motion and they derive their power from an external compressor that provides a high-pressure reservoir. Therefore, they have to be properly isolated or anchored down to solid ground. Currently, the Gifford–McMahon and the pulse tube cryocoolers are the most used for cryogenic optics applications because they do not require cascades of different types of cryocoolers to reach cryogenic temperatures. In general, the pulse tube is cheaper, smaller, more reliable, and has smaller vibrations than the Gifford–McMahon because it uses a standing wave in the gas instead of a mechanically oscillating displacer[16] (Tomaru *et al.*, 2004; Antonini, 2005). Still, it generates vibrations from the waves of helium gas at the fundamental frequency (typically 1–60 Hz) and its harmonics (Wang *et al.*, 1999; Lienerth *et al.*, 2001). Such vibrations can be suppressed by careful design (Caparrelli *et al.*, 2006; Ikushima *et al.*, 2008; Wang and Hartnett, 2010) as well as by external vibration-isolation mechanisms.

The second law of thermodynamics predicts that the efficiency of an ideal Carnot-cycle cryocooler becomes 0 at 0 K. Therefore, cooling power becomes smaller at lower temperature. For example, the typical power of a pulse tube cryocooler is about 40 W and 1 W at 45 K and 4.2 K, respectively. Evacuated ^{3}He or ^{3}He-^{4}He dilution cryocoolers have less than 1 mW of cooling power below $\sim$0.3 K. Currently, it is difficult to reach sub-Kelvin temperatures with a heat load of $\sim$0.1 mW (as discussed in Section 8.3.1) because of limited cooling power.

8.3.4 Contamination

One concern with cooled optics is the possible catastrophic optical power absorption from residual gas condensation on the coating surface. The gas molecules accumulate on the cold mirror surface, forming a layer. This layer leads to degradation of reflectance through optical absorption, scattering, and refractive index changes (see Chapters 10 and 11 for absorption and scattering effects, respectively). Through these effects, cavity finesse may degrade over time during cryogenic operations. Therefore, mirrors must be surrounded by other independently cooled and colder objects, namely, radiation shields. In general, a cooling cycle should be started only after a good vacuum level is obtained in order to avoid such effects.

Contamination effects have been investigated in detail at 1064 nm and at 10 K using a high finesse ($\mathcal{F} \sim 30\,000$) rigid cavity with SiO_2/Ta_2O_5 coatings (Miyoki *et al.*, 2001).

---

[16] When a pulse-tube is driven by a Gifford–McMahon-type compressor and rotary valve, it is called a GM-type pulse tube. In Stirling-type pulse tubes, valveless compressors or pressure oscillators are used.

During cooling, while molecules were absorbed, the mirror reflectance remained constant within ±5 ppm over a few weeks in a carefully operated cryostat.

8.4 Design examples

Cryogenic cooling of mirrors has been achieved in a number of fields. In this section, we overview cryogenic system designs and scientific results.

8.4.1 Rigid cavities as length reference

Cryogenic cavities are used to measure thermal expansion of materials and to lock laser frequencies to a stable length. At a low temperature T, the thermal linear expansion coefficient drops as $\alpha \propto T^3$, which leads to a reduced sensitivity to temperature changes. Examples of cryogenic cavity experiments and system designs are summarized below. General design considerations of frequency stabilization cavities are discussed in Chapter 15.

The use of cooled rigid cavities started in the 1970s to measure thermal expansion of materials, see Berthold and Jacobs (1976); Jacobs (1986); Richard *et al.* (1990).[17] Notcutt *et al.* (1995) operated a fused silica mirror cavity with a sapphire spacer at cryogenic temperatures in which differential thermal contraction between the mirror and the spacer was created. The cavity was mounted vertically in an evacuated copper can, which was housed inside the vacuum of a bottom-loading liquid helium dewar flask. The details of their cryo-system design are shown in Figure 8.8 (Notcutt *et al.*, 1996). They also tested an all-sapphire cavity between 5 K and 77 K and measured the thermal expansion of sapphire to be $\alpha \sim 10^{-12}(T/1 \text{ K})^3/\text{K}$ in that temperature region (Taylor *et al.*, 1996).

Cryogenic cavities were first intensively used as optical frequency references by groups led by Stephan Schiller at the Universität Konstanz. They first demonstrated laser frequency stabilization at around 3 K using two identical, all-sapphire, cryogenic resonators in two liquid helium cryostats (Seel *et al.*, 1997). Their cavities achieved the lowest short-term instability (fractional Allan deviation $\sigma_A \sim 3.5 \times 10^{-15}$ at 100 s) among any oscillator at that time. The drift of the cavity resonance was measured to be less than 3 kHz over 6 months (Storz *et al.*, 1998). The cavity could be cooled down to 1.3 K by pumping on the liquid helium in the pot. This system was used to test the foundations of relativity (isotropy of space) (Müller *et al.*, 2002; Braxmaier *et al.*, 2002; Müller *et al.*, 2003a),b) and to search for fluctuations in space (Schiller *et al.*, 2004). It was later modified by the implementation of a pulse-tube cryocooler (Antonini, 2005; Antonini *et al.*, 2005; Schiller *et al.*, 2006) which cooled the sapphire resonators down to 3.4 K. This technique was chosen to avoid disturbances from the periodic refilling of liquid-based cryostats.

Figure 8.9 shows this experimental setup. A copper radiation shield was connected to the first stage of the cooler (about 40 K), which enclosed the second stage and the experiment

---

[17] Cryogenic optical cavities without spacers were also implemented as readout transducers in resonant-bar gravitational-wave detectors, see Richard (1992).

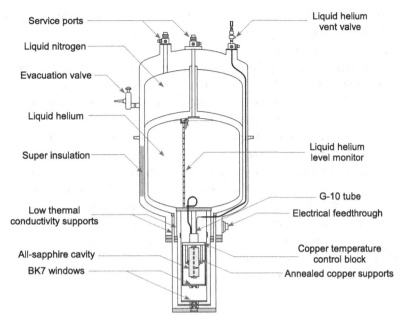

Figure 8.8 Cryostat for cooling a fully contained, small cavity. A sapphire cavity was vertically suspended below the temperature control block (Notcutt *et al.*, 1996).

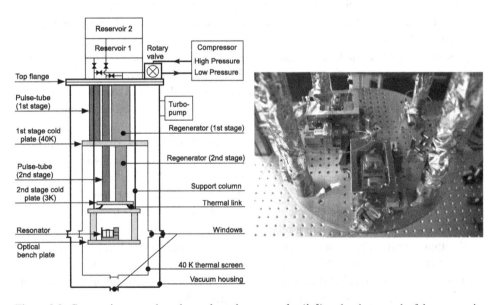

Figure 8.9 Cryogenic system based on pulse-tube cryocooler (*left*) and a photograph of the cryogenic sapphire cavity (*right*) for the test of the constancy of the speed of light (Antonini, 2005).

which consisted of two sapphire cavities oriented at right angle. The shield, the tubes of the cooler, and the second stage were wrapped by superinsulation aluminized foil to insulate them from radiation from the cryostat walls. The effects of vibrations created by the pulse-tube cryocooler were reduced by the use of mechanical vibration isolation (copper braids; thermal link in Figure 8.9) and by fiber-coupling of the laser beams to the resonators. This experiment set a limit on violations of Lorentz invariance from any anisotropy of space to $\delta c/c = 6.4 \times 10^{-16}$. In a later experiment, an upper limit for the long-term drift of a sapphire resonator was set using a frequency comb measurement and found to be less than 3 kHz over 140 days (Ernsting, 2009).

Currently, several groups are working on frequency stabilization experiments using cryogenic cavities based on the latest understanding of thermal noise and other technical advances, see Chapter 15 for a full discussion. The vibration-insensitive cavity support system (Chen *et al.*, 2006a) is very promising for cryogenic systems in which low-frequency vibration isolation cannot be implemented easily. Detailed mechanical structures, thermal noise, and heat flows can be calculated by numerical modeling. All-sapphire and all-silicon cavities have been investigated to achieve higher laser frequency stability, mainly for optical clocks.

8.4.2 Laser interferometer gravitational-wave detectors

In the field of gravitational wave detection using terrestrial laser interferometers, thermal noise of the mirrors is the sensitivity limiting factor in the center of their observation bands, typically around 100 Hz. The mirrors in gravitational-wave interferometers must be suspended as pendulums individually, so that they behave as free masses in the observation band and to provide isolation from mechanical disturbances (seismic noise). The use of cryogenics is currently being developed to minimize thermal noise in mirrors and in their pendulum suspensions for use in future detectors. Gravitational-wave interferometers are described in detail in Chapter 14.

A part of the budget was recently approved for LCGT (Large scale Cryogenic Gravitational wave Telescope) (Kuroda and The LCGT Collaboration, 2010) (see Section 14.5.2), an interferometer that will use cryogenic sapphire mirrors. The Einstein Telescope (Punturo and The Einstein Telescope Collaboration, 2007, 2010) (see Section 14.5.3) is in the early planning stage and is also considering cryogenic mirrors for its lower frequency detector. In LCGT, a high power laser with a relatively large beam radius ($\sim$3 cm) will be circulated within the cavity. The heat generated in the mirror is expected to be about 1 W and is planned to be extracted through heat conduction along the suspending fibers to cryocoolers. Ultimately, the suspending fibers have to have low mechanical loss, as well as high thermal conductivity, in order to minimize suspension thermal noise.

As a technology demonstrator for LCGT, the CLIO (Cryogenic Laser Interferometer Observatory) experiment has been built at Kamioka, in an underground site in

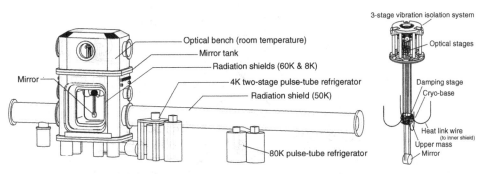

Figure 8.10 Schematic view of the cryogenic system in CLIO. Left: vacuum system containing the mirror suspension. Right: cryogenic mirror suspension system in the mirror tank. (Courtesy of Dr. Takashi Uchiyama at Institute for Cosmic Ray Research, the University of Tokyo.)

Japan (Uchiyama *et al.*, 2006; Miyoki and The CLIO and LCGT Collaboration, 2010). The CLIO interferometer has two Fabry–Perot cavities with 100 m arm length. The four main mirrors are made of sapphire with a diameter of 10 cm and a thickness of 6 cm. Figure 8.10 shows the cryogenic system in CLIO. The mirrors are housed in four cryogenic vacuum tanks, each of which has two radiation shields at 60 K and 8 K. The shields are cooled by a 4 K, two-stage, pulse-tube cryocooler in which vibration reduction techniques are applied (Tomaru *et al.*, 2004; Yamamoto *et al.*, 2006b; Ikushima *et al.*, 2008). In order to reduce radiation from the 300 K environment, each tank includes a 5 m long, 50 K cryogenic sleeve inside the beam vacuum pipe. This sleeve is cooled by a separate 80 K pulse-tube cryocooler. The sleeve has several baffles in order to reduce the transmission of 300 K radiation along the sleeve as well as to minimize its scattering of the main laser beam (Tomaru *et al.*, 2008a,b).

A three-stage vibration isolation system and alignment optical stages are set on top of the radiation shields in a room-temperature vacuum environment. Inside the radiation shield, a cryogenic suspension is suspended from the optical stage by 800 mm long amorphous metal wires whose thermal conductivity is small. The suspension is cooled from the shield through heat link wires made of pure aluminum, which is soft and pliant enough not to disturb the seismically isolated mirrors. The suspension consists of four masses, a damping stage, a cryo-base, an upper mass, and then the main mirror. The suspension is designed to attenuate vibrations from the heat link wires as well as the residual seismic noise from the room temperature stage while under cryogenic operation. Mechanical resonances of these systems are suppressed by passive magnetic dampers. The mirror is suspended at the bottom stage of the suspension by 99.999% pure aluminum wires, 400 mm in length and 1 mm in diameter. These aluminum wires will be replaced by sapphire in the future to reduce thermal noise. The mirror can be cooled down to 21 K within 7 days. CLIO is presently under cryogenic operation and will study the reduction of mirror thermal noise from cooling in the near future.

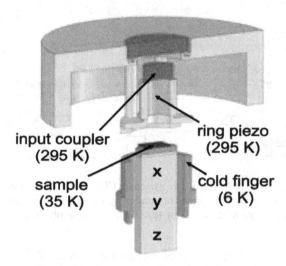

Figure 8.11 Cavity mounting inside the cryostat for the radiation-pressure self-cooling experiment (Gröblacher *et al.*, 2008).

8.4.3 Quantum ground state of opto-mechanical system

Quantum optomechanics is a rapidly growing field, which is discussed in detail in Chapter 16. In order to achieve the quantum ground state of a mechanical oscillator, its "effective" temperature is reduced by cold damping in addition to physical cooling by a cryocooler.[18] The effective temperature T_{eff} is defined as $T_{eff} = T Q_{eff}/Q_{mec}$, where $Q_{mec} (= 1/\phi)$ and Q_{eff} are unperturbed and effective Q values of the mechanical oscillator, while T and T_{eff} are physical and effective temperatures, respectively. Therefore, when a reflective coating is used in such a quantum experiment, it has to have low mechanical loss (and thus maintain a high Q_{mec}) and low absorption in order to achieve low T, and thus low T_{eff}.

Currently, the quantum ground state has been approached most closely (32 quanta) in a SiO_2/Ta_2O_5-coated Si_3N_4 membrane with $T_{eff} = 1$ mK, $Q_{mec} = 3 \times 10^4$, and $T = 5.3$ K at 945 kHz (Gröblacher *et al.*, 2009a,b). Compared to earlier experiments with a SiO_2/Ta_2O_5-coated silicon cantilever (2×10^5 quanta) (Kleckner and Bouwmeester, 2006; Kleckner *et al.*, 2006) and a SiO_2/Ta_2O_5 coating as an oscillator (10^4 quanta) (Gröblacher *et al.*, 2008), the use of low mechanical-loss cantilever materials ($Q > 2 \times 10^6$ around 4 K in Si_3N_4 (Zwickl *et al.*, 2008)) and low T has enabled near quantum-limited operation. This high Q was achieved even with a relatively high mechanical loss coating because the coating does not deform much at the resonance, thus minimizing the amount of elastic energy lost in the coating (see Chapter 1).

In the 32-quanta experiment, a Fabry–Perot cavity was set in a continuous ^{4}He flow cryostat (Figure 8.11). Vibration of the cryostat was not an issue because of the high

---

[18] This cold damping can be done passively via photons in a detuned cavity (Braginsky and Vyatchanin, 2002) or actively via an external control system (Milatz *et al.*, 1953).

Table 8.1 *Mechanical properties of mirror materials*

Material	Sapphire	Silicon	Calcium fluoride	Fused silica	Ta_2O_5
Young's modulus γ ($\times 10^{10}$ Pa)	40	16	9.0	7.2	14
Poisson's ratio σ,	0.29	0.22	0.28	0.17	0.23
Density (g/cm^3) ρ	4.0	2.3	3.2	2.2	—

observation frequency. The 25 mm cavity was formed by a small mechanical oscillator and a more massive mirror. The mechanical oscillator, 100 μm × 50 μm × 1 μm and coated with a 4 μm thick SiO_2/Ta_2O_5 coating, was cooled down to about 5 K through a cold finger. The massive mirror was attached to the outer radiation shield. Alignment of the mechanical oscillator was adjusted from outside of the cryostat using a positioner under the cold finger.

In order to get closer to the quantum ground state, efforts are being made to reduce the physical temperature, T, and the mechanical losses of cantilevers and coatings. Use of a cryocooler based on nuclear adiabatic demagnetization (Kleckner and Bouwmeester, 2006) and monocrystalline $Al_xGa_{1-x}As$ heterostructure coating (Cole *et al.*, 2008) are currently being implemented.

8.5 Appendix: Thermal and mechanical parameters of optical materials

8.5.1 Mechanical properties

Table 8.1 summarizes some mechanical properties of mirror materials. They are from Braginsky and Vyatchanin (2003a) for sapphire, Wortman and Evans (1965) and JAHM software (1998) for silicon, Nawrodt *et al.* (2007a) for calcium fluoride, and Braginsky and Vyatchanin (2003a) for fused silica and tantala (in deposited films). Crystalline materials can have different values for mechanical properties along different crystal axes; typical values are shown in Table 8.1. See the appropriate reference for more details.

Figure 8.12 shows the temperature dependence of mechanical loss in various substrate and coating materials. The data are from Uchiyama *et al.* (1999) for sapphire, Nawrodt *et al.* (2008) for silicon, Nawrodt *et al.* (2007a) for calcium fluoride, Schnabel *et al.* (2010) for fused silica, and Yamamoto *et al.* (2006a); Martin (2009); Martin *et al.* (2010) for SiO_2/Ta_2O_5 dielectric coatings.

8.5.2 Thermal properties

Figure 8.13 shows the thermal linear expansion coefficient of mirror substrate materials from Okaji *et al.* (1995) for fused silica, from Lyon *et al.* (1977) for silicon, and Touloukian and Ho (1970) for others.

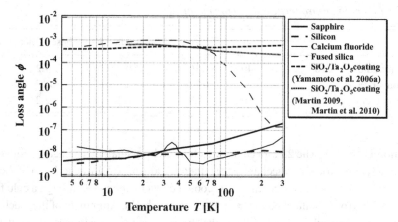

Figure 8.12 Temperature dependence of mechanical loss ϕ of mirror substrate and coating materials.

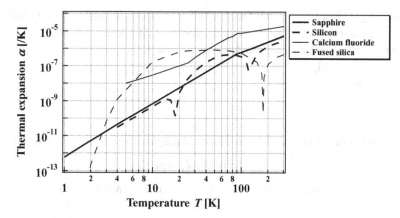

Figure 8.13 Temperature dependence of the thermal linear expansion coefficient, α, of mirror substrate materials. Absolute values $|\alpha|$ are plotted for fused silica below 180 K and for silicon between 20 K and 120 K.

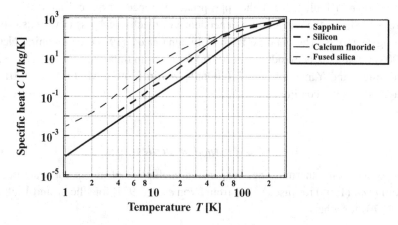

Figure 8.14 Temperature dependence of the specific heat per unit mass, C, of mirror substrate materials.

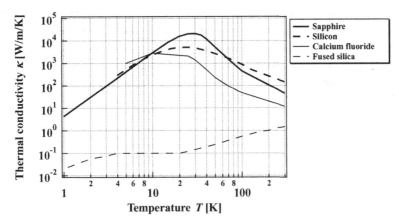

Figure 8.15 Temperature dependence of the thermal conductivity, κ, of mirror substrate materials.

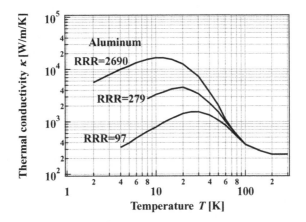

Figure 8.16 Temperature dependence of the thermal conductivity, κ, of aluminum with different residual resistivity ratios (RRR).

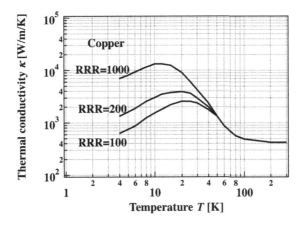

Figure 8.17 Temperature dependence of the thermal conductivity, κ, of copper with different residual resistivity ratios (RRR).

Figure 8.14 shows the specific heat of materials from Zeller and Pohl (1971) for fused silica, Jonscher (1964) for silicon, and Touloukian and Ho (1970) for others.

Figures 8.15, 8.16, and 8.17 show thermal conductivity of mirror substrate materials, aluminum, and copper, respectively, from Zeller and Pohl (1971) for fused silica, Ho *et al.* (1972) for silicon, Seeber and White (1998) for aluminum and copper, and Touloukian and Ho (1970) for others.

9

Thermo-optic noise

MATTHEW EVANS AND GREG OGIN

9.1 Introduction

The dielectric coatings typically used in high precision optical measurement consist of alternating layers of materials with different refractive indices (see Figure 9.1). For high-reflectivity, the coating layers form pairs of high and low refractive index materials, referred to as doublets, with each component layer a quarter wave thick.[1] In most cases, it is safe to assume that the coating reflects incident light from a mirror surface; a conceptually perfect plane at the boundary between the coating and the vacuum. However, a more precise picture is that of a reflection from each interface between coating layers, all of which add at the mirror surface to form the total reflected wave.

The displacement of the mirror surface, as interferometrically measured with the reflected wave, is given by the simple relation $\Delta z = \Delta\varphi\lambda/4\pi$, where $\Delta\varphi$ is the change in reflection phase of a field with wavelength λ, and Δz is the apparent change in position of the surface. In this chapter we will consider thermo-optic noise, which manifests itself as a change in the reflection phase of the coating resulting from thermal fluctuations in the coating materials.

9.2 How to change the reflection phase of a coating

9.2.1 The thermoelastic mechanism

Thermal expansion of the coating materials causes the thickness of a coating to change as a function of temperature. These temperature changes can be driven, or can result from statistical fluctuations in the coating material. Changes in the thickness cause the reflection phase of the coating to change with respect to a reference plane attached to the substrate or to the center of mass of the optic.

---

[1] Coating designs beyond the basic quarter wave doublet design are discussed in Chapter 12. The formalism in this chapter will be kept general so that it may be applied to any coating design, but quarter wave doublet design will be used in most examples.

Optical Coatings and Thermal Noise in Precision Measurement, eds. Gregory M. Harry, Timothy Bodiya and Riccardo DeSalvo.
Published by Cambridge University Press. © Cambridge University Press 2012.

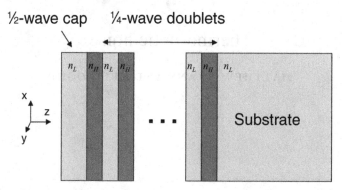

Figure 9.1 A high reflection coating made of quarter wave doublets with a half wave cap layer.

The variation in reflection phase as a function of temperature due to thermal expansion of the coating is given by the product of the average coefficient of thermal expansion (CTE) of the coating and the coating thickness

$$\frac{\partial \varphi_{cTE}}{\partial T} = 4\pi \bar{\alpha}_c \frac{d}{\lambda}. \tag{9.1}$$

Assuming that the substrate and coating have similar elastic coefficients, the substrate constrains the coating expansion such that the effective CTE is given by

$$\bar{\alpha}_c \sim 2\alpha_c (1 + \sigma_c), \tag{9.2}$$

where α_c is the average CTE of the coating, and σ_c the average Poisson's ratio.[2]

Thermal expansion of the coating, when driven by statistical fluctuations in coating temperature, is the basis for "thermoelastic" noise discussed later in Section 9.3. See Braginsky et al. (1999); Braginsky and Vyatchanin (2003a); Fejer et al. (2004) for more on coating thermoelastic noise.

9.2.2 The thermorefractive mechanism

A somewhat more subtle mechanism by which coating temperature affects the apparent position of an optic is related to the change in refractive index of the coating material, and the change in the thickness of each coating layer. This mechanism is referred to as the "thermorefractive" mechanism, though it includes both thermal expansion and changes in refractive index inside the coating (Braginsky et al., 2000).

In this case the coefficient used is an effective change in refractive index with temperature, and the length scale of interest is the wavelength of the reflected field,

$$\frac{\partial \varphi_{cTR}}{\partial T} = -4\pi \bar{\beta}. \tag{9.3}$$

---

[2] The definitions of average coating properties are given in Appendix A of Evans et al. (2008).

This relationship defines $\bar{\beta}$ such that

$$\frac{\partial \Delta z_{TR}}{\partial T} = -\bar{\beta}\lambda. \qquad (9.4)$$

The computation of $\bar{\beta}$ depends on the coating structure, but it is based on a combination of material properties which gives the fractional change in optical path length in each coating layer as a function of temperature,[3]

$$B_k = \beta_k + \frac{1 + \sigma_k}{1 - \sigma_k}\alpha_k n_k \qquad (9.5)$$

where $\beta_k = \partial n/\partial T$ for the kth coating layer, $\bar{\alpha}_k$ is the effective CTE for that layer, and n_k is the layer's refractive index at the operating temperature. Example calculations of $\bar{\beta}$ for a few simple coating structures are given later in Section 9.4.

To clarify the distinction between the thermoelastic (TE) and thermorefractive (TR) mechanisms; the TR mechanism changes the reflection phase of a coating as measured at its surface, while the TE mechanism changes the position of the coating surface relative to the substrate (Braginsky and Vyatchanin, 2003a). This distinction is somewhat arbitrary, and we will see later that a complete picture is only obtained when the two mechanisms are treated together.

9.3 Noise resulting from fluctuations in coating temperature

9.3.1 *Heat flow, dissipation and thermal noise*

As discussed in Chapter 1, noise resulting from thermal fluctuations can be related to dissipation resulting from thermal relaxation (e.g., heat flow) (Levin, 2008). The Fluctuation–Dissipation Theory approach outlined there addresses the general problem of thermal noise processes, and thus leaves open a problem specific "form factor" $q(\vec{r})$ which we will now derive for thermo-optic noise.

We start by expressing the impact of temperature on the measurement of interest,

$$\hat{z} = \oint \delta T(\vec{r}, t) q(\vec{r}), \qquad (9.6)$$

where $\delta T(\vec{r}, t)$ is the deviation of the temperature from the average at location $\vec{r}$ and time t. This is similar to Equation 1.7 except that we have defined a specific measurement variable $\hat{z}$, the apparent position of the mirror surface. We restrict further the formalism presented in the introduction by assuming that the measurement is made by a Gaussian beam with a normalized intensity profile given by

$$I(r_\perp) = \frac{2}{\pi w_m^2} e^{-2r_\perp^2/w_m^2}, \qquad (9.7)$$

---

[3] Equation 9.5 does agree with Equation 8.9, but does not agree with Evans *et al.* (2008). This is believed to be due to the use of incorrect boundary conditions in Evans *et al.* (2008), although work is still ongoing to understand this discrepancy.

where $r_\perp^2 = x^2 + y^2$ is the distance from the beam center on the mirror surface. Such a beam measures the reflection phase as

$$\hat{z} = \frac{\lambda}{4\pi} \oint \delta T(\vec{r}, t) I(r_\perp) \frac{\partial \varphi(\vec{r})}{\partial T} \qquad (9.8)$$

from which we can identify the overall form factor

$$q(\vec{r}) = \frac{\lambda}{4\pi} I(r_\perp) \frac{\partial \varphi(\vec{r})}{\partial T}. \qquad (9.9)$$

Continuing on the path described in Chapter 1, we consider a periodic injection of entropy, here expressed as energy in the form of heat per unit time or simply "heat power",

$$\frac{P(\vec{r})}{\delta V} = T F_P \sin(2\pi f t) q(\vec{r}) = T F_P \sin(2\pi f t) \frac{\lambda}{4\pi} I(r_\perp) \frac{\partial \varphi(\vec{r})}{\partial T}, \qquad (9.10)$$

where in this expression the scale factor F_P has SI units of $m^2 \times W/m^3 = N/s$. This equation is a direct result of Equation 1.7 from Chapter 1, through the replacement of entropy with heat power via $P(\vec{r}) = T \partial s(\vec{r})/\partial t$ and $F_P = 2\pi f F_0$.

In the case of *coating* thermo-optic noise we must remove the component of $q(\vec{r})$ which results from the substrate. To do this properly in the context of Fluctuation–Dissipation Theory, which we use by computing dissipation via heat-flow, we must remove the component of power injection which does not result in heat-flow between the coating and substrate

$$\frac{1}{\rho(\vec{r}) C(\vec{r})} \frac{P_c(\vec{r})}{\delta V} = \frac{1}{\rho(\vec{r}) C(\vec{r})} \frac{P(\vec{r})}{\delta V} - \frac{1}{\rho_s C_s} \frac{P_s(\vec{r})}{\delta V}, \qquad (9.11)$$

where C is the specific heat per unit mass of each material and ρ is the density. This expression can be rearranged to give the coating specific form factor

$$q_c(\vec{r}) = \frac{\lambda}{4\pi} I(r_\perp) \left(\frac{\partial \varphi(\vec{r})}{\partial T} - \frac{\rho(\vec{r}) C(\vec{r})}{\rho_s C_s} \frac{\partial \varphi_s}{\partial T} \right). \qquad (9.12)$$

Note that $q_c(\vec{r})$, the coating-only form factor, is zero in the substrate material where $\varphi(\vec{r}) = \varphi_s$ and $C(\vec{r}) = \rho_s C_s / \rho(\vec{r})$.

Next we will make a few simplifying assumptions. The first and least limiting of these is that the coating and substrate are uniform in the plane of the mirror surface, which allows us to replace $\vec{r}$ with z everywhere except in the beam intensity profile $I(r_\perp)$. We will also assume that the heat-flow is dominantly into the substrate from the coating, and that heat flow parallel to the surface can be ignored. This is equivalent to assuming that the measurement beam radius is large compared with the thermal diffusion length, $w_m \gg r_T$, where

$$r_T = \sqrt{\frac{\kappa}{2\pi f C \rho}}. \qquad (9.13)$$

Since typical mirror materials have $r_T \sim 10 \, \mu m \sqrt{\frac{100 Hz}{f}}$ at room temperature and higher values at cryogenic temperatures, this assumption will be violated in low-frequency measurements which utilize small beams, especially in cryogenic environments (see Chapter 8) such as cryogenic reference cavities (see Chapter 15). Lastly, we will assume that the coating is thin enough that heat flow in the coating itself can be ignored; $d \ll r_{Tc}$. A typical high reflection coating will have $d \sim 10\lambda$, so this assumption may be violated at high frequencies or in coatings with thermally conductive materials. This assumption will be explored in more depth in Section 9.3.2.

Finally we are in a position to compute the dissipation, and thus thermal noise, associated with the coating. The general expression for dissipation due to heat flow is

$$W_{diss} = \left\langle \oint \frac{\kappa}{T} (\vec{\nabla} \delta T)^2 \right\rangle, \tag{9.14}$$

where the average $\langle \dots \rangle$ is over the cycles of our injection of heat power. Given the above assumptions, we can integrate the form factor $q_c(\vec{r})$ over the coating

$$q_c(r_\perp) = \int_0^\infty dz \, \frac{\lambda}{4\pi} I(r_\perp) \left(\frac{\partial \varphi(z)}{\partial T} - \frac{\rho(z) \, C(z)}{\rho_s C_s} \frac{\partial \varphi_s}{\partial T} \right) \tag{9.15}$$

$$= I(r_\perp) \left(\bar{\alpha}_c d - \bar{\beta} \lambda - \frac{\rho_c C_c}{\rho_s C_s} \bar{\alpha}_s d \right) \tag{9.16}$$

so that it can be treated as a heat source at the surface of the mirror. Again using the assumption that heat-flow is dominantly into the substrate from the coating, we can solve the diffusion equation to find

$$\vec{\nabla} \delta T \simeq \frac{\partial \delta T}{\partial z} = \frac{-T F_P}{\kappa_s} e^{\frac{-z}{\sqrt{2}r_T}} \sin\left(2\pi f t - \frac{z}{\sqrt{2}r_T} \right) q_c(r_\perp). \tag{9.17}$$

We perform the average over injection cycles, leaving

$$
\begin{aligned}
W_{diss} &= \frac{-T F_P^2}{2\kappa_s} \oint e^{\frac{-2z}{r_T}} q_c(r_\perp)^2 \\
&= \frac{-T F_P^2}{2\kappa_s} \int_0^\infty dz \, e^{\frac{-2z}{r_T}} \int_0^\infty r dr \, q_c(r_\perp)^2 \\
&= \frac{T F_P^2 r_T}{4\kappa_s} \int_0^\infty r dr \, q_c(r_\perp)^2 \\
&= \frac{T F_P^2 r_T}{2\sqrt{2}\pi w_m^2 \kappa_s} \left(\bar{\alpha}_c d - \bar{\beta} \lambda - \frac{\rho_c C_c}{\rho_s C_s} \bar{\alpha}_s d \right)^2,
\end{aligned}
\tag{9.18}
$$

which is directly related to the measurement thermal noise by

$$S_{\hat{z}}(f) = \frac{8 k_B T \, W_{diss}}{F_P^2} \tag{9.19}$$

$$= \frac{2\sqrt{2} k_B T^2 r_T}{\pi w_m^2 \kappa_s} \left(\bar{\alpha}_c d - \bar{\beta} \lambda - \frac{\rho_c C_c}{\rho_s C_s} \bar{\alpha}_s d \right)^2 \tag{9.20}$$

$$= \frac{2 k_B T^2}{\pi w_m^2 \sqrt{\pi f \kappa_s \rho_s C_s}} \left(\bar{\alpha}_c d - \bar{\beta} \lambda - \frac{\rho_c C_c}{\rho_s C_s} \bar{\alpha}_s d \right)^2. \tag{9.21}$$

Taking a longer look at the equation for thermal noise, we can see that it easily breaks into two parts. The first part represents the temperature fluctuations averaged over by a

Gaussian beam profile,

$$S_{\hat{T}}(f) = \frac{2\sqrt{2}k_B T^2 r_{Ts}}{\pi w_m^2 \kappa_s}. \tag{9.22}$$

The second part is simply the sum of the TE and TR mechanisms that make up the coating thermo-optic mechanism, which can be thought of as an overall coefficient of thermal expansion,

$$\bar{\alpha}_{TO} = \bar{\alpha}_c - \beta \frac{\lambda}{d} - \rho_c C_c / (\rho_s C_s) \bar{\alpha}_s. \tag{9.23}$$

Further interrogation of $S_{\hat{T}}(f)$ allows us to break it again into parts,

$$S_{\hat{T}}(f) = \frac{\sqrt{2}k_B T^2}{2\pi f \rho_s C_s r_T^3} \frac{2r_T^2}{\pi w_m^2}, \tag{9.24}$$

where the first fraction represents the thermal fluctuations in a volume of size r_T^3, and the second is the averaging over this volume performed by a Gaussian beam of size w_m. Thus, the complicated expression for thermo-optic noise can be understood as the simple average over thermal fluctuations which are converted to apparent surface displacement via the two thermo-optic mechanisms.

9.3.2 Correction for thick coatings

In the previous section a number of simplifying assumptions were made, one of which is that the coating is thin compared with the thermal diffusion length, or $d \ll r_{Tc}$. Here we will relax this assumption, allowing diffusion in the coating to occur, as it may at high frequencies or with special coating materials where r_{Tc} is small.

In thick coatings, an important difference between the TE and TR mechanisms comes into play; the TE mechanism is the expansion of the entire coating, and is thus affected by heat deposition in any part of the coating, while the TR mechanism is dominated by changes in the first few layers of the coating where most of the laser field is reflected. In terms of the Fluctuation–Dissipation Theory noise calculation, the implication is that we can refine our model by injecting heat power associated with the TR mechanism at the surface of the coating, rather than evenly throughout the coating. The result is a more complicated solution to the heat diffusion equation which can be expressed as a change in $\bar{\alpha}_{TO}$, such that for thick coatings we have

$$\bar{\alpha}_{thick}^2 = \frac{p_E^2 \Gamma_0 + p_E p_R \xi \Gamma_1 + p_R^2 \xi^2 \Gamma_2}{R \xi^2 \Gamma_D} \tag{9.25}$$

$$\Gamma_0 = 2(\sinh(\xi) - \sin(\xi)) + 2R(\cosh(\xi) - \cos(\xi))$$

$$\Gamma_1 = 8\sin(\xi/2)(R\cosh(\xi/2) + \sinh(\xi/2))$$

$$\Gamma_2 = (1 + R^2)\sinh(\xi) + (1 - R^2)\sin(\xi) + 2R\cosh(\xi)$$

$$\Gamma_D = (1 + R^2)\cosh(\xi) + (1 - R^2)\cos(\xi) + 2R\sinh(\xi)$$

where we have made the expression more compact with TE and TR coefficients

$$p_E = \bar{\alpha}_c d - \frac{\rho_c C_c}{\rho_s C_s} \bar{\alpha}_s d \ , \ p_R = -\bar{\beta}\lambda \tag{9.26}$$

and the dimensionless scale-factors

$$R = \sqrt{\frac{\kappa_c \rho_c C_c}{\kappa_s \rho_s C_s}} = \frac{\kappa_c r_{Ts}}{\kappa_s r_{Tc}}, \tag{9.27}$$

$$\xi = \frac{\sqrt{2}d}{r_{Tc}} = \sqrt{\frac{4\pi f \rho_c C_c}{\kappa_c}} d. \tag{9.28}$$

The resulting TO noise is then given by

$$S_{\hat{z}}(f) = S_{\hat{T}}(f)\bar{\alpha}_{thick}^2. \tag{9.29}$$

We can connect this to the result in Section 9.3 by considering the case of $d \ll r_{Tc}$ or $\xi \ll 1$ in which the coating is again thin compared with the thermal diffusion length. In this case, we can use the Taylor expansion around $\xi = 0$ to find

$$\bar{\alpha}_{thick}^2 \simeq \bar{\alpha}_{TO}^2 + \frac{p_E^2 + 3\bar{\alpha}_{TO}(p_R - \bar{\alpha}_{TO}R^2)}{3R}\xi$$
$$- \bar{\alpha}_{TO}\frac{p_E - 3\bar{\alpha}_{TO}(1 - R^2)}{6}\xi^2 \tag{9.30}$$

which approaches $\bar{\alpha}_{TO}^2$ as ξ goes to 0. In the other extreme, a thick coating with $\xi \gg 1$, the TE and TR mechanisms contribute incoherently to the total thermo-optic noise such that

$$\bar{\alpha}_{thick}^2 \simeq \frac{2p_E^2}{R(1 + R)\xi^2} + \frac{p_R^2}{R}. \tag{9.31}$$

9.4 Example calculations

In the previous sections we have presented the theory of coating thermo-optic noise in a general way, without making reference to any particular coating structure or material choice. Here we will apply the formalism developed thus far to a few example coating structures in order to clarify and make more concrete the use of this formalism.

9.4.1 The classic high-reflector

As mentioned in Section 9.1, the most common coating structure used in high-precision optical measurements is the multi-layer dielectric high-reflector. In the following example calculation we will use a coating with a half wave cap layer of silica (SiO_2), followed by 20 quarter wave doublets of tantala (Ta_2O_5) and silica (SiO_2). The material parameters for the coating are given in Table 9.1.

Table 9.1 *These material parameter values are taken from Evans et al. (2008).*

symbol	Ta_2O_5	unit
α	3.6	$10^{-6}/K$
β	14	$10^{-6}/K$
κ	33	W/mK
C	290	$J/K\,m^3$
ρ	7200	kg/m^3
Y	140	GPa
σ	0.23	
n_H	2.06	

symbol	SiO_2	unit
α	0.51	$10^{-6}/K$
β	8	$10^{-6}/K$
κ	1.38	W/mK
C	745	$J/K\,ms$
ρ	2200	kg/m^3
Y	72	GPa
σ	0.17	
n_L	1.45	

The coating thickness d is the sum of the layer thicknesses d_k, which are *optically* a quarter wave, such that $d_k = \lambda/4n_k$. Thus, the 19 Ta_2O_5 layers have $d_{odd} = 129$ nm, and the SiO_2 layers have $d_{even} = 183$ nm. The exception is the cap layer, which is $d_0 = 367$ nm. The resulting total coating thickness is

$$d = \frac{\lambda}{4}(21/n_L + 20/n_L) = 6.4 \ \mu m. \tag{9.32}$$

We will also need to know the thermal diffusion length in the coating and substrate materials. For the substrate, this is simply a matter of taking values from Table 9.1,

$$r_{Ts} = 37 \ \mu m \sqrt{\frac{100\,Hz}{f}}. \tag{9.33}$$

While for the coating we must first compute the average thermal conductivity and heat capacity:

$$\kappa_c = \left(\sum_{k=1}^{N} \frac{1}{\kappa_k} \frac{d_k}{d} \right)^{-1} = 2.2 \ W/mK, \tag{9.34}$$

and

$$\rho_c C_c = \sum_{k=1}^{N} \rho_k C_k \frac{d_k}{d} = 1.8\,\text{MJ}/\,\text{K}\,\text{m}^3. \tag{9.35}$$

This results in a value similar to the substrate value,

$$r_{Tc} = 44\,\mu\text{m} \sqrt{\frac{100\,\text{Hz}}{f}}. \tag{9.36}$$

We are now in a position to compute the spectrum of thermal fluctuations, $S_{\hat{T}}(f)$, which makes up the first half of the measurement noise $S_{\hat{z}}(f)$. For a fused-silica substrate at room temperature, we have

$$S_{\hat{T}}(f) = 3 \times 10^{-17} \frac{\text{K}^2}{\text{Hz}} \sqrt{\frac{100\,\text{Hz}}{f}} \left(\frac{1\,\text{mm}}{w_m}\right)^2. \tag{9.37}$$

This leaves us with the second half of $S_{\hat{z}}(f)$, the overall thermo-optic coefficient $\bar{\alpha}_{TO}$, to compute. The coating thickness is small compared with the thermal diffusion length, $\xi = 0.2$ at $100\,\text{Hz}$, so we will use the simpler Equation 9.23 to compute $\bar{\alpha}_{TO}$, though we should use Equation 9.25 for frequencies above $100\,\text{Hz}$.

We will start with $\bar{\alpha}_c$, the effective coating thermal expansion coefficient. First, the effective CTE for each layer is

$$\bar{\alpha}_k = \alpha_k \frac{1 + \sigma_s}{1 - \sigma_k} \left[\frac{1 + \sigma_k}{1 + \sigma_s} + (1 - 2\sigma_s)\frac{Y_k}{Y_s}\right] \tag{9.38}$$

where Y and σ are the Young's modulus and Poisson's ratio of each coating layer, and of the substrate material. In our case we have $\bar{\alpha}_{odd} = 13 \times 10^{-6}/\,\text{K}$ for the Ta_2O_5 layers, and $\bar{\alpha}_{even} = 1.2 \times 10^{-6}/\,\text{K}$ for the SiO_2 layers. The average value for the coating is given by

$$\bar{\alpha}_c = \sum_{k=1}^{N} \bar{\alpha}_k \frac{d_k}{d} = 5.8 \times 10^{-6}/\,\text{K}. \tag{9.39}$$

Equation 9.38 simplifies for the substrate to $\bar{\alpha}_s = 2\alpha_s(1 + \sigma_s) = 1.0 \times 10^{-6}/\,\text{K}$.

The final ingredient in the calculation of $\bar{\alpha}_{TO}$ is the thermorefractive coefficient $\bar{\beta}$. A useful approximation for high-reflectors is

$$\bar{\beta} \simeq \frac{B_H + B_L(2(n_H/n_L)^2 - 1)}{4(n_H^2 - n_L^2)}, \tag{9.40}$$

where the subscripts H and L refer to the high- and low-index layers of the coating, Ta_2O_5 and SiO_2 in this example. Using Equation 9.5 we find $B_H = 40 \times 10^{-6}/\,\text{K}$ and $B_L = 10 \times 10^{-6}/\,\text{K}$, and $\bar{\beta} \simeq 8 \times 10^{-6}/\,\text{K}$, though we should note here that the value of β for both coating materials, is poorly constrained (see Section 9.5 below).

Putting these values into Equation 9.23, we find $\bar{\alpha}_{TO} = 3.3 \times 10^{-6}/\,\text{K}$. Combining this with our value for $S_{\hat{T}}(f)$, we finally arrive at the power spectral density of apparent displacement noise in our measurement of the position of a mirror with a high-reflection coating

$$S_{\hat{z}}(f) = S_{\hat{T}}(f)\,\bar{\alpha}_{TO}^2 d^2 = 1.3 \times 10^{-38}\,\frac{\text{m}^2}{\text{Hz}}\,\sqrt{\frac{100\,\text{Hz}}{f}}\left(\frac{1\,\text{mm}}{w_m}\right)^2. \tag{9.41}$$

9.4.2 Gedanken coating designs

This section will briefly explore a few single-layer coatings which serve only to demonstrate some of the features of coating thermo-optic noise. In a single layer coating, the computation of $\bar{\alpha}_{TO}$ can be done in only a few lines. The value of $\bar{\alpha}_c$ comes directly from Equation 9.38, since the coating consists of only a single layer. The value of $\bar{\beta}$ comes from Equation 9.5, combined with simple optics (see Appendix B of Evans *et al.* (2008)). The reflectivities of the vacuum-coating interface, and the coating substrate interfaces are

$$r_c = \frac{1 - n_c}{1 + n_c},\quad r_s = \frac{n_c - n_s}{n_c + n_s}. \tag{9.42}$$

Assuming that our coating layer is a quarter wave thick, the reflectivity of the coating–substrate combination is

$$r_{single} = \frac{r_c - r_s e^{-i\varphi_k}}{1 + r_c r_s e^{-i\varphi_k}}, \tag{9.43}$$

where φ_k is the deviation in the round-trip phase in the coating from π, as would be given by a perfect quarter wave layer. The change in overall reflection phase with respect to φ_k is

$$\frac{\partial \arg(r_{single})}{\partial \varphi_k} = \frac{n_c^2 - n_s^2}{n_c(1 - n_s^2)}. \tag{9.44}$$

This gives a $\bar{\beta}$ for a single layer coating of

$$\bar{\beta}_{single} = \frac{n_c^2 - n_s^2}{4n_c(1 - n_s^2)}B_c, \tag{9.45}$$

reminiscent of Equation 9.40 for a high-reflection coating. The above is derived using Equation 9.1 and the definition of B_c as the fractional change in optical path length such that

$$\frac{\partial \varphi_k}{\partial T} = \pi B_c. \tag{9.46}$$

The result is a single layer coating coefficient of

$$\bar{\alpha}_{single} = \bar{\alpha}_c - \frac{n_c^2 - n_s^2}{1 - n_s^2}B_c - \frac{C_c}{C_s}\bar{\alpha}_s, \tag{9.47}$$

where we have used the fact that for our single quarter wave coating layer $4n_c d = \lambda$.

We will now apply these results to a few cases. The first of these is the "conceptual coating" in which we simply identify a thin layer of the substrate as a "coating". Thermo-optic noise must be zero for a coating made from the substrate since it is equivalent to no coating at all, and this is borne out by Equation 9.47 since $n_c = n_s$, so the TR component vanishes, and $C_s = C_c$ such that the coating and substrate TE components cancel. Note, however, that even if the coating material is optically and mechanically identical to the substrate yet has a different heat capacity, such that $C_s \neq C_c$, $\bar{\alpha}_{single} \neq 0$ since heat flows between the coating and substrate.

Another "conceptual coating" can be made by taking a small layer of the vacuum outside of our mirror as a coating layer. Clearly, for true vacuum the heat capacity is zero, so there is no heat flow and no thermal noise. If, however, we allow the coating to remain invisible ($n_c = 1$ and $\beta_c = 0$), while having some heat capacity again we get $\bar{\alpha}_{single} \neq 0$ due to heat flow. Note that unlike the first example, this is true even if $\bar{\alpha}_c = \bar{\alpha}_s$ and $C_c = C_c$, since the TR component remains.

From these examples we can see that heat flow, and thus loss and thermal noise, will result from essentially any difference between the substrate and coating, even if the coating has no interaction with the optical field.

9.5 Measurement of thin film properties

Any discussion of thermo-optic noise would be incomplete without noting that critical materials coefficients are notoriously different from those of the same materials in bulk, and their measurement is currently an area of active research. In this section we will review various techniques being used to find these numbers.

9.5.1 Spectroscopic ellipsometry

One promising option is spectroscopic ellipsometry. Ellipsometry is a method of studying surfaces and thin layers by looking at how the polarization state of light changes on reflection from these surfaces. When light hits a surface at a non-zero angle of incidence, it undergoes partial reflectance and a phase shift, which are generally characterized by a complex reflection coefficient:

$$r_p = |r_p|e^{i\delta_p} \tag{9.48}$$
$$r_s = |r_s|e^{i\delta_s}. \tag{9.49}$$

These reflectance effects will be different for polarization in the plane of incidence (p-polarization) and perpendicular to the plane of incidence (s-polarization). Thus incident light which is linearly polarized and contains both s- and p-components will be reflected with some elliptical polarization that is a function of the index and structure of the surface.

Ellipsometry measures the ratio of these two quantities,

$$\rho \equiv \frac{r_p}{r_s} = \frac{|r_p|}{|r_s|} e^{i(\delta_p - \delta_s)}, \tag{9.50}$$

which is traditionally parameterized by the "ellipsometric quantities" Ψ and Δ, related to the magnitude ratio and phase difference respectively by

$$\tan(\Psi) = \frac{|r_p|}{|r_s|} \tag{9.51}$$

$$\Delta = \delta_p - \delta_s. \tag{9.52}$$

With a thin coating on top of a substrate, it is possible to use ellipsometric measurements at many wavelengths to determine the thickness of the coating independently of the index of refraction of the coating if there is a model for how the index of refraction changes with wavelength. A few such measurements at different temperatures will theoretically yield measurements of the coefficient of expansion of the film as well as β. These models are non-trivial though; simple Sellmeier coefficients are usually not enough, since the index of refraction across the visible spectrum is in many materials affected by the temperature-dependent position of near-UV features such as band gaps and atomic resonances. More complicated models (e.g. the Tauc–Lorentz model, see Jellison and Modine (1996) and subsequent erratum) involve more free parameters and are significantly more difficult to fit to. For more on ellipsometry see Fujiwara (2007); Azzam and Bashara (1987).

9.5.2 Cyclically heated Michelson interferometry

Work is currently under way to measure the combined thermorefractive effect of a dielectric mirror in a Michelson interferometer. A standard Michelson interferometer is set up, and length changes of the arm that contains the mirror under test are measured as the coating is cyclically heated by an amplitude modulated carbon dioxide laser.

At first glance, this should be the simplest experiment of all to understand, but in practice the dynamics of the thermal waves that penetrate from the coating into the substrate are key to understanding the response. Solutions to the heat equation in the body of a mirror with a time harmonic heat source at the surface take the form of traveling waves that decay exponentially over one wavelength,

$$T(x,t) = \frac{A}{\sqrt{2\pi f}} e^{-x/\lambda_t} e^{i(x/\lambda_t - 2\pi t)}. \tag{9.53}$$

Here A is a constant, f is the frequency of the thermal driving force, and λ_t is the thermal length scale – a function of f given by

$$\lambda_t = \sqrt{\frac{2\kappa}{2\pi f \rho C}} \tag{9.54}$$

in a material with thermal conductivity κ, density ρ, and specific heat C. For fused silica and a driving force at 100 Hz, the thermal wavelength is 55 µm.

When deriving the expected response of a mirror to this type of heating, there will be a coating response (including both thermoelastic and thermorefractive parts) proportional to $T(0, t)$, which will fall as $f^{-1/2}$ and have 45° of phase lag relative to the driving force. The substrate response will be an integral of this temperature profile multiplied by the substrate coefficient of thermal expansion (CTE), and will pick up another factor of $f^{-1/2}$ and another 45° of lag.

The results of such an experiment will reveal the overall thermo-optic coefficient of a coating relative to the known substrate CTE, including the coating volume-weighted CTEs as well as the thermorefractive coefficient. These numbers are a mix of the CTEs and βs of both materials. However, by (1) measuring coatings with different compositions, and (2) combining the results with an experiment that can isolate the CTEs (for example, cavity assisted photothermal displacement, see Section 9.5.3), both numbers could be measured for the case of interest – thin films in the geometry of a dielectric coating.

9.5.3 Cavity assisted photothermal displacement spectroscopy

A promising method for measuring the overall (volume-weighted) coefficient of thermal expansion of a coating has been demonstrated by Black *et al.* (2004b). A mirror with the coating to be tested is overcoated with a gold layer and used as the rear mirror in a Fabry–Perot cavity. A probe laser is then frequency-locked to the cavity (using the Pound–Drever–Hall, or "PDH" technique, see Section 15.2.1), and used to measure cavity length changes due to the cyclic heating of the test mirror by a pump beam. The pump beam is derived from the same laser as the probe beam (which means it also will be resonant in the cavity, increasing the power incident on the test mirror), but has an orthogonal polarization so it can be separated from the probe for length measurement purposes. The pump is then amplitude modulated at various frequencies, and the response of the cavity at the modulation frequency is measured with a lock-in amplifier.

At low pump modulation frequencies, thermal waves penetrate deep into the substrate, and the main observed effect is the substrate expanding and contracting over heating cycles. As the modulation frequency is increased, the thermal waves penetrate less deeply, and the effect of the coating expansion becomes more important, until the frequency where the coating thickness is on the same order as the thermal wavelength. At higher frequencies, the coating expansion dominates the response of the experiment, allowing the coefficient of thermal expansion for just the coating to be measured. A straightforward interpolation between these two regimes gives the expected response as a function of frequency

$$\delta L(f) \propto \left[\frac{\alpha(1+\sigma)}{f} + \left(\frac{\rho C}{\kappa \rho_c C_c} \right) \frac{\alpha_c(1+\sigma_c)}{f_t + f} \right], \tag{9.55}$$

where α is a coefficient of thermal expansion, σ is a Poisson's ratio, ρ is a density, C is a specific heat, and κ is a thermal conductivity with parameters with no subscript referring

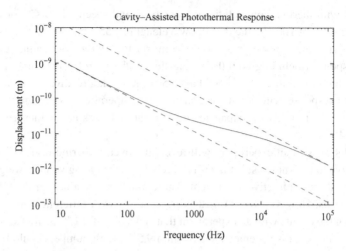

Figure 9.2 Predicted mirror expansion response for the cavity photothermal experiment assuming values similar to those in Black *et al.* (2004b). The asymptotic substrate and coating response behaviors are shown by the lower and upper dashed lines, respectively. The transition frequency $f_t \approx 10$ kHz.

to substrate values and parameters with a subscript c referring to volume averaged coating parameters. The transition frequency f_t is where the thermal wavelength is equal to the coating thickness, given by

$$f_t = \frac{\kappa_c}{\pi \rho_c C_c t^2},\tag{9.56}$$

where t is the thickness of the coating. There are two important pieces of information that can be determined from the measured response in this experiment. The lower frequency $1/f$ regime involves the substrate, whose properties are typically known – the amplitude of this regime can be used as an overall calibration. The amplitude of the $1/f$ coating dominated regime gives information about the coefficient of thermal expansion of the coating (multiplied by the appropriate expression involving the Poisson's ratio). The transition frequency between these two regions gives information about the thermal conductivity and heat capacity of the coatings, both of which are typically not well known. There is uncertainty in the transition frequency, since Equation 9.55 is just an interpolation, but an order of magnitude measurement can still be quite useful. The expected response of the cavity assisted photothermal experiment, showing the two $1/f$ regions, is shown in Figure 9.2.

This method has many benefits. Relative to most interferometric techniques, cavity assisted photothermal displacement increases the sensitivity by a factor of roughly the cavity finesse squared – one factor of the finesse for the increased PDH sensitivity, and another for the increased pump power incident on the test mirror. It also has the advantage that it can extract the coefficient of thermal expansion without "contamination" by changes in the index of refraction, since the gold layer blocks any thermorefractive effects. The down side is that this method will not give information about β. In addition, it gives a volume weighted average of the CTEs for both coating materials rather than the individual numbers themselves – a shortfall which can be theoretically overcome by testing mirrors with varying ratios of high- and low-index material.

The main experimental challenge of this method is maintaining the crossed polarization of the pump and probe beams. Since the two are overlapped and of the same wavelength, once a component of the pump has the same polarization as the probe, it will be read out directly as a signal.

9.5.4 Dielectric band edge shift

Another way to measure the thermo-optic coefficients of a coating has been discussed by Gretarsson (2008). This technique takes advantage of coating reflectivity as a function of wavelength having a "plateau" shape. The first order approach is to use a wavelength that is on the edge of the high-reflectance plateau, and then heat and cool the mirror while measuring the reflectance at that wavelength as the coating parameters change with temperature. From the change in the position of that edge with respect to a reference laser, the optical coefficients can be extracted from the shift in the plateau position. Experimentally, it is more straightforward to take a laser with a wavelength that the mirror can reflect, and have the laser light incident on the mirror at an angle for which the mirror was not designed. This allows the wavelength of the laser to be effectively increased with respect to the coating structure until it reaches an effective wavelength on the edge of the mirror's reflectance.

This measurement has the benefit of being non-interferometric (or at least single-beam, since the interference in the coating layers is what causes the reflectivity plateau), thus relatively simpler to perform. Its main drawback is that it also does not distinguish between coating layer expansion and change in index. However, as mentioned in Section 9.5.2, this technique could be combined with an experiment that measured only the CTE or β in order to extract the second value.

9.5.5 Measurements for films with thickness greater than a micron

There are a few additional techniques that deserve mention in passing that measure coefficients of films on the order of a few microns in thickness. The first is from Inci (2004) and uses a fiber interferometer in a Michelson configuration, with one fiber end uncoated (as the control "mirror"), and the other end coated with a 3 μm coating of tantala. This effectively forms an etalon reflector. The incident light is from a superluminescent diode with a bandwidth large enough to cover many free spectral ranges of the tantala etalon. By examining the shift in the interference pattern over wavelength between different temperatures, both the CTE and β can be inferred. This allows measurement of both quantities independently, unfortunately some modification is needed to extend this technique to optical sized thin films. A much broader light source would be necessary for the observation of multiple fringes in this case.

The second measurement is by Chu *et al.* (1997) and uses a Mach–Zender setup where one of the optical paths traverses a long (1.5 mm), thin (3.3 μm) waveguide of tantala. Results are compared at different temperatures. This method measures only β, since the physical expansion of the film is assumed to be determined by the known CTE of the substrate. As above, some modification would be needed to extend this technique to optical sized thin films, for which it would not be possible to build a suitably confining waveguide structure.

10

Absorption and thermal issues

PHIL WILLEMS, DAVID OTTAWAY, AND PETER BEYERSDORF

10.1 Overview

Light incident on high quality optical surfaces and substrates can be lost from the beam in two possible ways. The light can be scattered from the beam due to imperfections such as microroughness or point defects, which is discussed in Chapter 11. Alternatively, the light can be absorbed by the coating or the substrates. Experiments over many decades have shown that for high quality mirrors reflecting radiation in the IR range, the dominant loss mechanism is scatter by often more than an order of magnitude. This large bias towards loss due to scatter makes it very challenging to measure the absorption in such optics. The optical absorption of the mirror coatings in high precision applications like gravitational wave interferometers (see Chapter 14) typically ranges from a few tenths of parts per million (ppm) to several ppm. Given that the loss in mirrors is completely dominated by scatter, one may ask the question: why then is such a small level of absorption important? The reason is thermal aberrations.

Clearly, absorption at far larger than ppm levels is widely tolerated in many other types of optical instruments – metallic mirror coatings typically absorb a few percent of light incident upon them, for example. But most optical instruments are either of low power, or low precision, or both. Absorption does play a central role in setting the damage threshold intensity in high power optics. Precision measurements are typically not at such high optical power that laser damage is an issue. In the field of precision measurement, laser stabilization (see Chapter 15), cavity quantum electrodynamics (see Chapter 17), but especially gravitational wave interferometers (see Chapter 14) are among the most sensitive to thermal aberrations caused by coating absorption.

Astrophysically sensitive gravitational wave interferometers employ kilometer-scale optical cavity lengths and beam spot sizes on their mirrors of several centimeters. See Chapter 14 for more on gravitational wave detector design and Figure 14.3 in particular for the placement of optics. Such large spot sizes are required to reduce the impact of coating thermal noise on the performance of the interferometer (see Chapter 4). This requires the

Optical Coatings and Thermal Noise in Precision Measurement, eds. Gregory M. Harry, Timothy Bodiya and Riccardo DeSalvo. Published by Cambridge University Press. © Cambridge University Press 2012.

optical cavities to approach marginal stability – their cavity g-factors are close to ± 1 (see Chapter 13 for more on cavity stability). The recycling cavities for these interferometers, located before the longer cavities and which create and maintain stable beams, often maintain the same beam size, but over cavity lengths of tens of meters. These cavities can be very close to marginally stable. For example, the initial LIGO interferometers had power recycling cavities that are designed to be marginally unstable when operated at low power and rely on thermal lensing effects to make them marginally stable at their optimum operating points. Finally, the cavity finesses are high, of order 100, and the circulating power quite large, from tens to hundreds of kilowatts of stored power in the long arm cavities. These design choices enhance detector sensitivity: the long arms increase the effective displacement for a given spacetime strain (see Section 14.2), the large laser spot size averages over more thermal fluctuations in the mirror surfaces, the high finesse improves the interferometer sensitivity down into the 100 Hz region, and the high optical power reduces photon shot noise.

These choices all conspire to make the interferometers highly sensitive to thermally induced aberrations. Absorption in the arm cavity mirrors of even a part per million produces relatively large heat gradients in the mirrors. These heat gradients then produce thermorefractive phase gradients (or *thermal lenses*), thermoelastic deformations, and thermal depolarization, to which the interferometer is highly susceptible as a consequence of its nearly marginally stable cavities.

In this chapter we will discuss the levels of absorption typically found in high quality coatings and the techniques used to measure them. We will then elaborate how thermal aberrations can limit the performance of precision measurements like gravitational wave detectors with high circulating power. Finally we will discuss some methods that can prevent absorption from significantly degrading the performance of high power interferometers.

10.1.1 Measuring absorption

Low absorption levels are most commonly measured in pump-probe experiments by observing with a probe beam the effect of the thermal energy deposited in an optic by a modulated pump beam. Such measurements are linear over many orders of magnitude in absorption, allowing small loss levels to be accurately determined by comparing them to much higher (and therefore easier to directly measure) loss levels of calibration optics.

The *photothermal deflection* technique (Commandre and Roche, 1996; Jackson *et al.*, 1981) uses a high power pump beam incident on the optical coating to produce a temperature change in the substrate proportional to the absorbed power. This temperature change produces a refractive index gradient in the surrounding material as well as a deformation of the surface from the thermal expansion of the surrounding substrate. A nearby grazing probe beam deflects in the presence of the refractive index gradient and/or thermal bump on the surface, analogous to the mirage effect seen when looking along a hot road surface. The deflection, Δ, of the probe beam is measured and is related to the absorption by

$$\Delta = A_{abs} G, \tag{10.1}$$

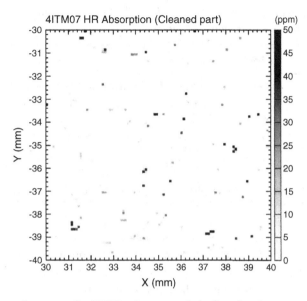

Figure 10.1 Absorption map of a LIGO mirror recorded after cleaning contamination that had developed after several years of use. This absorption map was measured using photothermal common path interferometry (Zhang *et al.*, 2008). Figure courtesy of L. Zhang.

where A_{abs} is the fractional absorption of the pump power and G depends on the thermal and optical properties of the coating and substrate as well as various experimental parameters such as the pump power.

The proportionality constant G is determined from the measurement of the photothermal deflection produced from an optic with similar thermal and optical properties, but known losses. Using a modulated 10 W pump beam with the amplitude of the deflection measured at the modulation frequency, Beauville and The Virgo Collaboration (2004) were able to measure absorption with a sensitivity of 20 ppb.

An alternative to the photothermal deflection technique, *photothermal common path interferometry* (Alexandrovski *et al.*, 2009) uses a modulated pump beam much smaller than the probe beam, and measures the amplitude change of the probe beam caused by interference between the portion traversing the thermal distortion and that traversing the undistorted part of the optic. At one Rayleigh length of the probe beam away from the sample, the Guoy phase shift is optimal for producing amplitude modulation from the interference. This technique also requires calibration of the signal to an optic with known loss. The spatial resolution provided by the small pump beam allows raster scanning of the optical surface to produce an absorption map, such as that shown in Figure 10.1. This technique has been used to measure losses in coatings with a sensitivity of 100 ppb with 10 W of pump power (Harry *et al.*, 2006a).

Other methods have been proposed for spatially mapping the absorption in low loss optical coatings using the surface thermal lens induced by absorption of a pump. Notably,

Table 10.1 *Typical absorption for tantala/silica* (Ta_2O_5/SiO_2) *coatings deposited via ion beam sputtering on superpolished substrates.*

Absorption loss	Scattering loss	Wavelength	Reference
0.6 ppm	4–8 ppm	1064 nm	Beauville and The Virgo Collaboration (2004)
0.5–13 ppm[a]	7–21 ppm	1064 nm	Zhang *et al.* (2008)
0.9–2.5 ppm	N/A	1064 nm	Harry *et al.* (2006a)
1.1 ppm[b]	N/A	850 nm	Rempe *et al.* (1992)

[a] A loss of 13.2 ppm was observed on a "contaminated" optic that had a measured loss of 1.2 ppm after cleaning.

[b] Loss here is total loss from a cavity (absorption plus scattering) but does not include small angle scattering that is captured by the cavity mirrors, which would likely be the dominant portion of the scattering loss

Hao and Li (2008) proposed a lock-in detection method that can be applied to a CCD sensor. This technique allows full frame capture of the spatial map of the absorption with acquisition speeds several orders of magnitude faster than raster scanning, but with a theoretical sensitivity comparable to photothermal deflection and photothermal common path interferometry.

The state-of-the-art low absorption multi-layer coatings are typically made from alternating layers of silica, SiO_2, and tantala, Ta_2O_5, see Chapter 2. These are the coatings used in first generation gravitational wave detectors. The absorption and scattering losses of various examples of these coatings are summarized in Table 10.1. It has been established that these coatings do not have sufficiently low mechanical loss to prevent coating thermal noise from limiting the performance of advanced gravitational wave detectors, see Chapter 4. To improve the mechanical loss various changes to the composition of the Ta_2O_5 layers have been investigated. Unfortunately, many of these alternative compositions have been measured to have greater absorption. Coatings using hafnia–silica (HfO_2–SiO_2) layers paired with silica layers have been measured to have an absorption of 6.6–7.2 ppm (Markosyan *et al.*, 2008), for example. However, titania doped tantala (TiO_2–Ta_2O_5) layers paired with silica layers have shown an absorption loss of <0.5 ppm at 1064 nm with proper coating design (see Chapter 12). This represents an improvement over the 0.9–2.5 ppm levels seen with undoped tantala (Harry *et al.*, 2006a). It is fortunate that this same titania doping into tantala also improves the mechanical loss (see Chapter 4), so TiO_2-doped Ta_2O_5/SiO_2 coatings are the preferred coatings for next generation gravitational wave detectors (Harry and The LIGO Scientific Collaboration, 2010). As an alternative route to achieving optical coatings for precision measurement, recent work has focused on the characterization of monocrystalline mirrors based on epitaxially grown compound semiconductor materials (Parker, 1985; Ludowise, 1985). Measurements of the optical absorption in multilayer mirrors based on AlGaAs heterostructures have revealed absorption values between 50–70 ppm in chemical vapor deposited epitaxial films, with the limiting absorption below 10 ppm for high purity films generated via molecular beam epitaxy (the details of this material system are covered in Section 16.2.3). The challenge will be for optimization of this materials system to yield absorption values competitive with state-of-the-art dielectric coatings.

The aforementioned methods have quantified the absorption in pristine optics, prior to their installation into working interferometers. Unfortunately, practical experience has demonstrated that optics frequently become contaminated once they are placed in a vacuum chamber. For example, an initial LIGO arm cavity mirror was discovered to have increased its absorption to over 10 ppm after a significant period within the LIGO vacuum envelope (Ottaway *et al.*, 2006; Zhang *et al.*, 2008). It is therefore necessary to have methods available that can measure the absorption of optics *in situ*. The laboratory techniques described above cannot do this. A number of techniques, described below, have been developed for directly measuring the thermal aberrations in mirrors of interferometric gravitational wave detectors. Such a measurement, coupled with knowledge of the circulating power incident on the interferometer optics, yields an absorption value.

One such technique is the white light interferometry developed by Zelenogorsky *et al.* (2006). A probe beam with a short coherence length (millimeters to centimeters) but well defined wavefront is injected into the interferometer vacuum chamber and reflected from two surfaces, one behind the thermal aberration and one in front (for example, the two faces of a thermally distorted mirror). The two return beams have different optical path lengths and propagation directions. These return beams are then reflected from a pair of reference mirrors outside the vacuum which match the wavefronts and optical paths of the two return beams to within the coherence length of the probe beam, allowing Fizeau fringes to be measured. Precision of wavefront measurement at the level of $\lambda/1000$ can be achieved.

An alternative *in situ* measurement technique utilizes a Hartmann sensor (Hartmann, 1900) that has been refined by Brooks *et al.* (2009). Using this method, it was demonstrated that optical distortions at the $\lambda/15\,000$ level are detectable. This latter method will be deployed on the Advanced LIGO interferometers to provide *in situ* absorption measurements and feedback to the thermal compensation system that will correct for the absorption-induced aberrations.

Such *in situ* measurements require sophisticated optics and cannot practically be retrofitted into existing gravitational wave interferometers. The initial LIGO and Virgo interferometers were not designed with remote wavefront sensing and so some creative methods were developed that made use of the infrastructure that was available. Betzweiser *et al.* (2005) developed a method that made use of the fact that the *g*-factor of the arm cavities changes when the radius of curvature of the optics changes due to surface heating. Ottaway *et al.* (2006) developed a method using the change in the spot size of the interferometer input beam as it reflects from a distorted arm cavity input mirror to an interferometer output port. This technique was calibrated by heating the mirror under test with a known amount of carbon dioxide laser power, and achieved sub-ppm absorption sensitivity. However, the technique cannot probe the arm cavity end mirrors, which are 4 km distant and viewable by the input beam only through the already thermally lensed input mirrors. The heating of a mirror by absorption is also detectable by the change in its acoustic resonant frequencies, which change with the temperature dependence of the density and elastic moduli of the mirror substrate. These acoustic resonances appear in the output signal of the interferometer, and by monitoring how they shift as the interferometer stored power is varied, the absorption of all the arm cavity mirrors can be measured, again to sub-ppm levels (McIvor *et al.*, 2007).

10.1.2 Treatments that affect coating absorption

Absorption can change during the life of the optic due to contamination, annealing or other treatments. The absorption properties of many materials can be improved by annealing (Alexandrovski *et al.*, 2001). For typical Ta_2O_5/SiO_2 optical coatings, annealing at 250 °C or less has been shown to increase absorption as new energy levels are formed in the band-gap due to a reduction in the degree of misorientation of the short-range ordering in the amorphous tantala (Volpyan and Yakovlev, 2002). Higher temperature annealing in N_2O can improve the stoichiometry of the oxide and reduce absorption (Milam *et al.*, 1982; Atanassova *et al.*, 2004). Increased absorption can result from surface contamination and may be improved by cleaning, and treatments of the optic may contribute to changes in absorption as well, such as the increased absorption seen after ultraviolet irradiation for discharging (see below).

It is well established that the cleanliness of the optical surface is paramount to maintaining low optical losses in an optical coating. The total integrated scatter from the mirrors of the initial LIGO interferometer were analyzed via a raster scan to map out the scattering losses on the mirrors. For mirrors with a scattering loss of 8–21 ppm averaged over the surface, localized scattering losses were as high as 1000 ppm. Additionally, the absorption losses were found to be a function of contamination on the optical surface. An absorption loss of 13.2 ppm on an optical coating was reduced to 1.2 ppm by cleaning (Zhang *et al.*, 2008). Different techniques can be used to keep the coating clean, or re-clean it after it has been contaminated. A protective cap can be kept affixed to the optic during installation, and removed just before closing the vacuum chambers. A film of First Contact[TM] can be applied to the surface of an optic before storage or transport and peeled off before use.[1] Measurements have shown no increased absorption in regions covered with the film for one week with a measurement resolution of 0.1 ppm, compared to pristine surfaces (Markosyan *et al.*, 2008).

Precision optical measurements that rely on dielectric test masses can also be adversely affected by charge that accumulates on the surface. This can be caused by friction with particulate matter or from free electrons produced by cosmic rays striking nearby matter (see Section 3.5). Accumulation of charge and its resultant attraction to external objects can create mirror displacements that may limit the sensitivity of the interferometer (Jafry and Sumner, 1997; Braginsky *et al.*, 2006b). It has been demonstrated that UV radiation can be used to remove the excess charge from mirror surfaces (Ugolini *et al.*, 2008; Pollack *et al.*,

---

[1] First Contact is available through www.photoniccleaning.com

2010; Sun *et al.*, 2006). However, the effect of ultraviolet radiation on optical coatings has been investigated and found to degrade optical losses. The absorption in Ta_2O_5/SiO_2 coatings exposed to 16 J/cm^2 of ultraviolet radiation over 250 h was observed to increase asymptotically from 0.4 ppm prior to irradiation to 4–6 ppm immediately after irradiation. Similar results were seen in coatings with other compositions; a TiO_2–Ta_2O_3/SiO_2 coating had its absorption loss increase from 1.2 ppm to 3–5 ppm. Two months after exposure to the UV radiation, however, the losses had decreased to 1–2 ppm (Sun *et al.*, 2008), showing that the degradation produced by UV exposure is temporary. However, this effect is not well understood and the mechanism responsible is uncertain.

10.2 Thermal compensation

10.2.1 Thermal aberration physics

Mirror heating leads to three physical effects that change the optical phase profile of the interferometer mirrors. Adopting the formalism of Lawrence (2003), we denote the optical path length through a transmissive optic as $OPL = nh$, where n is the refractive index of the optic and h is its thickness. A change in temperature of the optic will create a change ΔOPL in the optical path:

$$OPL + \Delta OPL(x, y) = (n + \Delta n_{TR}(x, y) + \Delta n_E(x, y))(h + \Delta h(x, y)) \quad (10.2)$$
$$\approx OPL + h\Delta n_{TR}(x, y) + n\Delta h(x, y) + h\Delta n_E(x, y).$$

The terms in this equation arise from the following.

(1) The thermorefractive effect (Δn_{TR}). The index of refraction of the mirror substrate will vary with temperature, and so a temperature gradient will produce a refractive index gradient, effectively turning the mirror into a graded index (GRIN) lens. This is sometimes referred to as *thermal lensing*. If $\Delta T(x, y)$ is the temperature profile in the mirror, then the change in OPL due to this effect is

$$\Delta OPL_{TR}(x, y) = \int_0^h \Delta n_{TR}(x, y)dz = \frac{dn}{dT} \int_0^h \Delta T(x, y)dz, \quad (10.3)$$

where we use z to label the axis through the mirror.

(2) The thermoelastic effect (Δh). The mirror substrate will expand as it is heated, so a thermal gradient will change the curvature of the mirror surface. If α is the thermal expansion coefficient of the mirror substrate, then the change in OPL due to this effect is

$$\Delta OPL_{TE}(x.y) = n\Delta h(x, y) \sim \alpha \int_0^h \Delta T(x, y)dz. \quad (10.4)$$

This formula is approximate and is most accurate in predicting surface deformation near the heated region of the mirror. Far from the heated region, thermal effects can be significant which can only be accurately calculated by a full thermoelastic model of the mirror. For example, heating of one side of a mirror will cause it to expand laterally, creating a bulk curvature of the mirror and of the unheated side, even though the heating and temperature gradient may be spatially uniform.

(3) The elasto-optic effect (Δn_E). The thermoelastic strain field in the substrate will also cause the refractive index to vary, adding to the thermal lens produced by the thermorefractive effect. Unlike the thermorefractive effect, the magnitude and shape of the lens depends upon the polarization of the light passing through it. This effectively turns the mirror into a spatially nonuniform waveplate and can cause depolarization of transmitted light. If p_{11} is the component of the elasto-optic tensor along the probe beam polarization axis, then the change in OPL due to this effect is

$$\Delta OPL_E(x, y) = -\alpha p_{11} \int_0^h \Delta T(x, y) dz. \tag{10.5}$$

The relative importance of these effects depends upon many factors. Different materials used for the mirror substrate will have different thermophysical parameters. Not to be neglected is the thermal conductivity of the material, which sets the magnitude of the thermal gradient, a higher conductivity giving a more spatially uniform temperature and therefore less aberration. The optical absorption of the substrate also contributes to the total heat load and its spatial distribution in the mirror; the same heat power deposited along the beam path in the mirror substrate as in the coating will produce approximately the same thermorefractive lens in the mirror, but much less thermal expansion of the mirror surface. Absorption, whether it occurs in the mirror coating or the mirror substrate, produces thermal gradients in the substrate that lead to the effects described above, although with different spatial profiles. The methods of compensation below are useful regardless of the location of the absorption in the coating, or substrate, or both. Current gravitational wave detectors all use ultrapure fused silica for their mirror substrates, but sapphire is also a likely material for future high power interferometers. See Chapter 7 for more on substrate materials.

The combination of these thermal aberration effects will be to produce a thermal phase profile φ across the mirror:

$$\varphi(x, y) = \frac{2\pi}{\lambda} \Delta OPL(x, y), \tag{10.6}$$

where x, y are the position coordinates transverse to the mirror surface.

Hello and Vinet have produced analytical solutions for the thermorefractive and thermoelastic aberrations in a circularly cylindrical mirror with axially symmetric heating as a Bessel function expansion (or *Dini series*) (Hello and Vinet, 1990a,b). These solutions are quite accurate for mirrors made of fused silica, for which the elasto-optic effect is negligible compared to the other two effects. Figure 10.2 shows the aberrations calculated using this method as seen from the front and rear surfaces of an Advanced LIGO input test mass mirror with spatially uniform 1 ppm optical absorption at the high reflectance coating and 10 kW of arm cavity optical power. At the front surface (the reflective, arm cavity side), the aberration is entirely due to thermoelastic deformation of the mirror, $n \Delta h(x, y)$. From the recycling cavity side, the aberration also includes the thermorefractive effect $h \Delta n_{TR}(x, y)$

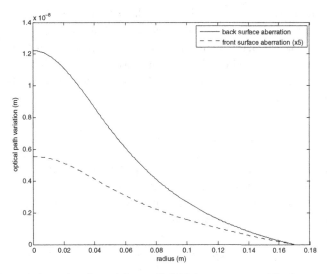

Figure 10.2 Thermal aberrations in an Advanced LIGO input test mass. The front surface aberration has been magnified $5\times$ for clarity.

and so is much larger. In fused silica, the thermorefractive coefficient is ~ 16 times the size of the thermoelastic coefficient and so dominates. Of course, the optical absorption of the coating need not be spatially uniform, nor the mirror be a perfect circular cylinder. For example, patches of contamination could lie on the surface. In this case the aberration will not have the simple profile of Figure 10.2, but can be calculated using finite element methods if the exact heating distribution is known.

Finally, although thermal aberrations are most important in the input arm cavity mirrors (we shall see why below), they also are present in other optics, such as the arm cavity end mirrors and the beamsplitter which divides the beam into two equal beams, one going to each arm. In the beamsplitter the thermal aberrations are necessarily astigmatic, due to the off-axis illumination of the optic. Thus the thermal lens is different between the two arms of the interferometer because the beam to one arm is transmitted through the beamsplitter and the beam to the other arm is not.

10.2.2 Thermal aberration and interferometer performance

An aberration profile such as that of Figure 10.2, inserted into the interferometer input arm cavity mirrors, will impair the interferometer performance in several ways. We still use the formalism of Lawrence (2003).

Distortion of the arm cavity mode

The thermoelastic "bump" in Figure 10.2 changes the effective radius of curvature of the arm cavity mirrors, thus changing the arm cavity spatial mode shape. As the arm cavity spatial mode differs from the spatial mode injected into it, the coupling into the arm cavity drops, and with it the stored power. With very high aberration the arm cavity can be made unstable, to where it has no stable resonant mode at all, if the nominal geometry is close to flat–flat. Consider a nearly flat–flat arm cavity in which the two end mirrors have the same radius of curvature R_0. The effect of the thermoelastic aberration in Figure 10.2 can be approximated as a thermal radius of curvature change R_t. The cavity will become flat–flat and therefore unstable when

$$\frac{1}{R_0} - \frac{1}{R_t^{ETM}} = -\left(\frac{1}{R_0} - \frac{1}{R_t^{ITM}}\right),\tag{10.7}$$

where ETM labels a cavity end mirror and ITM a cavity input mirror. Conversely, if the cavity is nearly concentric, the thermoelastic aberrations will make the cavity more stable. In either case, the hot cavity mode shape will differ from the cold cavity mode shape, decoupling the arm cavity from the input beam.

Distortion of the RF heterodyne sidebands in the recycling cavities

The RF sideband light that is used to control the interferometer does not resonate in the arm cavity (exceptions to this have been proposed (McClelland *et al.*, 1999), but not implemented), see Chapter 14. We show below how resonance in the arm cavity stabilizes the carrier light against thermal aberrations, but the RF sidebands' spatial mode will become distorted, especially if the recycling cavity is degenerate, as is commonly done. Since the heterodyne control signals for the interferometer rely on good overlap between the carrier and RF sideband light, this distortion will reduce the control servo gains and the impact of shot noise in the interferometer performance.

Referring to Figure 10.3, the carrier field reflected from the overcoupled arm cavity is the superposition of the field promptly reflected from the input mirror and the field leaking through the input mirror from inside the arm cavity. Each of these fields sees the thermal aberration $\varphi(x, y)$ in the input mirror substrate. However, the promptly reflected field suffers the aberration two times (as it passes through the substrate twice upon reflection), and the leakage field suffers the aberration only once. Thus,

$$E_r^{(c)} \approx r e^{2i\varphi(x,y)} E_i^{(c)} + it e^{i\varphi(x,y)} E_c^{(c)}\tag{10.8}$$

$$\approx r(1 + 2i\varphi(x, y))E_i^{(c)} - \frac{1 - r^2}{1 - r}(1 + i\varphi(x, y))E_i^{(c)}\tag{10.9}$$

$$\approx -E_i^{(c)},\tag{10.10}$$

where we have assumed that the incident field $E_i^{(c)}$ is matched to the arm cavity field $E_c^{(c)}$, that $r \approx 1$, and that the end mirror of the cavity is perfectly reflecting. Remarkably, the thermal aberration has no effect on the carrier mode inside the recycling cavity, so long as the arm cavity is overcoupled.

For the RF sideband fields, there is only the promptly reflected field, and so

$$E_r^{(sb)} \approx e^{2i\varphi(x,y)} E_i^{(sb)}.\tag{10.11}$$

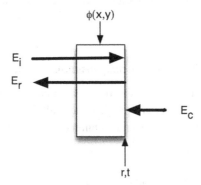

Figure 10.3 Optical fields to and from the input arm cavity mirror. E_i is the field incident from the recycling cavity, E_c is the field incident from the arm cavity, and E_r is the field leaving the mirror in the recycling cavity. The optical phase distortion in the mirror substrate is φ, and r and t are the amplitude reflectivity and transmissivity of the high-reflectance coating on the mirror's arm cavity face.

The RF sideband light is thus scattered into higher order transverse optical modes by the thermal aberration, and no longer overlaps completely with the carrier field. Moreover, the overall power of the RF sidebands in the power recycling cavity will drop as the aberrations distort the recycling cavity mode profile away from that of the field injected at the input mirror of the recycling cavity.

Distortion of the gravitational wave signal sidebands in the signal recycling cavity

Resonant sideband extraction (Mizuno *et al.*, 1993) causes the gravitational wave sidebands on the carrier light to be resonant in the signal recycling cavity on their way out of the interferometer; this tailors the frequency response of the detector. If the signal recycling cavity resonates at a different spatial mode than the arm cavities, the gravitational wave sidebands will be "bottled up" in the arm cavities, thereby limiting the efficiency with which they are transferred to the output detectors, hence reducing the interferometer sensitivity.

Reduction of the dark port contrast

The output port of the interferometer is ideally kept dark (or nearly so) to minimize shot noise. If the thermal aberrations are different between the two interferometer arms, the wavefronts returning to the beamsplitter will not match, with the result that some excess light will leak to the output port and increase the shot noise, reducing sensitivity.

Again, the relative importance of these various effects will change dramatically depending upon interferometer parameters. For example, an output mode cleaner can be used to filter out non-TEM00 light at the dark port due to differential thermal aberrations. The cavity lengths, finesses, and stabilities, presence or absence of power and signal recycling cavities, the control and readout scheme used, etc., all factor into the susceptibility of an interferometer to thermal aberrations.

10.2.3 Thermal compensation techniques

It is an unfortunate fact that the most sensitive gravitational wave interferometers now in operation or installation would be crippled by thermal aberrations if they could not be corrected in some way. Fortunately, several techniques have been developed to compensate for thermal aberrations.

Absorption compensated interferometer optical designs

It was recognized early in the initial LIGO design process that absorption in the optical coatings and substrates would have a detrimental impact on the performance of the instruments. For this reason the curvatures of the optics were set such that they were sub-optimal when the interferometer was operated in the cold state. It was then anticipated that with a known absorption the optics would thermally distort by a specified amount and the optimal optical state of the instrument would be achieved for a fixed input power level. Similar techniques are utilized to compensate for the thermal lenses that are always present in high power solid state lasers. However, the absorption in an individual optic's coating is affected by its environment and as a result is not easy to predict. Therefore, designing such a point design interferometer is not in itself reliable. Curvature offsets can be utilized to lessen the required compensation but methods such as discussed next still need to be employed to compensate individual optics.

Incandescent heaters

By positioning incandescent filaments and shields appropriately around a thermally distorted mirror, corrective thermal gradients can be added to the optic to minimize the net aberration along the optical axis. This technique was first proposed and tested by Lawrence *et al.* (2004) on a prototype interferometer mirror at the Massachusetts Institute of Technology (MIT). This is not to say that the thermal profile of the mirror is made uniform. For instance, a positive thermal lens produced in the high reflectance coating on one side of a mirror can be compensated by a negative thermal lens produced by an incandescent heater on the other side of the mirror.

Figure 10.4 shows the Lawrence incandescent heater prototype experiment at MIT. The mirror (at right) is thermally loaded by a carbon dioxide laser beam with a Gaussian spot. The heater (at left) is a coil of nickel-chromium wire wound around a ring of ceramic beads on a metal former. Surrounding the heater is a set of metallic shields which block the unwanted part of the heater's radiated power. The radiation that emerges from the shield array has a spatial profile at the mirror that compensates the thermal lens of the carbon dioxide laser beam. Figure 10.5 shows the degree of compensation that was achieved using this device.

Figure 10.4 The MIT prototype incandescent heater experiment. Figure courtesy of R. Lawrence.

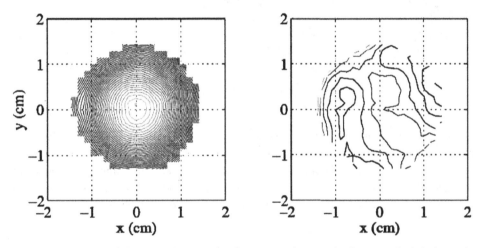

Figure 10.5 Results of the MIT prototype ring heater experiment. The figure on the left shows the transmitted phase profile contours for the thermal aberrations produced by the carbon dioxide laser alone. The figure on the right shows the contours with both the carbon dioxide laser and the ring heater. Figure courtesy of R. Lawrence.

Incandescent heaters are simple to build and operate. Because they thermally average input electrical power fluctuations before reradiating them to the mirror, they deliver very stable power. This is important, because the compensation works by adding a corrective optical path profile to the mirror, and fluctuations in the compensation power induce path length fluctuations in the interferometer, potentially decreasing its sensitivity. However, they generally require a large unimpeded solid angle around the mirror to produce the desired heating profile. As they can be practically installed only in vacuum, they do not permit much adjustment or variation of the spatial profile.

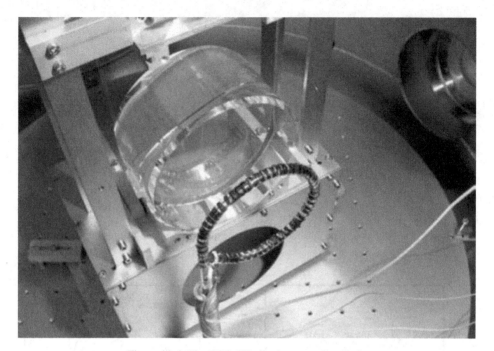

Figure 10.6 The GEO 600 ring heater as installed.

This technique is not limited to transmissive optics or to thermal aberrations. The GEO 600 gravitational-wave detector uses a radiant heater (shown in Figure 10.6) positioned behind a totally reflective folding mirror in order to correct the radius of curvature of its high-reflectance surface (Lück *et al.*, 2007). The function in this case is to correct a mismatch in the polished radius of curvature between mirrors in the two cavities.

Carbon dioxide laser projectors

Like incandescent heaters, carbon dioxide laser projectors add heat to the mirror with a spatial pattern that cancels the absorbed heat along the optical axis. These projectors are installed outside the vacuum and inject their heat via a laser beam through a viewport, and can produce a variety of different spatial profiles as needed should the absorption profile not be as expected or change over time. Projectors are either of a "staring" design, in which the spatial pattern is constant and produced by lenses, apertures and axicons (lenses with conical profiles), or of a "scanning" design, in which a beam is raster-scanned across the optic with a power or dwell time that varies from spot to spot. Compared to incandescent heaters, the carbon dioxide laser projector's convenience of pattern adjustability and operation outside the vacuum is offset by greater complexity and noisier delivered power.

Carbon dioxide laser projectors are currently in use on the LIGO and Virgo gravitational wave detectors and both are of the staring design. The optical layout of the LIGO carbon

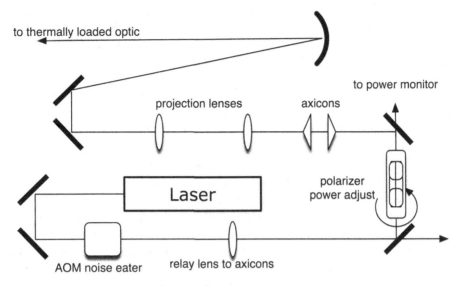

Figure 10.7 Schematic of the LIGO carbon dioxide laser projector.

dioxide laser projector is shown in Figure 10.7. The Virgo carbon dioxide laser projector is similar.

A scanning carbon dioxide laser projector design has been prototyped by Lawrence (2003). In this design, a pair of galvanometer mirrors sweeps the carbon dioxide laser beam in a spiral pattern on the thermally loaded mirror's face. The frequency of the pattern cycle is 6 Hz, slow compared to the frequency band of gravitational wave detectors but fast compared to the heat diffusion time of the thermal aberrations. At each point on the mirror, an acousto-optic modulator sets the intensity of the projected beam to achieve the desired thermal lens. The Lawrence design has the additional feature of using an active servo to determine the optimal projected spatial profile, using a real-time measurement of the thermal aberration taken by a Shack–Hartmann sensor.

A variant of the scanning carbon dioxide laser projector has been prototyped by Soloviev *et al.* (2006) in which the scanned beam, rather than repeating a raster scan on the face of the compensated mirror, maintains a constant intensity and sweeps through a fixed circular orbit on the mirror face. The pattern achieved is similar to that of a radiant ring heater.

Compensation plates

Correction of the arm cavity mirror's high reflective surface radius of curvature can only be done by acting directly on the deformed mirror. However, test mass radius of curvature deformations tend to be a small effect and less important than thermal aberrations in the recycling cavities. These recycling cavity aberrations can be corrected by applying the compensating heat pattern not only to the arm cavity input mirror directly, but also or even exclusively to an additional transmissive optic – a *compensation plate* – suspended just before the arm cavity input mirror, inside the recycling cavity. Because the wavefront of the gravitational wave interferometer beam has a radius of curvature of the order of kilometers, the compensation plate and arm cavity mirror together act as a thin lens, even though together they can be nearly half a meter thick.

The primary advantage of a compensation plate is that noise in the power applied by an incandescent heater or carbon dioxide projector will change only the apparent length of the recycling cavity, in which the compensation plate hangs, rather than that of the high-finesse arm cavity, in which the arm cavity mirror hangs and where the gravitational wave signal is generated. Since noise in the recycling cavity couples much less strongly to the output port of the interferometer, this allows more thermal compensation to be applied before its injected noise degrades the sensitivity of the detector.

The use of a separate optical element as an active or passive component in high power optical systems is not unique to gravitational wave interferometry. For example, Khazanov *et al.* (2004) have demonstrated a high power Faraday isolator using a compensation plate of opposite dn/dT to that of the Faraday crystal. In their design, the thickness of the compensation plate is chosen so that the thermal lens passively produced in the plate just corrects that of the Faraday crystal. A compensation plate that is conductively heated by a resistive element wrapped around its barrel has also been developed for thermal tuning of a laser beam profile to facilitate coupling into optical systems (Arain *et al.*, 2007), as has a design with a compensation plate heated by an auxiliary laser beam (Quetschke *et al.*, 2006).

An actively controlled, conductively heated compensation plate has been tested at the Gingin facility in Western Australia (Zhao *et al.*, 2006). This experiment was intended as a proof of principle of thermal compensation, rather than as a test of a prototype design, and so has many features that would not be applicable to an astrophysically sensitive interferometric detector. For example, their plate rested on a v-block inside the high-finesse cavity, and likely contributed significant absorption and displacement noise into the cavity. Nevertheless their experiment showed that the beam profile inside a high power optical cavity subject to thermal aberrations can be controlled.

Radiative cooling

A novel approach to thermal compensation has been demonstrated by Kamp *et al.* (2009). Rather than add extra heat to the mirror to compensate the heat absorbed from the interferometer beam, the unwanted heat can be radiated away by exposing the warmer section of the mirror to a cold surface, generally via reflective relay optics. This method has the advantage of very low injected noise, allowing direct action upon the arm cavity mirrors. Like the incandescent heater, a radiative cooler resides inside the vacuum chamber with the thermally loaded optic, and so has limited adjustability. A radiative cooler must cover a fairly large solid angle around the mirror surface to produce sufficient cooling for current high power gravitational wave interferometers.

10.2.4 Predicted impact of coating absorption on other experiments

The impact of absorption can be detrimental to other precision experiments in a number of ways, including photothermoelastic noise (see Sections 3.4 and 3.10) and reduction in achieved power build up in Fabry–Perot cavities. In this section, we will outline the limits that optical absorption can place on an experiment. The photon nature of light means that it is necessary to increase the circulating power within an optical cavity to increase the measurement sensitivity that can be achieved. This limit is known as the shot noise limit. The shot noise limit can be improved for a given circulating power by using non-classical states of light. Currently the state-of-art in non-classical light implementation allows these limits to be reduced by an amount equivalent to increasing the circulating power by a factor of 10, see also Section 11.3.

High circulating power in a Fabry–Perot cavity can cause physical displacement of the mirror surfaces by two mechanisms: radiation pressure and photothermoelastic noise. Radiation pressure is the dominant noise mechanism in the experiments that require suspended optics. This is because intensity fluctuations result in radiation pressure fluctuations which cause an oscillating force on the mirrors. Suspended mirrors are unconstrained above the resonant frequency of the suspension and hence move in response to this applied force. Hence radiation pressure is the dominant noise mechanism due to high power operation for suspended mirrors. However, most other high precision experiments use Fabry–Perot mirrors fixed by a spacer. These fixed mirrors are virtually insensitive to radiation pressure noise. Photothermoelastic noise results from the displacement of the mirror surface caused by absorbed circulating optical power. This absorbed optical power causes local heating which in turn causes the mirror surface to expand slightly due to non zero thermal expansion coefficients. In this section we will explore how coating absorption can make this noise source dominant. The level of photothermoelastic noise is directly proportional to the circulating power and the level of shot noise is inversely proportional to the square root of the circulating power. If an experiment is limited by these two noise sources then there is an optimum circulating power in which the combination of these noise sources is minimized.

The shot noise limit of a Fabry–Perot cavity in terms of the minimum power spectral density of the displacement of one of its mirror surfaces is given by

$$S_d(f) = \left(\frac{\lambda}{2\pi}\right)^2 \frac{h\nu}{8P_{circ}}, \tag{10.12}$$

where h is Planck's constant, ν is the frequency of light, λ is the wavelength of light and P_{circ} is the power circulating within the cavity.

In many cases Fabry–Perot mirrors are used as high sensitivity frequency references (see Chapter 15). Equation 10.12 can be re-written in terms of the minimum frequency noise spectral density that can be sensed:

$$S_f(f) = \left(\frac{FSR}{\pi}\right)^2 \frac{h\nu}{8P_{circ}}, \tag{10.13}$$

where FSR is the free spectral range of the cavity.

The displacement spectral density caused by photothermal noise is given by Braginsky *et al.* (1999) (see also Section 3.4)

$$S_d(f) = 2\alpha^2(1+\sigma)^2 P_{abs}^2 \frac{S_I(f)}{\pi^4 \rho^2 C^2 w_m^4 f^2}, \tag{10.14}$$

where σ is the Poisson's ratio of the mirror substrate, $P_{abs}(f) = aP_{circ}$ is the power absorbed at an oscillating frequency f, ρ is the density of the substrate, C is the specific heat of the substrate, w_m is the radius of the mode in the cavity, and a is the absorption coefficient of the coating.

The optimal circulating power is found by equating Equations 10.14 and 10.12,

$$P_{circ} = \left(\sqrt{h\lambda c} \frac{\pi f \rho C w_m^2}{8\alpha(1+\sigma)S_I(f)a} \right)^{2/3}. \tag{10.15}$$

It is interesting to consider what sort of limit this sets on the available displacement sensitivity and hence frequency stability given some currently achievable values. Consider a 30 cm long fused silica reference cavity with $a = 10^{-6}$, a 10 Hz measurement frequency, a cavity mode radius of 0.1 mm, and a relative intensity noise of 5×10^{-9} (Kwee *et al.*, 2009). With the optimum circulating power of 375 Watts, a displacement noise of 1.3×10^{-18} m/$\sqrt{\text{Hz}}$ and frequency noise of 1.3 mHz/$\sqrt{\text{Hz}}$ can be achieved. These numbers are intended as a point of reference only because there is a large possible parameter space that can be explored. There are currently no high precision experiments limited by photothermal noise.

11

Optical scatter

JOSHUA R. SMITH AND MICHAEL E. ZUCKER

11.1 Introduction

Many high precision optical measurements require high laser power, low optical loss or a combination of both to achieve their sensitivity requirements. Optical loss in an apparatus is often dominated by light scattering from coatings while optical absorption sets limits on the advances that can be made with higher optical power. In some cases, the achievable sensitivity given these limits is not sufficient and experiments require more advanced techniques such as a manipulation of the quantum behavior of light or avoiding any transmissive optics. In this chapter we give an overview of our current understanding of scattered light from high performance coatings. We also present simple overviews on squeezed light, in Section 11.3, and diffractive optics, in Section 11.4.

11.2 Optical loss

Optical power loss in experiments can result from a variety of processes; scattering, absorption, transmission, polarization effects, and imperfect quantum efficiency in detection. Some precision optical measurements are limited by the standard quantum limit, discussed in Section 1.4 and below in Section 11.3, where optical power becomes a key parameter in sensitivity. When position of a test mass is measured interferometrically, the signal will scale linearly with the laser power while the quantum shot noise increases as the square-root of power. In these situations, the signal-to-shot-noise ratio will increase proportionally to the square-root of the power. Optical loss can limit the achievable power in a system, and degrade this signal-to-noise ratio. It is also a limiting factor in the quantum noise reduction achievable with squeezed states of light, as discussed below in Section 11.3.

Absorption and thermal effects are discussed in Chapter 10. The direct optical loss due to absorption is often negligible when compared to scatter for high quality optics, but the heat generated by absorption can cause local temperature changes that lead to changes in index of refraction or material expansion, creating so-called thermal lensing. Scattering in

Optical Coatings and Thermal Noise in Precision Measurement, eds. Gregory M. Harry, Timothy Bodiya and Riccardo DeSalvo.
Published by Cambridge University Press. © Cambridge University Press 2012.

high quality optics is a direct result of surface roughness and any irregularities or impurities in the optical surfaces. In addition to reducing the overall light power, scattering can also lead to spurious noise in the experiment readout.

11.2.1 Causes of scattering

In real optical systems, surface imperfections scatter light from the nominal paths defined by ideal specular surfaces and homogeneous media. Scattered energy may simply be lost, or it may couple to the external environment and return as interference.

Local defects or foreign particles can produce a range of scattering effects, depending on their sizes and compositions. In principle, Mie theory (Mie, 1908) can be used to estimate the scattered field from each isolated anomaly. In many cases of interest, particles (or inclusions) are distributed randomly and with random heights (or depths). In such cases, the independent ensemble scatters incoherently. Nevertheless, local diffraction may still enhance the effective cross-section of each particle or defect. Perhaps counterintuitively, a particle bigger than about λ/π in radius may actually scatter several times the incident energy intersecting its geometric profile (Born and Wolf, 1999).

Particle and local defect scattering normally yield broad angular light distributions, particularly when the offenders are comparable to or smaller than the wavelength in size. We can characterize the differential fraction of energy scattered into a small element of solid angle $d\Omega$ situated at angle θ from the surface normal by the *Bidirectional Reflectance Distribution Function* or BRDF,

$$\text{BRDF} = \frac{1}{P_i} \frac{dP_s}{d\Omega \cos\theta}, \tag{11.1}$$

where P_i is the incident power and P_s is the scattered power. In the case of uniform hemispherical scattering, also known as *Lambertian* scattering, normalizing the integral of this function over 2π steradians yields

$$\text{BRDF}_{Lambertian} = \frac{R}{\pi}, \tag{11.2}$$

where $0 < R < 1$ is the scattering reflectance. The distribution of light from collections of particles and from rough, diffuse surfaces is generally Lambertian or nearly so.

For smooth surfaces, residual topography or coating index perturbations spatially modulate the optical phase, giving rise to *topographic scattering*. Most interferometric and imaging optics maintain small root-mean-square height variations, that is, $\sigma_z \ll \lambda$. As a result, all parts of the illuminated surface may contribute coherently. A comparison of the BRDF of a Lambertian scatterer and a smooth reflecting surface is shown in Figure 11.1.

Thinking of the perturbed optic as a weak phase grating, the scattered field at some angle θ from the normal will have constructive contributions from those features varying over baseline d according to the grating equation, $d \sin\theta = \lambda$. Integrating over the surface, the total energy scattered to this angle is thus determined by the surface *autocorrelation function* evaluated at d, or equivalently, by its Fourier transform, the surface *power spectral density*

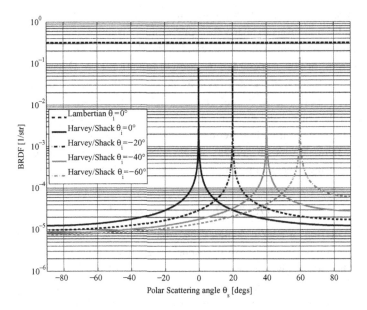

Figure 11.1 Theoretical BRDF curves for a Lambertian diffuser and a smooth reflector, plotted versus polar scatter angle θ_s for fixed azimuthal incident α_i and scatter α_s angles. Both samples have $R = 1$.

evaluated at the corresponding spatial frequency $f = d^{-1}$. Given surface perturbations $z(x, y)$ the two-dimensional power spectral density for an isotropic surface will be

$$S_2(f) = \lim_{L \to \infty} \frac{1}{L^2} \left| \int_{-L}^{L} dx \int_{-L}^{L} dy\, z(x, y) \exp(-2\pi i f\,(x + y)) \right|^2. \tag{11.3}$$

As discussed below, practical limits of integration L depend on the application. The diffractive scattering from this topography at angle θ is then given by (Stover, 1995; Elson and Bennett, 1979)

$$\mathrm{BRDF}(\theta) = \left(\frac{4\pi}{\lambda^2}\right)^2 S_2\left(\frac{\theta}{\lambda}\right). \tag{11.4}$$

Optically polished surfaces and coatings often adopt fractal surface topographies (Church, 1988), leading to inverse power-law power spectral densities of the form

$$S_2(f) = A f^{-m} \tag{11.5}$$

where m is a positive index.[1] For example, measured topography of 250 mm diameter precision mirrors deployed in gravitational wave interferometers (Abbott and The LIGO Scientific Collaboration, 2009)(see also Chapter 14) is described closely by this form, with

---

[1] Many authors use 1-D power spectral densities, which are more easily derived from profilometer measurements. Under certain assumptions, a 2-D power-law power spectral density with index m is equivalent to a 1-D power-law spectrum with index $n = m - 1$; see, for example, Yamamoto (2007).

$m \approx 2.4$, over spatial frequencies ranging from $\sim 10^6\, \mathrm{m}^{-1}$ to $\sim 10\, \mathrm{m}^{-1}$ (that is, on spatial scales ranging from microns to nearly the optic radius) (Walsh *et al.*, 1999).

There is a small problem in that the integral of the BRDF (the *total integrated scattering* or "TIS") diverges for power spectra of the form of Equation 11.5. Of course, all real optics have finite size, so there is in fact a "lowest frequency" to the power spectral density. Also, at a certain point light "scattered" into narrow angles merges with the specular fluence, say within the relevant Fresnel divergence angle or the imaging aperture, and is not really "lost". It's often better to treat low-frequency topographic phase perturbations as classical aberrations or figure errors. In general, however, appropriate angular limits for what is defined as "scattering" depend on how a given application is affected by it.

11.2.2 Effects of scattered light

In imaging, scattering of course depletes available intensity. It can also limit achievable contrast and feature detectability, spilling spurious energy from bright features into unrelated fields. The BRDF gives a straightforward measure of this interfering light flux. In particular, the typically strong angular dependence of topographic scatter (Equation 11.5) confounds the capacity to resolve dim features nearby a bright object, or to measure an object's size.

Net power loss is also an issue in laser interferometry. Unlike absorption (see Chapter 10), scattering losses generally do not also cause heating or distortion. However for quantum-limited applications (see Section 11.3), scatter losses are equally effective at introducing unwanted vacuum fluctuations, and must therefore be minimized.

Scattering may also introduce spurious phase or displacement noise in precision interferometry. An instrument resolving optical phase differences at the level of 10^{-11} radian will be degraded by infiltration of less than a part in 10^{-22} of the sensing beam's power. To illustrate the problem, we'll consider two typical examples of phase noise infiltration: pseudo-specular backscatter, and reciprocal coupling.

Consider an interferometer which compares optical paths traversed by two matched Gaussian laser beams, interfered at a beamsplitter. To derive the phase, the residual intensity must be detected somehow; this may involve relay mirrors, windows, and imaging optics to deliver the interference fringe to a photosensor and to match its sensitive area. With careful design, none of these elements will have specular surfaces aligned normal to the beam. However, each will produce a finite degree of backscatter.

The fraction spatially matched to the beam mode can be approximated by the amount falling within the mode's characteristic solid angle $\Omega_e \equiv \lambda^2 / \left(\pi w_m^2 \right)$, where w_m is the local Gaussian waist radius (Siegman, 1986). We can thus define a scattering *pseudo-reflectance*

$$\mathcal{R}_{sc} \approx \mathrm{BRDF}(\theta) \cdot \Omega_e, \qquad (11.6)$$

where θ is the angle between the beam and the surface normal. From the viewpoint of the interferometer, the backscattering optic behaves like a mirror aligned and curved to match the beam, having effective reflectance $\mathcal{R}_{sc}$.

Notice that the pseudo-reflectance varies inversely with the square of the Gaussian beam waist as projected in the vicinity of the scatterer. Reducing the beam size thus incurs the penalty of casting a wider net for backscatter. Indeed, the detector or other aperture-limiting element, such as an electro-optic crystal, may itself have high intrinsic BRDF, compounding the problem.

The interferometer phase error $\delta\varphi_0$ caused by the backscattered field can be estimated by superimposing the retroreflected field on other fields present at the beamsplitter. The total phase may wander through many radians, but presuming the motion is stochastic, the typical error will be proportional to the motion of the scattering surface δz_{sc} (Flanagan and Thorne, 1995; Fritschel, 2006):

$$\delta\varphi_0 \approx \frac{2\pi\,\delta z_{sc}}{\lambda}\sqrt{\frac{\mathcal{R}_{sc}}{2}}. \tag{11.7}$$

A second scatter-induced phase noise mechanism invokes three separate scattering processes. In this scenario, beam power impinging on an interferometer optic is scattered and strikes surrounding objects such as supports, chamber walls, or baffles. These objects rescatter the light, impressing their motion on its phase. A portion of this rescattered light falls back on the original mirror, and (since the process is symmetric) is reciprocally scattered back into the beam, contaminating its phase. In this case the effective pseudo-reflectance $\mathcal{R}_{sc}$ is proportional to the wall or environmental object BRDF as well as *two* powers of the optic BRDF, integrated over relevant angles. This mechanism is important for several applications explored in Flanagan and Thorne (1995) and Fritschel and Zucker (2010).

11.2.3 Measurement of scattering

Because of the close relationship between BRDF and the topography of optical surfaces, experimental characterization of scattering can be used to gain insight into polishing, coating and material technology, even where instrument performance is not directly impacted.

Instrumentation to measure scatter falls into two broad classes. Integrating instruments attempt to capture all energy leaving a target surface in order to assign the *total integrated scatter* (TIS) (sometimes also called *total hemispherical scatter* or "THS"). The precise method of excising the specular beam involves practical judgment based on the application and the required sensitivity. Modern sub-Ångstrom polishing methods may yield TIS in the parts per million, making reliable integrating measurement a challenge.

More information can be derived by resolving the BRDF as a function of angle. A scatterometer for this type of measurement is shown in Figure 11.2. In the variant shown, a camera or imaging array is used as a detector to enable mapping the sample surface under broad illumination. This allows one to distinguish the scattered energy due to local defects and particles from that arising from intrinsic topographic scattering. In both types of scatterometer, intrinsic scattering from the probe beam relay optics forms an irreducible measurement background, sometimes referred to as the *instrument signature*.

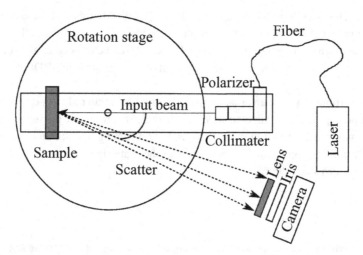

Figure 11.2 Diagram of an imaging scatterometer. The sample is illuminated by a linearly polarized and collimated laser beam at normal incidence. The azimuthal scattering angle is fixed and scattered light from the sample surface can be imaged by the lens, iris and camera. The sample and laser output are fixed to a rotation stage that allows imaging over a set of polar scattering angles from a few degrees to ninety degrees from the normal.

For simplicity, our overview has so far assumed isotropic surfaces and polar symmetry. General surfaces will not necessarily conform, and azimuthal variability is a critical indicator. The instrument shown may be generalized by moving the detector out of the plane of incidence. Polarization of scattered light is another important indicator for general surfaces, and may be probed by introducing an analyzer in front of the detector. Such augmented instrumentation is especially critical in characterizing surfaces with *deliberate* anisotropy, such as diffractive optics (see Section 11.4).

11.2.4 Reducing light scattering

For precision optical measurements that are sensitive to light scattering, care must be taken in the experiment design, the handling, and cleaning of optics. As described above, a lower limit on the BRDF of a smooth reflector is given by its surface roughness. However, the role of defects and contaminants can easily dominate this scatter. Even a small number of defects with characteristic size greater than the diffraction limit will increase the BRDF significantly (Lequime *et al.*, 2009). To avoid excess scatter, extreme care must be taken in the handling of substrates, the uniformity and purity of coatings, and the cleanliness of the optics in the experimental setup. The optics should be manufactured, polished and coated in a clean environment and with the appropriate specifications for roughness, contaminants and scratches.

After manufacture, the optics must be handled cleanly. This can be aided by specially designed cases for transportation and storage that keep the optics isolated under a light vacuum to avoid contamination by organic compounds, dust, and moisture which can penetrate the optical surface via micropores and cracks. Should the optics require cleaning, a commonly used method is drag-wiping the surfaces with a non-abrasive clean room-quality optical tissue wetted with high grade methanol (which is hydroscopic and should thus be used from a fresh bottle when possible). Another effective cleaning method is to use First Contact™, a polymer that is spread onto a surface while still liquid, and then attaches itself to surface particles as it dries.[2] Once the polymer has solidified it is peeled off, taking the surface contaminants with it. Removal of First Contact can generate significant electric charge, however, which can re-attract nearby dust. So care must be used when removing this polymer. See Section 10.1.2 for more on charging of optics.

The pernicious effects of stray light in an instrument can be reduced by using materials with high absorption and low scatter to baffle or trap the diffuse radiation. In addition, secondary beams, such as reflections off anti-reflection coatings should be identified and dumped to avoid them scattering off an auxiliary surface. In the experimental setup, baffles and beam dumps constructed from highly absorptive materials with low BRDF and installed in key locations at Brewster's angle, when possible with linearly polarized light, can effectively reduce the intensity of stray light that is re-reflected into the main beam or any measurement photodiodes. By using large optics with rigid optical mounts, scattering due to beam clipping and optic motion, respectively, can be reduced. Finally, beam waists efficiently couple scattering back into the main beam (Hild, 2007), so placing components in them should be avoided.

There are a number of ways to identify and eliminate scattered light in a functioning optical experiment. To check for dominant scattered light noise from components in an auxiliary scattering beam path, such as a monitored beam transmitted by an end mirror in an optical cavity, an optical attenuating filter (such as a neutral density filter) inserted in the path between the cavity and the scatterer can give a double-path reduction of the scattered light field re-entering the cavity. Alternatively, a transparent opto-mechanical phase-shifting device can be used in the same situation to shift the scattered light noise to a higher (out-of-band) frequency (Lück *et al.*, 2008). In both of these situations, the reduction of dominant scattered light noise can be seen as an improvement in the main measurement output. In rare cases where the scattered light source, once identified, cannot be improved or eliminated, absorptive filters or optical mechanical phase-shifting can be used to permanently reduce the coupling of the noise.

11.3 Squeezed states of light

The Heisenberg Uncertainty principle sets a fundamental limit on the minimum uncertainty product of any two quantum observables that do not commute, such as the phase and

---

[2] First Contact is available through www.photoniccleaning.com

amplitude of light. This quantum limit manifests itself as noise in precision optical measurements. In general, in un-squeezed electromagnetic fields, noise is equally distributed between amplitude and phase quadratures, and the minimum uncertainty state is given by zero-point, or vacuum, fluctuations. The noise of this un-squeezed state is referred to as the standard quantum limit, see Section 1.4. In a squeezed light state (Caves, 1981) the noise is reduced below the vacuum state for one quadrature and consequently enhanced for its conjugate. Thus squeezed light states can be used to reduce the uncertainty in the measurement quadrature and thereby improve the sensitivity of an optical measurement. Squeezed light has been demonstrated in several proof of principle experiments to reduce quantum noise in optical applications, see for example the list presented in Goda *et al.* (2008).

Squeezed states of light can be produced by creating quantum correlations between individual photons (Vahlbruch *et al.*, 2008). Often non-linear optics such as optical parametric oscillators are used to correlate the upper and lower quantum sidebands (The LIGO Scientific Collaboration, 2010) about the carrier for a given light field. State of the art squeezing has reached 10 dB quantum noise reduction (Vahlbruch *et al.*, 2008), limited by optical loss in the squeezing setup, which is dominated by imperfect photodetection.

An example of applying squeezed light states to precision optical measurement is in current and future interferometric gravitational-wave detectors whose astrophysical range is, or will be, limited by quantum noise (Caves *et al.*, 1980; Braginsky *et al.*, 1999). Phase and amplitude uncertainty manifest themselves as shot noise and radiation pressure noise, respectively, in the interferometer output signals, see Chapter 14. Shot noise, present at high frequencies, scales as the $1/\sqrt{P}$. Unfortunately, there are many technical challenges present, including thermal lensing and thermal loading for cryogenic detectors (see Chapters 8 and 10, respectively) which impose practical limits on the achievable laser power.

The use of squeezed light to surpass the quantum limit in interferometric gravitational-wave detectors was proposed by Caves (1981). Over the past several years, the injection of squeezed vacuum states has been demonstrated to reduce the quantum shot noise level in both prototype (Goda *et al.*, 2008) and kilometre-scale laser interferometric detectors (The LIGO Scientific Collaboration, 2010). The amount of noise reduction due to squeezing, F_{sqz}, is limited by the amount of squeezing produced, and the losses in the interferometer system,

$$F'_{sqz} = F_{sqz}\eta + (1 - \eta). \qquad (11.8)$$

One of the main components that contributes to the loss η is scatter and absorption from the optics. In addition, a scatterer can couple back to where the squeezing is produced and mask the quantum correlations. This effect limits the amount of squeezing present while possibly replacing quantum noise at low frequencies with amplified scattered noise.

11.4 Diffractive optics

Transmissive optics such as beamsplitters, input-coupling mirrors for optical cavities, and partially transmitting end mirrors can negatively affect precision measurements. First, coating transmissive optical substrates with stacks of dielectric materials increases the thermal noise of the optic because the mechanical loss of the dielectric coating materials is higher in general than that of the bulk materials. For a discussion of coating thermal noise and diffractive gratings, see Section 6.5.2. Second, absorption of laser power by the substrate material couples via temperature changes to the bulk index of refraction or expansion coefficient to form a so-called thermal lens as described in Chapter 10. Although higher light power is desirable from a shot-noise standpoint, the thermal deformations of the interferometer optics lead to changed beam parameters and potentially instability, setting a practical upper limit for the light power (Punturo and The Einstein Telescope Collaboration, 2010). These effects make choosing substrate and coating materials with both high optical transparency and desirable mechanical and thermal properties a significant challenge.

Diffraction gratings offer a reflective alternative to these transmissive optics. By carefully designing grating structures, reflective optics can be produced with similar optical properties to partially transmissive beamsplitters. Diffractive beamsplitters have the advantage of allowing the choice of substrate material to be optimized for thermal performance, independent of optical performance. So materials such as silicon, which can tolerate more than an order of magnitude higher circulating optical power than fused silica, can be used despite being opaque at visible and near IR wavelengths (Beyersdorf, 2001). Diffractive beamsplitters have been experimentally demonstrated in tabletop Michelson, Sagnac and Fabry–Perot interferometers (Sun and Byer, 1998; Friedrich *et al.*, 2008). Input couplers to optical cavities can also be replaced by diffraction gratings (Bunkowski *et al.*, 2004). These may be realized with either very high or very low diffraction efficiency, depending on if the reflected or diffracted beam is intended to provide the coupling to the cavity (Clausnitzer *et al.*, 2005). Low efficiency gratings formed by depositing a standard high-reflectivity dielectric stack on top of a corrugated optical substrate have been shown to have significantly less optical loss than high diffraction efficiency gratings formed by corrugating the top layer of the reflective dielectric stack (Clausnitzer *et al.*, 2005). They have been shown to have efficiencies as low as 0.02%. Grating input couplers in various configurations have been demonstrated in tabletop and suspended Fabry–Perot optical cavities with finesse as high as 1500 (Bunkowski *et al.*, 2006) and 1100 (Edgar *et al.*, 2010). Finally, gratings based on optical waveguides for use as normal incidence mirrors have recently been demonstrated with reflectivity of 99% (Brückner *et al.*, 2009) and 99.8% (Brückner *et al.*, 2010).

Figure 11.3 shows a standard transmissive beamsplitter, input coupler, and high reflectivity mirror, and their all-reflective diffraction grating equivalents. These diffractive optics allow for the construction of all-reflective optical experiments of high complexity (Barr and Burmeister, 2009). While these all-reflective technologies may reduce thermal lensing and thermal noise, they bring along other conditions that must be dealt with in the experimental setup. Beam geometry and noise coupling of alignment and translation to phase noise via

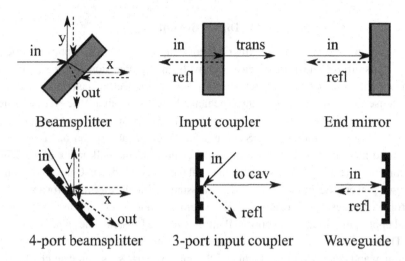

Figure 11.3 Standard transmissive optics and their all-reflective counterparts. The transmissive optics are, from upper left, a partially transmitting beamsplitter with four ports, a partially transmitting mirror used as a cavity input coupler, and a highly reflective end mirror. The diffractive optics from lower left are a four-port beamsplitter diffraction grating, a three-port input coupler diffraction grating, and a (monolithic) highly reflective waveguide.

the structured surfaces are areas where diffractive optics place more stringent requirements on the experimental setup (Freise *et al.*, 2007; Hallam *et al.*, 2009). Another concern is whether the scatter of diffractive optics is low enough for precision measurements (Siegman, 1986), although recent advances in design and manufacturing have greatly decreased optical loss in diffractive gratings.

12

Reflectivity and thickness optimization

INNOCENZO M. PINTO, MARIA PRINCIPE, AND
RICCARDO DESALVO

12.1 Introduction

This chapter is focused on design strategies for minimizing Brownian (see Chapter 4) and, more generally, thermal noises (see Chapters 3 and 9) in high-reflectivity optical coatings. It is organized as follows: in Section 12.2 we review the basic formulas needed to describe the optical properties of dielectric coatings (an *ab-initio* derivation of these formulas is included in the Appendix). Brownian noise formulas are the subject of Section 12.3. Section 12.4 presents the key ideas of coating thickness optimization. Thermo-optic noise issues are reviewed in Section 12.5, together with a discussion of pertinent minimization criteria. Section 12.6 contains a few comments on material characterization, and touches the important topic of glassy mixture modeling and optimization.

12.2 Coating formulas

In this section we summarize the basic coating formulas on which the subsequent analysis is based. A compact *ab-initio* derivation of these results is given in the Appendix.

Optical coatings are modeled as stacks of planar layers terminated on both sides by homogeneous halfspaces; the relevant geometry and notation is sketched in Figure 12.1. Layers are identified by an index $i = 1, 2, \ldots, N_L$. It is understood that $i = 0$ and $i = N_L + 1$ correspond to the left halfspace and the substrate, respectively. It is convenient to introduce a local coordinate system (x, y, z_i) for each layer, so that the internal layers $i = 1, 2, \ldots, N_L$ correspond to $-d_i \leq z_i \leq 0$, the left halfspace is defined by $-\infty < z_0 \leq 0$, and the substrate by $0 \leq z_{N_L+1} < \infty$. Plane wave incidence from the leftmost halfspace is assumed.[1] An $\exp(\imath 2\pi f t)$ time dependence of the field is understood and omitted.

---

[1] This is the usual (optical limit) approximation. The general case of an incident Gaussian beam, and its optical limit, are discussed in Hillion (1994).

Optical Coatings and Thermal Noise in Precision Measurement, eds. Gregory M. Harry, Timothy Bodiya and Riccardo DeSalvo. Published by Cambridge University Press. © Cambridge University Press 2012.

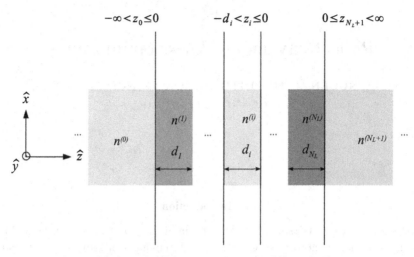

$$-\infty < z_0 \le 0 \qquad\qquad -d_i < z_i \le 0 \qquad\qquad 0 \le z_{N_L+1} < \infty$$

Figure 12.1   A stack of plane layers between different halfspaces.

12.2.1 Singlet matrix

Consider normal incidence first, where the field in each layer is a linearly polarized plane wave whose wave vector is normal to the (planar, parallel) interfaces. The properties of a *singlet*, i.e. a planar homogeneous layer with finite thickness d_i and (complex) refractive index $n^{(i)}$ are described by the transmission matrix

$$\mathbf{T}_i = \begin{bmatrix} \cos\psi_i & \iota(n^{(i)})^{-1}\sin\psi_i \\ \iota n^{(i)}\sin\psi_i & \cos\psi_i \end{bmatrix}, \tag{12.1}$$

where

$$\psi_i = \frac{2\pi}{\lambda_0} n^{(i)} d_i \tag{12.2}$$

is the phase-thickness of the layer, λ_0 being the light wavelength in vacuum. The matrix in Equation 12.1 relates the nonzero transverse components of the electric and magnetic field at the singlet terminal planes as follows

$$\begin{bmatrix} E^{(i)} \\ Z_0 H^{(i)} \end{bmatrix}_{z_i=-d_i} = \mathbf{T}_i \cdot \begin{bmatrix} E^{(i)} \\ Z_0 H^{(i)} \end{bmatrix}_{z_i=0}. \tag{12.3}$$

A vacuum characteristic impedance $Z_0 = \sqrt{\mu_0/\epsilon_0}$ factor is inserted in front of the magnetic fields in Equation 12.3 so as to make the transmission matrix dimensionless. The singlet matrix in Equation 12.1 is unimodular, i.e. $\det \mathbf{T}_i = 1$.

Maxwell's equations require the transverse field components to be continuous across the layer interfaces. Hence

$$\begin{bmatrix} E_t^{(i)} \\ Z_0 H_t^{(i)} \end{bmatrix}_{z_i=-d_i} = \begin{bmatrix} E_t^{(i-1)} \\ Z_0 H_t^{(i-1)} \end{bmatrix}_{z_{i-1}=0}. \tag{12.4}$$

Accordingly,

$$\begin{bmatrix} E^{(1)} \\ Z_0 H^{(1)} \end{bmatrix}_{z_1=-d_1} = \mathbf{T} \cdot \begin{bmatrix} E^{(N_L)} \\ Z_0 H^{(N_L)} \end{bmatrix}_{z_{N_L}=0}, \qquad (12.5)$$

where

$$\mathbf{T} = \mathbf{T}_1 \cdot \mathbf{T}_2 \cdots \cdot \mathbf{T}_{N_L} \qquad (12.6)$$

is the whole multi-layer transmission matrix.

In the substrate only a forward propagating plane wave exists, so that

$$E^{(N_L+1)} = Z^{(N_L+1)} H^{(N_L+1)} = (n^{(N_L+1)})^{-1} Z_0 H^{(N_L+1)}, \qquad (12.7)$$

$Z^{(N_L+1)} =: Z_S$ and $n^{(N_L+1)} =: n_S$ being the substrate characteristic impedance and refraction index, respectively. From Equations 12.5 and 12.7 it is possible to compute the input impedance Z_c of the whole substrate-terminated coating, and its effective refractive index n_c

$$\frac{E^{(1)}}{Z_0 H^{(1)}} =: \frac{Z_c}{Z_0} = (n_c)^{-1} = \frac{T_{11} + n_S T_{12}}{T_{21} + n_S T_{22}}. \qquad (12.8)$$

The (complex) coating reflection coefficient is thus

$$\Gamma_c = \frac{n^{(0)} - n_c}{n^{(0)} + n_c}, \qquad (12.9)$$

and the power transmittance is

$$\tau_p = 1 - |\Gamma_c|^2. \qquad (12.10)$$

Alternatively,[2] the coating reflection coefficient can be retrieved from the Airy–Schur formula:

$$\Gamma_{i-1} = \frac{\gamma_{i-1,i} + \Gamma_i \exp(-2\iota \psi_i)}{1 + \gamma_{i-1,i} \Gamma_i \exp(-2\iota \psi_i)}, \qquad (12.11)$$

where Γ_i is the ratio between the (complex) forward and backward wave amplitudes at $z_i = 0$, and

$$\gamma_{i-1,i} = \frac{n^{(i-1)} - n^{(i)}}{n^{(i-1)} + n^{(i)}} \qquad (12.12)$$

is the reflection coefficient for a wave incident at the interface from a halfspace with $n = n^{(i-1)}$ to a halfspace with $n = n^{(i)}$. The coating reflection coefficient is obtained by iterating Equation 12.11 from $i = N_L$, where $\Gamma_{N_L} = \gamma_{N_L,N_L+1}$ down to $i = 1$, where $\Gamma_1 = \Gamma_c$.

---

[2] Equations 12.6–12.9 and 12.11 have different error propagation properties. When dealing with truncated-periodic multi-layers, using Equations 12.6–12.9 yields better accuracy.

12.2.2 Oblique incidence

The normal-incidence results in the previous section are readily extended to the case of a general elliptically polarized obliquely incident plane wave. Such a wave can be always decomposed into the superposition of linearly polarized transverse-electric (TE) and transverse-magnetic (TM) plane waves, where, respectively, the electric field and the magnetic field is orthogonal to the incidence plane, defined by the normal to the interface(s) and the incident field wave vector (the xz plane in Figure 12.1). Equations 12.1–12.3 are still valid, provided the transverse fields components in Equation 12.3 are identified as

$$(E_t, H_t) = \begin{cases} (E_y, -H_x), & \text{for TE incidence} \\ (E_x, H_y), & \text{for TM incidence} \end{cases}, \qquad (12.13)$$

once the following formal substitutions are made,

$$n^{(i)} \longrightarrow n_T^{(i)} = \begin{cases} n^{(i)} \cos \theta_i, & \text{for TE incidence} \\ n^{(i)}/\cos \theta_i, & \text{for TM incidence} \end{cases}, \qquad (12.14)$$

and the phase thickness, Equation 12.2, is replaced by

$$\psi_i = \frac{2\pi}{\lambda_0} n^{(i)} d_i \cos \theta_i. \qquad (12.15)$$

The quantities n_T and $Z_T = Z_0/n_T$ are referred to as the *transverse* refraction index and characteristic impedance, respectively.

The angles θ_i ($i = 1, 2, \ldots, N_L$) and θ_{N_L+1} needed to compute the transverse indexes, Equation 12.14, and the phase thickness, Equation 12.15, are obtained from the incidence angle θ_0 by repeated application of Snell's law,

$$n^{(0)} \sin \theta_0 = n^{(1)} \sin \theta_1 = \ldots = n^{(N_L)} \sin \theta_{N_L} = n^{(N_L+1)} \sin \theta_{N_L+1}. \qquad (12.16)$$

For normal incidence $\theta_0 = 0$, the TE and TM cases are physically equivalent (except for an irrelevant rotation around the z-axis), all θ_i are zero, and $n_T^{(i)} = n^{(i)}$.

12.2.3 Optical losses

Optically lossy materials are characterized by complex refraction indices

$$\tilde{n}^{(i)} = n^{(i)} - \iota \kappa^{(i)}, \qquad (12.17)$$

with $\kappa > 0$ being known as the extinction coefficient. All Equations 12.1–12.16 remain valid, after the analytic continuation $n \to \tilde{n}$.

Snell's law, Equation 12.16, now yields *complex* values for $\sin \theta_i$, and the root sign in

$$\cos \theta_i = (1 - \sin^2 \theta_i) = \left[1 - \left(\frac{\tilde{n}^{(i-1)}}{\tilde{n}^{(i)}} \sin \theta_{i-1} \right)^2 \right]^{1/2} \qquad (12.18)$$

should be chosen so that

$$\Im(\cos\theta_i) \le 0. \tag{12.19}$$

The average power (per unit area) absorbed in all coating layers with $i \ge m$, including the substrate, is

$$\mathcal{P}_{[i \ge m]} = \frac{1}{2}\Re\left(\vec{E}_t^{(m-1)} \times \vec{H}_t^{(m-1)*} \cdot \hat{z}\right)_{z_{m-1}=0}, \tag{12.20}$$

where the asterisk denotes complex conjugation. The power absorbed in layer $i = m$ is thus

$$\mathcal{P}_m = \mathcal{P}_{[i \ge m]} - \mathcal{P}_{[i \ge m+1]}. \tag{12.21}$$

For $m = 1$

$$E_t^{(0)} = E_t^{(0),+}(1 + \Gamma_c), \quad Z_0 H_t^{(0)} = \tilde{n}_T^{(0)} E_t^{(0),+}(1 - \Gamma_c), \tag{12.22}$$

at $z_0 = 0$, and hence (assuming $\kappa^{(0)} = 0$)

$$\mathcal{P}_{[i \ge 1]} = n_T^{(0)}\mathcal{P}^{(0),+}(1 - |\Gamma_c|^2), \tag{12.23}$$

where $E_t^{(0),+}$ is the known (complex) amplitude of the incident plane wave at the coating interface, and

$$\mathcal{P}^{(0),+} = \frac{1}{2Z_0}\left|E_t^{(0),+}\right|^2 \tag{12.24}$$

is the related (known) power per unit area in vacuum. Using the inverse of Equation 12.3,

$$\begin{bmatrix} Et^{(m)} \\ Z_0 H_t^{(m)} \end{bmatrix}_{z_m=0} = \mathbf{T}_m^{-1} \cdot \begin{bmatrix} E_t^{(m-1)} \\ Z_0 H_t^{(m-1)} \end{bmatrix}_{z_{m-1}=0}. \tag{12.25}$$

Equations 12.20 and 12.21 allow recursive computation, starting from $m = 1$, the power (per unit cross section) dissipated in each layer of the coating.

12.3 Coating Brownian noise

For coatings using only two different materials (binary coatings), the (frequency-dependent) power spectral density (henceforth PSD) of Brownian noise induced random fluctuations of the mirror front face in the normal ($\hat{z}$) direction was first deduced via the Fluctuation–Dissipation Theorem (Callen and Welton, 1951) (see Chapter 1), and reads

$$S_z^{(B)}(f) = \frac{2k_B T}{\sqrt{\pi^3}f}\frac{1 - \sigma^2}{w_m Y}\phi_c, \tag{12.26}$$

where

$$\phi_c = \frac{d_1 + d_2}{\sqrt{\pi}\,w_m} \frac{1}{Y_\perp} \left\{ \left[\frac{Y}{1 - \sigma^2} - \frac{2\sigma_\perp^2 Y Y_\parallel}{Y_\perp(1 - \sigma^2)(1 - \sigma_\parallel)} \right] \phi_\perp \right.$$

$$+ \frac{Y_\parallel \sigma_\perp (1 - 2\sigma)}{(1 - \sigma_\parallel)(1 - \sigma)} (\phi_\parallel - \phi_\perp) \tag{12.27}$$

$$\left. + \frac{Y_\parallel Y_\perp (1 + \sigma)(1 - 2\sigma)^2}{Y(1 - \sigma_\parallel^2)(1 - \sigma)} \phi_\parallel \right\}$$

is the effective coating loss angle. This is Equation 4.37, reproduced here for convenience. In Equations 12.26 and 12.27, k_B is Boltzmann's constant, T the absolute temperature, w_m the laser beam width,[3] d_1 and d_2 are the total thicknesses of the two coating materials, ϕ, Y and σ the mechanical loss angle, Young's modulus and Poisson's ratio, pertinent to the substrate (no suffix), and to the coating under parallel (suffix $\parallel$) and perpendicular (suffix $\perp$) stresses. The quantities in Equation 12.27 are[4]

$$\begin{cases} Y_\perp = \dfrac{d_1 + d_2}{Y_1^{-1} d_1 + Y_2^{-1} d_2} \\[2ex] Y_\parallel = \dfrac{Y_1 d_1 + Y_2 d_2}{d_1 + d_2} \\[2ex] \phi_\perp = Y_\perp \left(\dfrac{Y_1^{-1} \phi_1 d_1 + Y_2^{-1} \phi_2 d_2}{d_1 + d_2} \right) \\[2ex] \phi_\parallel = Y_\parallel^{-1} \left(\dfrac{Y_1 \phi_1 d_1 + Y_2 \phi_2 d_2}{d_1 + d_2} \right) \end{cases} \tag{12.28}$$

These are Equations 4.33–4.36, also reproduced here for convenience. It is readily seen that letting

$$\tilde{Y}_i = Y_i(1 - \iota \phi_i), \tag{12.29}$$

where $\tilde{Y}_i$ is the complex Young's modulus, Y_i its real part, and ϕ_i the mechanical loss angle (assumed $\ll 1$), Equations 12.28 are obtained taking the real part and the argument of

$$\tilde{Y}_\perp = \langle \tilde{Y}_i \rangle_R, \quad \tilde{Y}_\parallel = \langle \tilde{Y}_i \rangle_V, \tag{12.30}$$

---

[3] The normalized beam intensity profile for a Gaussian beam is

$$\mathcal{I}(r) = \frac{2}{\pi w_m^2} \exp\left(\frac{-2r^2}{w_m^2} \right).$$

Occasionally, the beam radius $r_m = w_m/\sqrt{2}$ is used in the literature.

[4] Note that the equation for $\phi_\parallel$ in Equation 12.28 is mis-written in Harry *et al.* (2006a), whose Equation 8 entails an obvious dimensional error.

where $\langle \cdot \rangle_{R,V}$ denote the Reuss (isostress) and Voigt (isostrain) mixture-averages, respectively (see, e.g., Lakes, 2009).

The Poisson's ratios, on the other hand, may be computed from

$$\sigma_\perp = \frac{\sigma_1 Y_1 d_1 + \sigma_2 Y_2 d_2}{Y_1 d_1 + Y_2 d_2} \tag{12.31}$$

and taking the (only) positive root of the quadratic equation

$$\frac{\sigma_1 Y_1 d_1}{(1+\sigma_1)(1-2\sigma_1)} + \frac{\sigma_2 Y_2 d_2}{(1+\sigma_2)(1-2\sigma_2)} = -\frac{Y_\parallel(\sigma_\perp^2 Y_\parallel + \sigma_\parallel Y_\perp)(d_1+d_2)}{(\sigma_\parallel+1)[2\sigma_\perp^2 Y_\parallel - (1-\sigma_\parallel)Y_\perp]}. \tag{12.32}$$

The limiting form of Equations 12.27 for vanishingly small Poisson's ratios is remarkably simple [5]

$$\phi_c = \frac{d_1 + d_2}{\sqrt{\pi}\, w} \left(\frac{Y}{Y_\perp}\phi_\perp + \frac{Y_\parallel}{Y}\phi_\parallel \right). \tag{12.33}$$

Somiya (2009a) derived $\phi_\parallel$ and $\phi_\perp$ independently and showed that the formulas in Equations 12.28 are valid in the limit of vanishing Poisson's ratios. This suggests that Equation 12.33 could be a more consistent choice for use together with Equation 12.28.

12.4 Minimum noise coatings

It is seen from Equation 12.26 that coating Brownian noise can be reduced, in principle, by a number of techniques. Attention will be focused here on coating layers' thickness optimization, which has been experimentally demonstrated (Villar *et al.*, 2010b) (see also Section 5.4.3), and offers perhaps the best tradeoff to date between technological challenges/cost, and noise reduction. Other thermal noise reducing strategies are discussed in Chapters 4, 6, 8, and 13.

12.4.1 Blind coating thickness optimization

Adopting a controlled ignorance attitude, genetic optimization[6] was used to find the structure of minimal noise coatings under a prescribed transmittance constraint, treating the total number of layers and the thicknesses of each layer as free independent parameters (Agresti *et al.*, 2006). Genetic optimization was chosen in view of its known ability to handle non-convex optimization problems (Charbonneau, 2002).

The main result of this investigation was that optimized coatings display a tendency (as the number of optimization cycles is increased) toward configurations consisting of stacked identical low/high index doublets, except for the terminal (first/last) layers (Agresti *et al.*, 2006). These doublets are, however, *not* quarter wavelength, as in the standard design.

---

[5] This limiting form appears in Harry *et al.* (2002), Equation 23, with a misprint: an omitted $\perp$ suffix on the denominator of the second term in round brackets.

[6] In particular, we used PIKAIA, a public domain genetic optimization engine. The PIKAIA software is freely available at http://www.hao.ucar.edu/modeling/pikaia/pikaia.php.

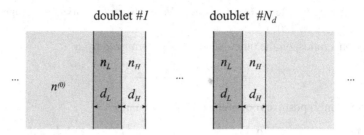

Figure 12.2 Stacked-doublets between different halfspaces.

In the optimized coatings, the amount of tantala (the noisier material, see Chapter 4) is reduced, while the number of doublets is slightly increased. On the basis of these findings, attention was focused on the analysis and optimization of coatings consisting of identical stacked-doublets.

12.4.2 Stacked-doublet coatings

Stacked-doublet coatings (sketched in Figure 12.2) are the simplest case of truncated, periodic multi-layers using only two different dielectrics. Each elementary doublet is described by the transmission matrix

$$\mathbf{D} = \mathbf{T}_L \cdot \mathbf{T}_H, \tag{12.34}$$

where

$$
\begin{cases}
D_{11} = \cos \psi_L \cos \psi_H - (n_H/n_L) \sin \psi_L \sin \psi_H \\[2mm]
D_{12} = \iota[(n_L)^{-1} \cos \psi_H \sin \psi_L + (n_H)^{-1} \cos \psi_L \sin \psi_H] \\[2mm]
D_{21} = \iota[n_L \cos \psi_H \sin \psi_L + n_H \cos \psi_L \sin \psi_H] \\[2mm]
D_{22} = \cos \psi_L \cos \psi_H - (n_L/n_H) \sin \psi_L \sin \psi_H
\end{cases}
\tag{12.35}
$$

The whole coating consisting of N_d doublets is accordingly described by the transmission matrix

$$
\mathbf{T} = \mathbf{D}^{N_d} =
\begin{bmatrix}
D_{11} - \dfrac{\Psi_{N_d-2}(\Theta)}{\Psi_{N_d-1}(\Theta)} & D_{12} \\[4mm]
D_{21} & D_{22} - \dfrac{\Psi_{N_d-2}(\Theta)}{\Psi_{N_d-1}(\Theta)}
\end{bmatrix}
\Psi_{N_d-1}(\Theta),
\tag{12.36}
$$

where

$$
\Psi_N(\Theta) = \frac{\sin[(N+1)\Theta]}{\sin \Theta}, \quad \Theta = \cos^{-1}\left[\frac{\mathrm{Tr}(\mathbf{D})}{2}\right],
\tag{12.37}
$$

and

$$\text{Tr}(\mathbf{D}) = 2 \cos \psi_1 \cos \psi_2 - \left(\frac{n_H}{n_L} + \frac{n_L}{n_H} \right) \sin \psi_1 \sin \psi_2 \tag{12.38}$$

is the trace of the doublet matrix.

The second equality in Equation 12.36 follows from a well known property of unimodular matrices (see, e.g. Born and Wolf, 1999). For computational purposes it helps to recognize that[7]

$$\Psi_N(\Theta) = U_N \left[\frac{\text{Tr}(\mathbf{D})}{2} \right], \tag{12.39}$$

with U_N being the Chebychev polynomial of the second kind.

12.4.3 Stacked-doublet coating loss angle

Equation 12.33 for the coating loss angle can be easily rewritten in terms of the low/high index medium parameters. Letting

$$z_{L,H} = \frac{n_{L,H}}{\lambda_0} d_{L,H}, \tag{12.40}$$

the layer thicknesses in units of the local wavalength, we may write

$$\phi_c = \phi_0 (z_L + \gamma z_H), \tag{12.41}$$

where

$$\phi_0 = \frac{N_d \lambda_0}{\sqrt{\pi} w} \frac{\phi_L}{n_L} \left(\frac{Y_L}{Y} + \frac{Y}{Y_L} \right), \tag{12.42}$$

and

$$\gamma = \frac{\phi_H}{\phi_L} \frac{n_L}{n_H} \frac{\left(\dfrac{Y_H}{Y} + \dfrac{Y}{Y_H} \right)}{\left(\dfrac{Y_L}{Y} + \dfrac{Y}{Y_L} \right)}. \tag{12.43}$$

According to Equation 12.41, the loss angle per unit thickness (scaled to the local wavelength) of the high-index material is γ times larger than that of the low-index material. For silica/tantala doublets, $\gamma \approx 7$ (see Chapter 4).

12.4.4 Minimal noise stacked-doublet coatings

Figure 12.3 shows a number of constant-loss-angle contours (straight lines), together with a number of iso-reflectance contours (closed curves), in the (z_L, z_H) plane.[8] It is seen

---

[7] It also helps to note that $\exp(\pm \iota \Theta)$ are the Bloch eigenvalues of $\mathbf{D}$, the corresponding eigenvectors being $\{D_{12}, \exp(\pm \iota \Theta) - D_{11}\}$. The reflection bands of the coating thus correspond to the forbidden bands of the infinite periodic structure.

[8] The former are obtained from Equations 12.41–12.43, the latter from Equations 12.8–12.10 and 12.36–12.38.

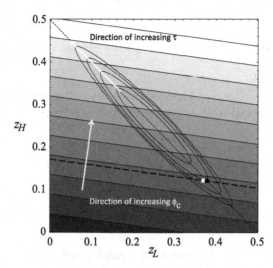

Figure 12.3 Constant transmittance (closed curves) and constant loss angle (straight lines) curves for a 10-doublet silica/tantala coating. The dotted line is $z_L + z_H = 1/2$, the dashed line is the tangent, the white marker is the minimal noise design, and the black marker is the approximate minimal noise design.

that the quarter wave design (the point $z_L = z_H = 0.25$) yields the largest reflectance at the corresponding fixed coating loss angle level. On the other hand, the design yielding the minimal coating loss angle for a prescribed transmittance corresponds to the point where the pertinent iso-reflectance contour is tangent (from above) to a constant ϕ_c contour. Such a point is the white marker in Figure 12.3. When the number of doublets becomes large (as implied by the large reflectances in order), the iso-reflectance contours squeeze unto the $z_L + z_H = 1/2$ line (the dashed line in Figure 12.3), and little error is made by taking the intersection between this latter and the iso-reflectance contour (the black marker in Figure 12.3) as the minimal noise design. Adopting this reasonable approximation[9] amounts to letting

$$z_H = \frac{1}{4} - \xi, \quad z_L = \frac{1}{4} + \xi, \quad \xi \in (0, 1/4), \tag{12.44}$$

which leaves only *two* free coating design parameters, namely, the number of doublets N_d and the quantity ξ in Equation 12.44. This leads to the simple optimization algorithm below (Agresti *et al.*, 2006; Villar *et al.*, 2010b).

(1) Start from the quarter wave design getting closest to the desired transmittance, for which $\xi = 0$ and $N_d = N_d^{(min)}$.
(2) Add one doublet, and adjust ξ until the same transmittance is recovered.

---

[9] Kondratiev *et al.* (2011) have derived a simple, implicit formula yielding the lowest order correction in the dielectric contrast to this approximation. The added accuracy, however, may be easily blurred by uncertainties and tolerances in the material parameters values.

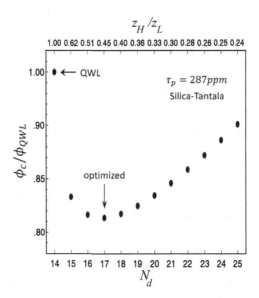

Figure 12.4 Loss angle reduction vs. number of doublets and layer thickness in a stacked-doublet coating.

(3) Calculate ϕ_c of the current coating configuration using Equations 12.41–12.44.
(4) Repeat steps (2) and (3) until the minimum of ϕ_c is reached.

The results of this procedure are illustrated in Figure 12.4 for the special case[10] of a silica/tantala coating with a transmittance of 287 ppm at 1064 nm. Increasing the number of doublets (and, in parallel, increasing ξ, to keep the transmittance fixed) has a twofold effect: the fraction of higher-index (lossier) material (tantala) is reduced, while the fraction of lower-index (less lossy) material (silica) is increased. The coating loss angle ϕ_c in Figure 12.3 has a minimum for $N_d = 17$, corresponding to the tradeoff between these competing effects.

Note that the minimum in Figure 12.4 is *shallow*. This is beneficial, both in view of uncertainties about the actual value of γ, and of possible variations due to manufacturing tolerances. The ϕ_c reduction factor featured by the optimal design with $N_d = 17$, compared to the quarter wave design, changes by less than 2% when γ is allowed to vary between 5 and 10.

The genetically optimized coatings differ from pure stacked-doublet configurations because the terminal layers are different. This suggests modifying the optimization procedure by implementing a final step which consists of *tweaking* the terminal layers as follows (Villar *et al.*, 2010b).

---

[10] These are the coating design figures for the Caltech Thermal Noise Interferometer mirrors (Black *et al.*, 2004a) discussed in Chapter 5.

(1) Tweak the top layer thickness so as to maximize the coating reflectance.
(2) Tweak the bottom layer thickness so as to bring back the reflectance to the design value.

Numerical experiments indicate that tweaking successive layers beyond the terminal ones does *not* yield any appreciable further noise reduction. There is also no advantage in seeking a minimum of the coating loss angle, at prescribed reflectance, using N_d, ξ, *and* the thicknesses of the top and bottom layers as free parameters, compared to the simplest sequential strategy where one first finds the optimal values of N_d and ξ for a pure stacked-doublet geometry, and subsequently tweaks the terminal layers, keeping N_d and ξ unchanged (Villar *et al.*, 2010b).

Optimized coating prototypes designed according to the above recipe were manufactured at the *Laboratoire des Matériaux Avancés* in Lyon, France. The noise PSD was measured using the Caltech Thermal Noise Interferometer (Black *et al.*, 2004a). See Section 5.4.3 for details and results of this experiment. Extension to dichroic coatings, featuring different transmittances at two different frequencies is possible, and relatively straightforward (Principe *et al.*, 2008).

Note that simpler coating optimization strategies, based on a single "compensating" layer, either within (Kimble *et al.*, 2008) or on top (Gorodetsky, 2008) of the coating do not yield significant reductions in the coating loss angle (Kondratiev *et al.*, 2011).

12.5 More coating thermal noises

In this section we shall briefly overview additional coating thermal noise terms. Within the limits of present day technologies, Brownian noise is by far the dominant coating noise term in most applications, but progress in optical materials and coating design may change the situation in a few years.

12.5.1 Coating thermo-optic noise

Fluctuations in the coating temperature may have a twofold origin; thermodynamic and photothermal. The latter stem from fluctuations in the laser intensity, resulting into fluctuations of the power dissipated in the coating. The power spectral densities of both kinds of temperature fluctuations have been derived in Braginsky *et al.* (2000) and Braginsky *et al.* (1999) (see also Chapter 3), and are, respectively,

$$S_{\Delta T}^{(\Theta)}(f) = \frac{2k_{\mathrm{B}}T^2}{\sqrt{\pi^3}w_m^2\sqrt{f\kappa_S C_S\rho_S}},$$ (12.45)

$$S_{\Delta T}^{(\Phi)}(f) = \frac{4P_{\mathrm{abs}}E_\lambda}{4\pi^3 w_m^4 \kappa_S\rho_S C_S f},$$ (12.46)

where k_{B} is Boltzmann's constant, T the temperature, w_m the beam amplitude radius, κ_S, ρ_S and C_S are the thermal conductivity, density, and heat capacity of the substrate, P_{abs} is the power dissipated in the coating due to optical losses, E_λ the beam photon energy, and f the frequency.

The underlying mechanisms being independent, the thermodynamic and photothermal fluctuations are uncorrelated, and their power spectral densities add incoherently to form the power spectral density of the coating temperature fluctuations,

$$S_{\Delta T}(f) = S_{\Delta T}^{(\Theta)}(f) + S_{\Delta T}^{(\Phi)}(f). \tag{12.47}$$

Temperature fluctuations in cavity mirrors imply fluctuations in the cavity phase length, via thermal expansion (thermoelastic noise), and temperature dependence of the refraction index of the coating materials (thermorefractive effect). See Chapter 9 for a discussion of coating thermoelastic and thermorefractive noises. It is expedient to define the quantities[11] α_{eff}, β_{eff}

$$\frac{\Delta z^{(TE)}}{d_{\text{tot}}} = \alpha_{\text{eff}} \Delta T \tag{12.48}$$

$$\frac{\Delta z^{(TR)}}{\lambda_0} = -\beta_{\text{eff}} \Delta T \tag{12.49}$$

relating, respectively, the (actual or equivalent) displacement of the mirror front face to a temperature change ΔT, with $d_{\text{tot}} = N_d(d_L + d_H)$ the total thickness of the coating, and λ_0 the beam wavelength. The superscripts TE and TR in Equations 12.48 and 12.49 identify the thermoelastic and thermorefractive displacement components, and the quantities α_{eff} and β_{eff} are the coating thermoelastic and thermorefractive coefficients.

The power spectral densities of the thermoelastic and/or thermorefractive displacements can be simply obtained thereafter, by the Wiener–Khinchin Theorem,

$$S_z(f) = \mathcal{F}_{\tau \to f} \langle \Delta z(t) \Delta z(t + \tau) \rangle_t = |H_{\Delta z}|^2 S_{\Delta T}(f), \tag{12.50}$$

where $\mathcal{F}$ is the Fourier transform operator, the angle brackets denote (integral) time averaging, and $H_{\Delta z}$ denotes the spectral transfer function connecting Δz to ΔT.

12.5.2 Thermoelastic coefficient

A formula for α_{eff} was first derived by Braginsky and Vyatchanin (2003a) (the post-publication version of this paper (v5) available in ArXiv (Braginsky and Vyatchanin, 2003b) contains important fixes and additions) and independently obtained by Fejer *et al.* (2004). Merging those results, it is possible to write

$$\alpha_{\text{eff}} = \alpha_L \frac{d_L}{d_L + d_H} + \alpha_H \frac{d_H}{d_L + d_H} \tag{12.51}$$

where

$$\alpha_{L,H} = 2(1 + \sigma_S) \left\{ \frac{\alpha_{L,H}}{2(1 - \sigma_{L,H})} \left[\frac{1 + \sigma_{L,H}}{1 + \sigma_S} + (1 - 2\sigma_S) \frac{Y_{L,H}}{Y_S} \right] - \alpha_S \frac{C_{L,H}}{C_S} \right\}.$$

$$\cdot \left(\frac{g(f)}{g(0)} \right)^{1/2} \Xi_{\text{fsm}}^{1/2}. \tag{12.52}$$

---

[11] In general, Equations 12.48 and 12.49 should be understood as being written in the spectral domain, α_{eff} and β_{eff} representing frequency-dependent complex transfer functions.

The frequency-dependent factor $g(f)$ was introduced in Fejer *et al.* (2004), and is

$$g(f) = \Im\left\{ -\frac{\sinh[(\iota 2\pi f \tau_f)^{1/2}]}{(\iota 2\pi f \tau_f)^{1/2}\left[\cosh[(\iota 2\pi f \tau_f)^{1/2}] + R\sinh[(\iota 2\pi f \tau_f)^{1/2}]\right]} \right\} \tag{12.53}$$

with

$$\tau_f = \frac{(d_L + d_H)C_f}{\kappa_f}, \quad R = \left(\frac{\kappa_f C_f}{\kappa_s C_s}\right)^{1/2}, \tag{12.54}$$

and

$$\kappa_f = (d_L + d_H)\left(\frac{d_L}{\kappa_L} + \frac{d_H}{\kappa_H}\right)^{-1} = \langle\kappa_i\rangle_R,$$

$$\tag{12.55}$$

$$C_f = \frac{d_L C_L + d_H C_H}{d_L + d_H} = \langle C_i\rangle_V.$$

The factor $\Xi_{\text{fsm}}(f)$, accounting for the finite size of the mirror, was first derived by Braginsky and Vyatchanin (2003a), and can be written

$$\Xi_{\text{fsm}} = \bar{w}_m \frac{(\Pi/2)^{1/2}}{\sqrt{2}\Lambda(1 + \sigma_S)}. \tag{12.56}$$

where $\bar{w}_m = w_m/R_m$, R_m being the mirror radius. The quantities Λ and Π in Equation 12.56 are given by Braginsky and Vyatchanin (2003a);

$$\Lambda = \sum_{i=L,H} \frac{d_i}{d_L + d_H}\left[-\frac{C_i}{C_S} + \frac{\alpha_i(1 + \sigma_i)}{2\alpha_S(1 - \sigma_i)(1 + \sigma_S)} + \frac{Y_i(1 - 2\sigma_S)}{Y_S(1 - \sigma_i)} \right] \tag{12.57}$$

and

$$\Pi = [U + VS_1]^2 + S_2 \tag{12.58}$$

where[12]

$$U = \sum_{i=L,H} \frac{d_i}{d_L + d_H}\left[\frac{\alpha_S}{\alpha_i}\left(\frac{1 + \sigma_i}{1 - \sigma_i} - \frac{2\sigma_S Y_i}{E_S(1 - \sigma_i)}\right) - \frac{C_i}{C_S} \right], \tag{12.59}$$

$$V = \sum_{i=L,H} \frac{d_i}{d_L + d_H}\left(\frac{C_i}{C_S} - \frac{\alpha_i Y_i(1 - \sigma_s)}{\alpha_S Y_s(1 - \sigma_i)} \right), \tag{12.60}$$

$$S_1 = 12\bar{h}^{-2}\sum_{m=0}^{\infty} \frac{\exp(-\zeta_m^{(1)}\bar{r}_0^2/4)}{(\zeta_m^{(1)})^2 J_0(\zeta_m^{(1)})}, \tag{12.61}$$

$$S_2 = \sum_{m=1}^{\infty} \Lambda_m^2 \frac{\exp[-(\zeta_m^{(1)}\bar{r}_0)^2/2]}{J_0^2(\zeta_m^{(1)})}. \tag{12.62}$$

---

[12] The infinite sums in $S_{1,2}$ converge rapidly, 30 terms are sufficient to achieve 16 figure precision.

In Equations 12.61 and 12.62, $\zeta_m^{(1)}$ is the mth zero of Bessel function J_1, $\bar{h} = H_m/R_m$, with H_m being the mirror thickness, and

$$\Lambda_m = A + BY_m, \tag{12.63}$$

with

$$A = \sum_{i=L,H} \frac{d_i}{d_L + d_H} \left[\frac{\alpha_i(1+\sigma_i)}{\alpha_S(1-\sigma_i)} - \frac{1+\sigma_S}{1-\sigma_i} \right], \tag{12.64}$$

$$\dot{B} = \sum_{i=L,H} \frac{d_i}{d_L + d_H} \left[\frac{\alpha_i Y_i(1-2\sigma_S)}{\alpha_S Y_S(1-\sigma_i)} + \frac{1}{1-\sigma_i} - 2\frac{C_i}{C_S} \right], \tag{12.65}$$

and

$$Y_m = \frac{(1+\sigma_S)[1 - \exp(-\sqrt{2}\zeta_m^{(1)}\bar{w}_m)]}{[1 - \exp(-\sqrt{2}\zeta_m^{(1)}\bar{w}_m)] - 4(\zeta_m^{(1)}\bar{h})^2 \exp(-\sqrt{2}\zeta_m^{(1)}\bar{w}_m)}. \tag{12.66}$$

12.5.3 Thermorefractive coefficient

A simple closed form expression for the thermorefractive coefficient β_{eff} valid for quarter wave coatings was given in Braginsky *et al.* (2000). This was based on a self-consistency argument valid for high reflection coatings, for which the addition of a further doublet does not change appreciably the coating input impedance, yielding

$$\beta_{\text{eff}} = \frac{n_H^2 \beta_L + n_L^2 \beta_H}{4(n_H^2 - n_L^2)}. \tag{12.67}$$

The same formula was obtained in Principe *et al.* (2007) using a different route (complete induction), disproving an alternative formula in Braginsky and Vyatchanin (2003b). The argument in Braginsky *et al.* (2000) can be extended to a general stacked-doublet coating to give (Principe *et al.*, 2007)

$$\beta_{\text{eff}} = \frac{1}{2\pi\iota} \frac{\dot{\overline{Y}}_c}{1 - \overline{Y}_c^2} \tag{12.68}$$

where $\overline{Y}_c$ is the coating input admittance (normalized to the vacuum), and the dot denotes derivative with respect to temperature. Both $\overline{Y}_c$ and $\dot{\overline{Y}}_c$ can be written in terms of the doublet matrix elements in Equation 12.35 as

$$\overline{Y}_c = -\frac{(D_{11} - D_{22}) + \sqrt{(D_{11} - D_{22})^2 - 4D_{12}D_{21}}}{2D_{12}}, \tag{12.69}$$

$$\dot{\overline{Y}}_c = \frac{\dot{D}_{21} + \overline{Y}_c(\dot{D}_{22} - \dot{D}_{11}) - \overline{Y}_c^2 \dot{D}_{12}}{D_{11} - D_{22} + 2\overline{Y}_c D_{12}}. \tag{12.70}$$

Both the thermoelastic and thermorefractive coefficients can be minimized by reducing the relative amount of high index (more noisy) material in the coating. This is similar to the Brownian case, and suggests that even if thermoelastic and/or thermorefractive noise were comparable to Brownian noise, one may still use the optimization strategy discussed in Section 12.4.4 to minimize the *total* coating noise.

12.5.4 Thermo-optic noise cancelation

It is expected on physical grounds (coating thickness and field build-up time are much smaller than the corresponding decorrelation scales of the temperature fluctuations) that thermoelastic and thermorefractive displacements should be added coherently[13] so that

$$|H_{\Delta z}|^2 = |d_{tot}\alpha_{eff} - \lambda_0\beta_{eff}|^2 \tag{12.71}$$

in Equation 12.50. As shown in Evans *et al.* (2008) and Chapter 9, adding coherently the thermoelastic and thermorefractive terms with the right relative signs entails partial cancelation between the two terms. Experimental checks of this cancelation are under way at the California Institute of Technology (see Section 9.5.2). Exact cancelation may occur (assuming positive values for the α_i and the β_i) for some specific, non-quarter wave coating configurations, in a frequency-dependent way, due to the f-dependent factor in the thermoelastic coefficient, Equation 12.52.

However, coating designs optimized for minimal Brownian noise turn out to be also *nearly* optimized, at least throughout the spectral band of interest for gravitational wave observations (see Chapter 14), when thermo-optic noise is included.

12.6 Material parameters and related uncertainties

Noise calculations and coating design optimization are dependent on the availability of reliable values for the pertinent material parameters. In this section we shall limit our discussion to the coating materials tantala and silica, as being common in precision measurements (see Chapter 2). Similar results will hold for other materials. The coating loss angle of silica thin films has been measured to high accuracy, yielding $\phi = (5 \pm 3) \times 10^{-5}$. The loss angle for tantala and titania-doped tantala (see Chapters 2 and 4) has been recently obtained from a direct thermal noise measurement (see Chapter 5) using different geometries and materials. The estimated values are[14] $(4.7 \pm 0.5) \times 10^{-4}$ for tantala and $(3.7 \pm 0.3) \times 10^{-4}$ for 16% titania-doped tantala. These values differ somewhat from those obtained by measuring the damping constant of cantilever or membrane shaped specimens (see Section 4.2.1), for reasons yet to be understood. However, being consistently estimated *directly* from measured

---

[13] In the extreme opposite case where thermoelastic and thermorefractive displacements were totally uncorrelated,

$$|H_{\Delta z}|^2 = |d_{tot}\alpha_{eff}|^2 + |\lambda_0\beta_{eff}|^2.$$

[14] These values are obtained using Equation 12.33. If Equation 12.27 is used instead, the estimated values are $(6.3 \pm 0.4) \times 10^{-4}$ for tantala, and $(4.9 \pm 0.6) \times 10^{-4}$ for titania-doped tantala.

noise power spectral densities and from the noise model in Section 12.3, these numbers are perhaps the most reliable ones obtained so far for coating optimization. The values presently in use for all other relevant parameters (elastic modulus Y, Poisson's ratio σ, etc.) are fiducially assumed as being equal to their bulk counterparts.

Thermoelastic and thermorefractive coefficients are presently known with much less accuracy. For tantala, values of α ranging from $-(4.43 \pm 0.05) \times 10^{-4}$ K^{-1} (Inci and Yoshino, 2000) to $(5 \pm 2) \times 10^{-6}$ K^{-1} (Braginsky and Vyatchanin, 2003a) and values for β ranging from $(2.3 \pm 2) \times 10^{-6}$ K^{-1} (Braginsky and Vyatchanin, 2003a) to $(1.2 \pm 2.) \times 10^{-4}$ K^{-1} (Inci, 2004) have been reported, possibly due to different manufacturing technologies. These uncertainties, however, have almost no impact on total coating noise, and coating design optimization for minimal noise.

Reliable numbers are expected from ongoing direct measurements on actual coating prototypes, see Section 9.5.4. These measurements face the basic difficulty of disentangling the thermoelastic and thermorefractive contributions (Gretarsson, 2008). Measurements taken at different wavelengths could be effective, provided a suitable model for the wavelength dependence of the α, β parameters is assumed.

Concerning titania–tantala mixtures, it is worth noting that sputtered titania films are known to exhibit a *negative* thermorefractive coefficient β (Xie *et al.*, 2008). Co-sputtered mixtures involving titania may thus be expected to exhibit positive or negative β, depending on the titania concentration. This has been observed, e.g., in silica/titania mixtures (Hirota *et al.*, 2005).

12.6.1 Beyond pure glasses – mixtures

The most successful attempt to date, in the search for optical materials featuring low mechanical loss angle and high refractive index, has been the introduction of co-sputtered titania–tantala mixtures (Harry *et al.*, 2007). The relevant design criteria, however, are to a large extent proprietary and undisclosed. On the other hand, the use of glassy mixtures in optics is by no means new, and is relatively well studied (see, e.g., Stenzel *et al.*, 2011).

Titania is perhaps the most interesting mixture candidate,[15] in view of its high refractive index and low mechanical loss angle (Scott and MacCrone, 1968). The optical and mechanical properties of co-sputtered mixtures based on both silica–titania (Chao *et al.*, 2001; Netterfield and Gross, 2007) and titania–tantala (Chao *et al.*, 2001; Harry *et al.*, 2007) have been investigated by several groups.

Modeling efforts aimed at understanding the properties of mixtures are also under way. Both a microscopic approach (see Chapter 4) and a macroscopic one (effective-medium theory), have been proposed and are being pursued. Preliminary results based on effective-medium theory reproduce nicely the measured properties of titania-doped tantala co-sputtered mixtures, and suggest that mixtures consisting of successively sputtered

---

[15] Crystallization, occurring already at thicknesses $\sim$50 nm and entailing an increase in the optical and mechanical losses, limits the use of pure titania for optical coatings (Wang and Chao, 1998).

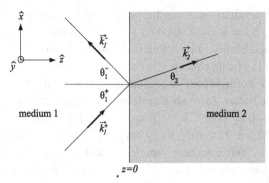

Figure 12.5 Plane waves at the interface between different halfspaces.

alternating sub-wavelength layers of silica and titania may feature an even better tradeoff between high optical index and low mechanical loss angle (Pinto *et al.*, 2010).

12.7 Appendix: Coating formulas primer

In this Appendix useful coating formulas are derived from first principles.

12.7.1 Plane waves

The electric and magnetic field of a monochromatic plane wave can be written:

$$\vec{e}(\vec{r}, t) = \Re[\vec{E}(\vec{r})\exp(\imath\Omega t)], \ \ \vec{h}(\vec{r}, t) = \Re[\vec{H}(\vec{r})\exp(\imath\Omega t)], \tag{12.72}$$

where the complex field vectors (i.e. phasors) are given by

$$\vec{E}(\vec{r}) = \vec{E}_0\exp(-\imath\vec{k}\cdot\vec{r}), \ \ \vec{H}(\vec{r}) = \vec{H}_0\exp(-\imath\vec{k}\cdot\vec{r}), \tag{12.73}$$

Ω being the (angular) frequency, and $\vec{k}$ the wave vector.

The source-free Maxwell equations in a time-invariant, spatially homogeneous and isotropic medium with constitutive parameters ϵ and μ yield

$$\begin{cases} k^2 = \Omega^2\epsilon\mu, \\ \hat{k}\cdot\vec{E}_0 = \hat{k}\cdot\vec{H}_0 = 0, \\ \hat{k}\times\vec{E}_0 = Z\vec{H}_0 \end{cases} \tag{12.74}$$

where $Z = \sqrt{\mu/\epsilon}$ is the medium characteristic impedance, and $\hat{k} = \vec{k}/k$ is the unit wave vector.

12.7.2 Interface between different halfspaces

Consider a plane wave impinging on the planar interface between two halfspaces with different time-invariant, spatially homogeneous and isotropic constitutive parameters ϵ and μ.

The problem's geometry is sketched in Figure 12.5. We adopt a Cartesian reference system with $z=0$ at the interface and $\hat{z}$ pointing toward medium 2.

The pertinent solution of Maxwell equations consists of three plane waves. Besides the incident wave, a reflected wave exists in the $z \leq 0$ halfspace, and a transmitted wave exists in $z \geq 0$. The relevant field phasors will be denoted as $(\vec{E}^{(1)+}, \vec{H}^{(1)+})$, $(\vec{E}^{(1)-}, \vec{H}^{(1)-})$ and $(\vec{E}^{(2)+}, \vec{H}^{(2)+})$, respectively, with the corresponding wave vectors

$$\begin{cases} \vec{k}_1^+ = k_1(\hat{x} \sin \theta_1^+ + \hat{z} \cos \theta_1^+) \\ \vec{k}_1^- = k_1(\hat{x} \sin \theta_1^- - \hat{z} \cos \theta_1^-) \,, \\ \vec{k}_2 = k_2(\hat{x} \sin \theta_2 + \hat{z} \cos \theta_2) \end{cases} \tag{12.75}$$

where the angles θ_1^+, θ_1^- and θ_2 are defined in Figure 12.5. The plane spanned by the vectors $\hat{z}$ and $\hat{k}_1^+$ (the zx plane in Figure 12.5) is referred to as the incidence plane.

Equations 12.74 imply that,

$$E_{0z} = -\hat{k}_z^{-1}(\hat{k}_x E_{0x} + \hat{k}_y E_{0y}), \quad H_{0z} = -\hat{k}_z^{-1}(\hat{k}_x H_{0x} + \hat{k}_y H_{0y}), \tag{12.76}$$

so that, in general, knowledge of the (x, y)-components alone (also referred to as *transverse*) is sufficient to reconstruct the whole $\vec{E}_0$ and $\vec{H}_0$ field vectors in Equation 12.73.

In the following, for the sake of simplicity, $\mu_{1,2} = \mu_0$, so that

$$Z_{1,2} = \frac{Z_0}{n_{(1,2)}}, \quad k_{1,2} = n_{(1,2)}k_0, \tag{12.77}$$

$n_{(1,2)} = \sqrt{\epsilon_{1,2}/\epsilon_0}$ being the refraction index, $Z_0 = \sqrt{\mu_0/\epsilon_0}$ the vacuum characteristic impedance, and $k_0 = \Omega\sqrt{\epsilon_0\mu_0} = 2\pi/\lambda_0$ the wavenumber in vacuum. A general (elliptically polarized) incident plane wave can be written as a superposition of a transverse-electric (TE) and transverse-magnetic (TM) plane wave, where, respectively, the electric or magnetic field is orthogonal to the incidence plane.

The nonzero transverse field components are, in view of Equation 12.74,

$$\begin{cases} E_{0y}, \ H_{0x} = -Z^{-1}\hat{k}_z E_{0y} & \text{(TE)} \\ H_{0y}, \ E_{0x} = Z\hat{k}_z H_{0y} & \text{(TM)} \end{cases} . \tag{12.78}$$

From Equations 12.78 and 12.76 it is seen that $\vec{E}_0$ and $\vec{H}_0$ can be derived from E_{0y} (or H_{0x}) alone in the TE case, and from H_{0y} (or E_{0x}) alone in the TM case.

Using a superscript $+$ $(-)$ to identify waves propagating in the forward (backward) z-direction, Equations 12.78 and 12.75 yield

$$Z_T^\pm := \begin{cases} \dfrac{E_{0y}^\pm}{H_{0x}^\pm} = \mp\dfrac{Z}{\cos\theta} = \mp\dfrac{Z_0}{n_T} & \text{(TE)} \\[3mm] \dfrac{E_{0x}^\pm}{H_{0y}^\pm} = \pm Z\cos\theta = \pm\dfrac{Z_0}{n_T} & \text{(TM)} \end{cases}, \tag{12.79}$$

where the $Z_T^\pm$ is referred to as the (forward, backward) *transverse impedance*, and

$$n_T = \begin{cases} n\cos\theta & \text{(TE)} \\ n/\cos\theta & \text{(TM)} \end{cases}, \tag{12.80}$$

is referred to as the (TE, TM) *transverse refraction index* of the medium.

Maxwell equations entail continuity of the transverse components of the *total* electric and magnetic fields across the interface at $z = 0$. Enforcing such continuity using Equations 12.75 and 12.77 in Equation 12.78 yields Descartes's and Snell laws,

$$k_{1x}^+ = k_{1x}^- = k_{2x} \iff \theta_1^+ = \theta_1^-, \ n_{(1)} \sin \theta_1^+ = n_{(2)} \sin \theta_2, \tag{12.81}$$

(we shall accordingly henceforth drop the $+, -$ superscript in θ_1), together with the Fresnel formulas

$$\Gamma := \begin{cases} \dfrac{E_{0y}^{(1),-}}{E_{0y}^{(1),+}} = \dfrac{n_{(1)} \cos \theta_1 - n_{(2)} \cos \theta_2}{n_{(1)} \cos \theta_1 + n_{(2)} \cos \theta_2} = -\dfrac{H_{0x}^{(1),-}}{H_{0x}^{(1),+}} \quad \text{(TE)} \\[2em] \dfrac{E_{0x}^{(1),-}}{E_{0x}^{(1),+}} = \dfrac{n_{(1)}/\cos \theta_1 - n_{(2)}/\cos \theta_2}{n_{(1)}/\cos \theta_1 + n_{(2)}/\cos \theta_2} = -\dfrac{H_{0y}^{(1),-}}{H_{0y}^{(1),+}} \quad \text{(TM)} \end{cases}, \tag{12.82}$$

where θ_2 is related to θ_1 by Equation 12.81. Clearly, it is possible to write

$$\Gamma = \frac{n_T^{(1)} - n_T^{(2)}}{n_T^{(1)} + n_T^{(2)}} =: \gamma_{12}, \tag{12.83}$$

for both the TE and TM case, using the appropriate formula for n_T from Equation 12.80, with $n = n_{(i)}$ and $\theta = \theta_i$. The quantity γ_{12} is referred to as the halfspace (or interfacial) plane-wave reflection coefficient from material 1 to material 2. Also,

$$\left.\begin{array}{l} \dfrac{E_{0y}^{(2),+}}{E_{0y}^{(1),+}} \quad \text{(TE)} \\[2em] \dfrac{E_{0x}^{(2),+}}{E_{0x}^{(1),+}} \quad \text{(TM)} \end{array}\right\} = 1 + \Gamma = \frac{2n_T^{(1)}}{n_T^{(1)} + n_T^{(2)}} =: \tau_{12}, \tag{12.84}$$

and

$$\left.\begin{array}{l} \dfrac{H_{0x}^{(2),+}}{H_{0x}^{(1),+}} \quad \text{(TE)} \\[2em] \dfrac{H_{0y}^{(2),+}}{H_{0y}^{(1),+}} \quad \text{(TM)} \end{array}\right\} = 1 - \Gamma = \frac{2n_T^{(2)}}{n_T^{(1)} + n_T^{(2)}} = \frac{n_T^{(2)}}{n_T^{(1)}} \tau_{12}, \tag{12.85}$$

using the appropriate transverse indexes from Equation 12.80. The quantity τ_{12} is referred to as the halfspace (or interfacial) plane-wave transmission coefficient from material 1 to material 2.

Summing up, the transverse fields in $z \leq 0$ can be written

$$\begin{cases} \vec{E}_T = \vec{E}_0^{(1),+} \left[\exp(-\imath \vec{k}_1^+ \cdot \vec{r}) + \gamma_{12} \exp(-\imath \vec{k}_1^- \cdot \vec{r}) \right] \\[1.5em] Z_0 \vec{H}_T = n_T^{(1)} (\hat{z} \times \vec{E}_0^{(1),+}) \left[\exp(-\imath \vec{k}_1^+ \cdot \vec{r}) - \gamma_{12} \exp(-\imath \vec{k}_1^- \cdot \vec{r}) \right] \end{cases}. \tag{12.86}$$

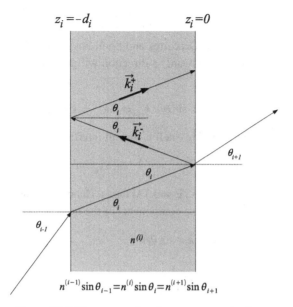

Figure 12.6 Plane waves in a homogeneous plan layer.

The fields in $z \geq 0$ can be written

$$
\begin{cases}
\vec{E}_T = \tau_{12} \vec{E}_0^{(1),+} \exp(-\imath \vec{k}_2^+ \cdot \vec{r}) \\
\\
Z_0 \vec{H}_T = n_T^{(2)} (\hat{z} \times \tau_{12} \vec{E}_0^{(1),+}) \exp(-\imath \vec{k}_2^+ \cdot \vec{r})
\end{cases}
\tag{12.87}
$$

As written, Equations 12.86 and 12.87 hold true for both the TE and TM case, where, respectively, $\vec{E}_0^{(1),+} = E_{0y}^{(1),+} \hat{y}$ and $\vec{E}_0^{(1),+} = E_{0x}^{(1),+} \hat{x}$, and $n_T^{(i)}$ is given by Equation 12.80 with $n = n_{(i)}$ and $\theta = \theta_i$.

For normal incidence $\theta_1 = \theta_2 = 0$, the TE and TM cases are physically equivalent (differing by an irrelevant $\pi/2$ rotation around the z-axis) and $n_T^{(i)} = n_{(i)}$.

12.7.3 *Plane layered media*

Consider a stack of planar layers, sandwiched between two different half spaces. The relevant geometry and notation is sketched in Figure 12.6. Layers will be identified by an index $i = 1, 2, \ldots, N_L$. It is understood that $i = 0$ and $i = N_L + 1$ correspond to the left halfspace and the substrate, respectively.[16]

A plane wave incident from the left halfspace will produce a cascade of plane waves propagating in the forward and backward z-direction, originating at the interfaces.

---

[16] The more realistic case where the refraction index is not piecewise constants, and partial inter-diffusion between successive layers exists, has been analyzed in Ignatchenko and Laletin (2004).

Plane waves in a layer

It is easy to prove that, in view of Descartes and Snell laws, all plane waves propagating in layer-i in the forward (resp. backward) z-direction will have the *same* wave vector $\vec{k}_i^+$ (resp. $\vec{k}_i^-$), with (see Figure 12.6)

$$k_{ix}^+ = k_{ix}^- = k_i \sin \theta_i, \ \ k_{iz}^+ = -k_{iz}^- = k_i \cos \theta_i, \tag{12.88}$$

where the θ_i are obtained by applying Snell law at all interfaces,

$$n^{(0)} \sin \theta_0 = n^{(1)} \sin \theta_1 = \ldots = n^{(N_L)} \sin \theta_{N_L} = n^{(N_L+1)} \sin \theta_{N_L+1}. \tag{12.89}$$

It is again convenient to focus on the TE and TM cases. Hence, in view of Equations 12.75, 12.77, 12.78 and 12.81, the transverse components of the field in layer-i can be written

$$
\begin{cases}
E_t^{(i)}(x, z) = F^{(i)}(z) \exp(\imath k_i x \sin \theta_i) \\[2mm]
Z_0 H_t^{(i)}(x, z) = G^{(i)}(z) \exp(\imath k_i x \sin \theta_i) \\[2mm]
\text{with} \\[2mm]
F^{(i)}(z) = E_0^{(i),+} \exp(-\imath k_i z \cos \theta_i) + E_0^{(i),-} \exp(\imath k_i z \cos \theta_i) \\[2mm]
G^{(i)}(z) = n_T^{(i)} [E_0^{(i),+} \exp(-\imath k_i z \cos \theta_i) - E_0^{(i),-} \exp(\imath k_i z \cos \theta_i)]
\end{cases}
\tag{12.90}
$$

where $(E_t^{(i)}, H_t^{(i)})$ represent, respectively, $(E_y^{(i)}, -H_x^{(i)})$ for the TE case, and $(E_x^{(i)}, H_y^{(i)})$ for the TM case, $n_T^{(i)}$ is given by Equation 12.80 with $n = n_{(i)}$ and $\theta = \theta_i$, and the superscripts $+, -$ identify the forward and backward propagating plane wave in the layer.

Layer transmission matrix

It is expedient to introduce a *local* reference system (x, y, z_i) for each layer, so that the layer corresponds to $-d_i \leq z_i \leq 0$ (see Figure 12.6). Letting $z_i = 0$ in Equation 12.90 it is possible to write $E_0^{(i),\pm}$ in terms of $F(0)$ and $G(0)$,

$$E_0^{(i),\pm} = \frac{1}{2} \left[F^{(i)}(0) \pm (n_T^{(i)})^{-1} G^{(i)}(0) \right]. \tag{12.91}$$

Using Equation 12.91 in Equation 12.90, and letting $z_i = -d_i$, yields

$$
\begin{bmatrix} E_t^{(i)} \\ Z_0 H_t^{(i)} \end{bmatrix}_{z_i = -d_i} =
\begin{bmatrix} \cos \psi_i & \imath (n_T^{(i)})^{-1} \sin \psi_i \\ \imath n_T^{(i)} \sin \psi_i & \cos \psi_i \end{bmatrix}
\begin{bmatrix} E_t^{(i)} \\ Z_0 H_t^{(i)} \end{bmatrix}_{z_i = 0},
\tag{12.92}
$$

where

$$\psi_i = \frac{2\pi d_i}{\lambda_0} n_{(i)} \cos \theta_i \tag{12.93}$$

is referred to as the *phase thickness* of the layer. Equation 12.92 relates the transverse field components at the terminal planes of a single homogeneous layer (or *singlet*).

Note that Maxwell equations imply continuity of the transverse field components across the interfaces, whence

$$
\begin{bmatrix} E_t^{(i)} \\ Z_0 H_t^{(i)} \end{bmatrix}_{z_i=-d_i} = \begin{bmatrix} E_t^{(i-1)} \\ Z_0 H_t^{(i-1)} \end{bmatrix}_{z_{i-1}=0} .
\tag{12.94}
$$

Airy–Schur formula

Define the reflection coefficient in layer-i as the (complex) ratio

$$
\Gamma_i = \frac{E_0^{(i),-}}{E_0^{(i),+}}
\tag{12.95}
$$

between the transverse components of the forward and backward propagating waves at $z_i = 0$. This definition is consistent with Equation 12.82.

Using Equation 12.95 in Equation 12.90 and enforcing Equation 12.94 at the interface $z_{i-1} = 0$ also known as $z_i = -d_i$ between layer-$(i-1)$ and layer-i yields

$$
\begin{cases}
E_0^{(i-1),+}(1+\Gamma_{i-1}) = E_0^{(i),+}\left[\exp(\imath\psi_i) + \Gamma_i \exp(-\imath\psi_i)\right] \\
n_T^{(i-1)} E_0^{(i-1),+}(1-\Gamma_{i-1}) = n_T^{(i)} E_0^{(i),+}\left[\exp(\imath\psi_i) - \Gamma_i \exp(-\imath\psi_i)\right]
\end{cases},
\tag{12.96}
$$

where ψ_i is given by Equation 12.93.

The solutions of this homogeneous system of equations in $E_0^{(i-1),+}$ and $E_0^{(i),+}$ must be non-zero, hence its determinant must vanish, yielding

$$
\Gamma_{i-1} = \frac{\gamma_{i-1,i} + \Gamma_i \exp(-2\imath\psi_i)}{1 + \gamma_{i-1,i} \, \Gamma_i \exp(-2\imath\psi_i)},
\tag{12.97}
$$

where

$$
\gamma_{i-1,i} = \frac{n_T^{(i-1)} - n_T^{(i)}}{n_T^{(i-1)} + n_T^{(i)}}
\tag{12.98}
$$

is the interfacial reflection coefficient between two half spaces with the same refraction index as layers $(i-1)$ and i. Equation 12.97, known as Airy–Schur formula, relates the reflection coefficients pertinent to neighboring layers.

13

Beam shaping

ANDREAS FREISE

13.1 Introduction

In this chapter we review recent research on using alternative beam shapes for reducing thermal noise and other thermal effects in mirrors. High-precision laser interferometry experiments typically make use of the fundamental Gaussian beam, which can be generated with the great spatial stability that is important for achieving low-noise signal readouts. However, the Gaussian beam might not be the optimal choice for all high-precision measurements. The idea of using an alternative beam geometry, specifically a flat-top intensity profile, to reduce mirror thermal noise in optical cavities was discussed first by Kip Thorne and his research group at Caltech in 2000 (Thorne *et al.*, 2000). Since then several groups have made progress in taking this idea closer to reality. Despite these efforts, no high-precision interferometric measurement actually showing lower thermal noise has so far been undertaken with alternative beam shapes.

Alternative beam shapes improve thermal noise, and specifically coating thermal noise, by effectively increasing the w_m parameter of Equation 4.9 and subsequent related equations. Straightforward increasing of the beam width w_m of a Gaussian beam will work up to a point, but the optical loss from light spilling over the mirror edge generally will be unacceptable at some level. Thus, with purely Gaussian beams, there is a tradeoff between lower thermal noise (larger w_m) and lower shot noise (higher optical power, thus smaller w_m). Using beam shapes other than Gaussian is a way to change this tradeoff, because alternative beam shapes can be effectively larger while keeping the light mostly contained on the face of the optic.

Most research so far has concentrated on the theoretical understanding of alternative beam shapes under the constraints of high-precision instrumentation. The theory involved is straightforward classical optics but used with a slightly new twist. New beam shapes have been proposed and new mathematical frameworks developed to describe their propagation and interaction with optics. In addition, numerical simulations have been used to verify the performance of alternative beam shapes in the presence of defects and deviations

Optical Coatings and Thermal Noise in Precision Measurement, eds. Gregory M. Harry, Timothy Bodiya and Riccardo DeSalvo. Published by Cambridge University Press. © Cambridge University Press 2012.

from a simplified theoretical optical system. Furthermore, first experiments with prototype interferometers have been performed or are under way. Vinet (2009) provides an in-depth, mostly analytical, review on beam shaping and its prospects for reducing thermal noise and thermal effects. In this chapter we review the ongoing research on alternative beam shapes for reducing thermal noise with a special focus on the practical design criteria for finding the optimal beam shape.

13.1.1 Interferometry with optical resonators

With the availability of ultra-stable laser sources, the use of laser interferometers for precision measurements has become commonplace. Laser interferometers are intrinsically well suited to comparing lengths and optical frequencies; in this type of application, typically the entire cross section of the laser beam can contribute to the interference signal. Thus, not spatial but *temporal* interference fringes are used to observe the optical signal.

Current interferometric gravitational wave detectors are Michelson interferometers enhanced with optical resonators. In order to improve the signal-to-shot-noise ratio, modern gravitational wave detectors include optical resonators in the arms of the Michelson interferometer. These and other techniques have led to a reduction of various noise sources so that in current state-of-the-art detectors (which are under construction) the thermal noise of the mirrors of the optical resonators will be a limiting noise source in a large section in the measurement band, see Chapters 4 and 14.

At the same time developments in other areas of precision measurement have also led to thermal noise limitations: the frequency stabilization of lasers with compact optical resonators, especially in the context of optical clocks or frequency standards as well as cavity optomechanical and quantum-electrodynamic experiments. Ongoing research is trying to improve the linewidth of lasers to reach a frequency stability beyond the level of one part in 10^{17}, see Chapter 15, improve the measurements of the quantum properties of macroscopic objects, see Chapter 16, and improve the measurements of the quantum nature of interactions between light and atoms, see Chapter 17.

13.1.2 Alternative beam shapes

The most common type of laser beam is the Gaussian beam, so called because of its radial Gaussian intensity profile. This type of beam is generated as the fundamental mode shape in an optical resonator using spherical (concave, convex or flat) mirrors. One of the key features of a Gaussian beam is that its beam shape and its propagation through optical systems can be fully described analytically. In addition, the Gaussian beam retains its shape over propagation, i.e. it remains Gaussian after passing a spherical lens or after being reflected by flat or spherical mirrors.

However, in many applications the use of beams with a different intensity profile can be beneficial or even essential. The most common examples are the use of so-called flat beams in which the light power is more evenly distributed. Several analytical descriptions of such beams have been suggested, such as the super-Gaussian beam (SGB) by de Silvestri et al. (1988), the flattened Gaussian beam (FGB) by Gori (1994) or the flat-topped multi-Gaussian beam (FMGB), proposed by Tovar (2001). The flat variants of the Gaussian beam are often used in unstable, high-power laser resonators in order to maximize the filling of the active medium for the maximum power output of the laser. However, they can also be useful for the optimized application of laser energy in many types of application; a more exotic type of usage of flat-top beams, for example, is laser paint-stripping (Forbes et al., 2009).

Another now quite common application of alternative beam shapes is the use of ring shaped beams (also called *doughnut modes*) in the field of cold atoms and quantum optics. In this field, beams with an intensity minimum at the center can be used as waveguide for cold atoms or for optical traps, see Bongs et al. (2001) and Kuga et al. (1997). More recently it has been recognized that certain types of Laguerre–Gauss modes carry orbital angular momentum and can be utilized for new applications such as quantum cryptography and quantum communications, see, for example, Gibson et al. (2004) and references within.

The following sections will introduce the concept of spatial modes for describing beam shapes and will recall the main features of the Gaussian beam and a number of alternative beam shapes discussed in the literature. Currently, nearly all research into thermal noise reduction with alternative beam shapes concentrates on reducing mirror thermal noise in nearly symmetric linear optical cavities. In Section 13.2 we discuss what constraints the choice of optical setup puts on the type of beam shape that can be used.

13.1.3 Optical resonators and eigenmodes

Laser optics with optical resonators is based on research on laser resonators, originally reviewed by Kogelnik and Li (1966). A very good resource for laser optics is Siegman (1986) and a recent review on interferometer techniques for gravitational wave detectors has been published by Freise and Strain (2010).

Optical cavities are employed in precision experiments to resonantly enhance circulating fields in order to increase the signal-to-noise ratio; in many cases with respect to the shot noise of the photo detection process. In order to achieve the resonant enhancement, the cavity must be constructed such that the incoming light field can interfere constructively with the field after one round-trip in the cavity. This is achieved when the phases of the fields inside the cavity are well matched, i.e. when the following conditions apply.

- The round-trip length of the cavity is set to be an exact multiple of the laser wavelength.
- The wave front, i.e. the transversal phase distribution, of the field after one round-trip is identical to that of the incoming field. In the case of linear cavities with Gaussian beams this is achieved by using spherical mirrors whose reflective surfaces are shaped

identically to the wave front of the beams, so that each beam is reflected back exactly into the same shape. Bélanger and Paré (1991) showed that the same method also works for non-Gaussian beam shapes and non-spherical wave fronts.

With the phase matched as described above, the beam interferes constructively and also remains the same shape over time; it can thus be considered an eigenmode of the system.[1]

The requirement on the wave front matching is one of the reasons why Gaussian beams are simple to generate and to use: their wave front is spherical (or flat at the beam waist) and simply requires mirrors with a spherical surface of the correct radius of curvature. In comparison, beams with other intensity patterns typically develop much more complex wave fronts and thus their use in optical resonators would require new mirror shapes which are currently more difficult to produce with the same precision as spherical mirrors. Mirror requirements in some precision measurements can be so tight that post machining processes like ion milling and/or corrective coatings are required. Once these processes are introduced, the gap in production difficulty between traditional spherical and numerical controlled shaped mirrors is greatly reduced. See Section 2.7 for more on the production of coatings for alternative beam shapes. Deviations from the required surface cause light to be lost through scattering, or may even produce phase noise in the cavity through back scattering from external surfaces, potentially spoiling any high-precision measurement. See Chapter 11 for more on scattering. Therefore, in designing a resonator for alternative beam shapes it is important to pay special attention to its robustness against the likely reduction in mirror surface quality and/or dynamic deviations from the optimal profile like thermal lensing (see Chapter 10) and gravitational sag.

We will later use some specific features of optical cavities, in particular those related to beam sizes. It is useful to derive the basic parameters for such a cavity with spherical mirrors using Gaussian optics. In the case of a two-mirror cavity the size of the beam can be computed conveniently from the stability parameters g_1, g_2, given as

$$g_{1,2} = 1 - \frac{L}{R_{C\,1,2}}, \tag{13.1}$$

with L the length of the cavity and $R_{C\,1,2}$ the radii of curvature of the input and end mirror, respectively. In the case of a symmetric cavity ($g = g_1 = g_2$) we can compute the beam size on the mirrors as

$$w_m^2 = \frac{L\lambda}{\pi} \sqrt{\frac{1}{1-g^2}} = \frac{\lambda}{\pi} \sqrt{\frac{R_C L}{2 - \frac{L}{R_C}}}. \tag{13.2}$$

In order to have a stable resonator the magnitude of the g-parameter must be smaller than 1. Using Equation 13.2 we see that the minimal beam size at the mirrors occurs at $L = R_C$ and is given by $w_{\min} = \sqrt{L\lambda/\pi}$, which corresponds to a minimal waist size in the center

---

[1] We ignore the possibility of using cavity mirrors with a non-uniform reflectivity for shaping the intensity pattern. This method has been used in connection with flat beams, however, it imposes extra optical losses and does not seem well suited for precision interferometric measurements.

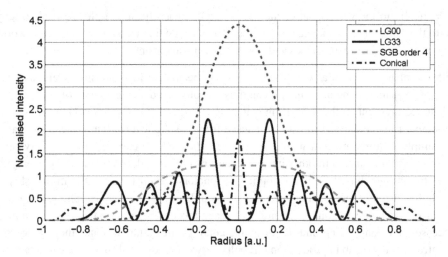

Figure 13.1 Beam profile of several different types of beams. Shown are the normalized intensity patterns (for 1 ppm clipping loss) of a fundamental Gaussian beam (LG00) a higher-order Laguerre–Gauss mode (LG33), a super-Gaussian beam of order 4 (SGB order 4) and a specific multiple-Gauss flat-top beam, the "Conical beam".

of the cavity of

$$w_{0,\text{min}} = \sqrt{\frac{\lambda L}{2\pi}}.$$
(13.3)

We will make use of this fact when discussing multi-Gauss flat beams.

13.2 Designing the right beam shape

In the following sections we will describe various alternative beam shapes. In particular, the mathematical details and several graphs are presented for a few selected alternative shapes which can be considered exemplary for different families of beam shapes.

Figure 13.1 shows four different example beam profiles. The reference beam shape is the fundamental Gaussian beam LG_{00} and the three "flat" beams are a higher-order Laguerre–Gauss mode LG_{33}, a Conical beam (which is a particular example from the family of flattened multi-Gauss beam) and a super-Gaussian beam (de Silvestri *et al.*, 1988). One can see clearly that in comparison to the fundamental LG_{00} mode, the alternative beams have a much more widely spread intensity distribution for a given mirror radius (and for a given clipping loss, see Section 13.2.2).

When designing a beam shape which can reduce the mirror thermal noise in a realistic optical experiment, one has to take into account several aspects.

- Thermal noise reduction: the main question is how much a different beam can reduce the thermal noise.

- Optical loss: the beam size of any beam must be chosen such that the beam fits onto the finite-size cavity mirror.
- Cavity stability: the high-precision experiments discussed here rely on stable and stationary setups. Alternative beams must be stable eigenmodes of the optical cavity and they should be robust against small optical defects such as misalignment.
- Practical considerations: the availability of special mirrors to match the wave front shape, how to generate the alternative beam in question, whether there exists an analytic description of beam propagation (which can significantly help in the design and commissioning phase of an experiment), and any mode degeneracies that would allow other modes in the cavity.

Some of these requirements are discussed in more detail below.

13.2.1 Thermal noise reduction with alternative beam shapes

Historically, mirror thermal noise is divided into different categories, all of which are covered in detail in this book, see Chapters 3, 4, 7, and 9. In this section we show briefly how to compute the thermal noise reduction for the coating Brownian noise when moving from a fundamental Gaussian beam to a higher-order Laguerre–Gauss mode. The same principle applies for computing thermal noise reductions for other beam shapes and other thermal noises. The interested reader is referred to Vinet (2009) where detailed calculations of various thermal noises such as substrate Brownian, coating Brownian, and thermoelastic thermal noise with alternative beam shapes are presented. Franc *et al.* (2010) provide a good example of how these equations are applied to accurately compute the noise reduction factors regarding all mirror thermal noises for a possible implementation of higher-order Laguerre–Gauss modes in the Einstein Telescope (see Section 14.5.3). The detailed equations used in this example are published in an accompanying article, Franc *et al.* (2009).

An interferometer can be set up such that it has a linear response to the longitudinal motion of one or more mirrors. The optical signal should be computed as the weighted average of the mirror surface shape and position. For an axisymmetric beam reflecting off a cylindrical mirror with radius R this can be written as

$$s(t) = \int d\phi \int_0^R dr \, r I(r) m(r, \phi, t), \tag{13.4}$$

with $I(r)$ the normalized intensity profile of the beam and $m(r, \phi, t)$ the two dimensional information about the mirror surface shape and its evolution over time. Mirror thermal noise will dynamically distort the mirror shape. This shows already that, for a random surface distortion m, a flat-top intensity distribution $I(r)$ would provide the best averaging and effectively reduce the imprint of the noise on the optical signal. However a perfect flat-top intensity pattern cannot be used due to diffraction effects, as described in Section 13.2.3.

The current models for estimating the impact of thermal noise on optical signals are based on a method introduced by Levin (1998), see Chapter 1. Traditionally the amount of noise is quantified as an equivalent displacement, which specifies the amount of motion of the whole mirror (surface) that creates the same optical signal as the surface deformation through thermal noise. According to Levin (1998), the power spectral density of displacement equivalent thermal noise is given by (Equations 1.2 and 1.3)

$$S_x(f) = \frac{4 k_B T}{\pi f} \phi U, \qquad (13.5)$$

with ϕ being the loss angle and U the strain energy of the static pressure profile on the mirror surface normalized to 1 N. The pressure profile is identical to the intensity profile I of the beam.

The most limiting thermal noise in most precision experiments is the coating Brownian thermal noise, see Chapter 4. In the case of a semi-infinite mirror the coating Brownian thermal noise induced by a LG$_{pl}$ mode can be calculated using the strain energy

$$U_{p,l,\text{coating}} = \delta_c \frac{(1+\sigma)(1-2\sigma)}{\pi Y w_m^2} g_{p,l}. \qquad (13.6)$$

Here δ_c is the thickness of the coating, σ is the Poisson's ratio, Y is the Young's modulus, w_m is the beam size at the mirror and $g_{p,l}$ is a numerical scaling factor depending on the LG$_{pl}$ mode used.[2] In the case of the fundamental LG$_{00}$ mode this scaling factor is $g_{0,0} = 1$, whereas for a LG$_{33}$ mode $g_{3,3} = 0.14$ must be used. Hence the power spectral density of displacement equivalent coating Brownian thermal noise is more than a factor of seven smaller for a LG$_{33}$ mode in comparison to the fundamental LG$_{00}$ mode.

A number of theoretical works have been published with predictions for the possible thermal noise reduction using alternative beam shapes, see Table 13.1. It can be seen that with a carefully designed alternative beam shape the limiting thermal noise could be reduced by a factor of 2 or more.

13.2.2 Beam size and clipping losses

We have introduced above the notion that wider beams reduce the thermal noise contribution in an optical cavity. However, there is an obvious limit to the size of the beam: the wider the beam becomes the more light power falls outside the mirror diameter and is consequently lost from the cavity. The light loss due to this effect is often referred to as *clipping loss*. In high-precision experiments the limits for allowable optical loss are typically very small, as low as 1 ppm for some applications. The clipping loss, l_{clip}, on a cylindrical mirror can be approximated as

$$l_{\text{clip}}(w_m, \rho, z) = 1 - \int_0^{2\pi} d\phi \int_0^{\rho} dr \cdot r$$
$$\times u(w_m, r, \phi, z) u^*(w_m, r, \phi, z), \qquad (13.7)$$

---

[2] The factor $g_{p,l}$ is just a convenient way of writing a short form of an integral equation, see Vinet (2009).

Table 13.1 *A selection of predicted reduction factors of the thermal noise equivalent displacement linear spectral density. Numbers larger than 1 represent a reduction in noise. Note that these factors have been computed for different optical setups and different materials and cannot be compared directly. They are listed to illustrate the order of magnitude of the effect.*

	Brownian		thermoelastic		
	coating	substrate	coating	substrate	
LG_{00}	1	1	1	1	
LG_{33}	1.67 to 1.93	1.88 to 2.13	1.08 to 1.45		Franc *et al.* (2010)
LG_{55}	1.64 to 2.11	1.99 to 2.49			Franc *et al.* (2010)
LG_{55}		4.84	3.56	5.13	Vinet (2009)
Mesa		2.78	17.18	5.93	Vinet (2009)
Mesa				1.17 to 6.17	O'Shaughnessy *et al.* (2004)
Mesa	1.7	1.7	1.55	1.9	Agresti (2008)
Conical	5.45	2.33		11.38	Bondarescu *et al.* (2008)

where w_m is the beam radius at the mirror, ρ is the radius of the mirror coating and $u(w, r, \phi, z)$ is the transversal field distribution of the beam shape of interest. Note that the parameter beam radius w_m refers to differently defined beam sizes for different types of beam shapes. For example, within the family of Gaussian modes, w_m is always a measure of the beam size of the fundamental Gaussian mode (LG_{00} or HG_{00}). Higher-order LG or HG modes of the same beam radius actually are more spatially extended, in the sense that a significant amount of light power can be detected at distances away from the optical axis larger than the beam radius. In Figure 13.2 the clipping losses for the fundamental Gaussian and three alternative beam shapes are plotted over the mirror-radius to beam-radius ratio.

It should be noted that all optimizations of beam sizes described here, in particular those mentioned in Section 13.3.3, are based on the effectively arbitrary requirement that the clipping losses should be less than 1 ppm. In practise, many cavities feature optical scattering losses which can be a hundred times larger than that. In such a case the requirements for clipping losses can be relaxed. An optimization of the beam size and profile should be performed specifically for each case, taking into account the actual requirement for optical losses.

13.2.3 Beam propagation and cavity stability

When mirror thermal noise is limiting the performance of an optical cavity the optimal setup is that of a symmetric linear cavity with two equal mirrors with beam and mirror shape optimized so that the impact of mirror thermal noise at each mirror is the same. For practical reasons, slightly asymmetric (or near symmetric) systems are often used, for example, to be able to detect the optical signal in reflection of the cavity. However, the

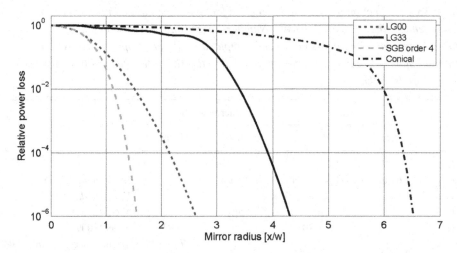

Figure 13.2   The relative power loss in reflection from a finite-sized mirror due to clipping loss for four different incident beam shapes over the mirror to beam radius ratio. The different curves shown are: dotted curve = TEM_{00} mode; solid curve = LG_{33} mode; dot-dashed curve = Conical beam; dashed curve: super-Gaussian beam of order 4.

principal design criteria can be described with the simpler symmetric system and we restrict the discussion to this type of cavity geometry.

The term "cavity stability" refers to the question of whether an eigenmode of the optical cavity exists (and is equal to the desired beam shape). This is exactly the same requirement as those stated in Section 13.1.3. We have seen in the previous section that the beam size on the mirrors is limited by requirements on the optical loss. However, this still leaves the freedom to change the wave front of the beam. The combination of intensity profile and wave front defines the beam. For Gaussian beams the cavity stability is well understood. For alternative beam shapes, however, a simple rule typically does not exist. Numerical simulations based on the Fourier transform, as proposed by Vinet *et al.* (1992), are a very powerful tool for modeling the behavior of an optical cavity or simply the propagation of a beam with an alternative shape. Such simulations are often referred to as *FFT propagation* and have, for example, been used to create Figures 13.3 and 13.4. Figure 13.3 illustrates the diffraction of a given beam shape over propagation; four different beam shapes are created at their waist such that they fit onto the same sized mirror at that location. Then the numerical simulation has been used to compute the beam shape after propagation, as shown in the right plot. It can be seen that all beams have widened through the unavoidable diffraction. The Gaussian modes (LG_{00} and LG_{33}) have retained their shape while the super-Gaussian beam has been transformed into an almost perfect Gaussian and the beam with an exact flat-top intensity profile shows strong distortions, which result from the sharp edges of the original flat-top profile. Figure 13.4 illustrates the beam shapes and mirror surfaces for three example cavity designs. In all three cases the cavity mirrors are assumed to be 2.5 cm in diameter and the length of the cavity is approximately 22 cm. The cavity setup

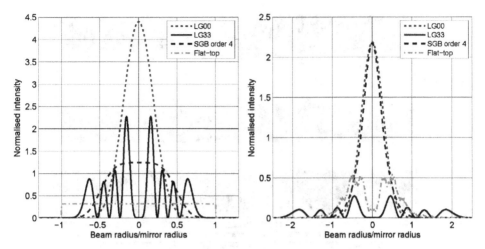

Figure 13.3 The left plot shows the intensity profile of four different beams at their respective beam waist. The size of the beam has been adjusted so that they fill the same area (with 1 ppm clipping losses). The right plot shows the same beams after propagation: the super-Gaussian becomes a Gaussian, the LG_{33} mode is more divergent than the LG_{00} mode and the flat-top beam distorts rapidly.

is symmetric and the beam size has been adjusted for a 1 ppm clipping loss per reflection. The three plots show the examples of an LG_{00} beam, a LG_{33} mode and a Conical beam (see Section 13.3.3). The required mirror surface shape is indicated by a vertical trace (the surface height changes are exaggerated and the beam intensity patterns are shown with a high contrast for clarity).

Comparing the higher-order LG mode with the fundamental Gaussian mode we see that in order to fit a higher-order mode optimally on the same mirror as the fundamental mode, the beam radius of the higher-order mode must be different from that of the fundamental mode. This corresponds to a different wave front curvature and consequently to a different spherical curvature of the cavity mirrors. This can be seen clearly in the two top plots of Figure 13.4. Therefore, changing an existing optical experiment such as an interferometer from a configuration using, e.g., the TEM_{00} mode to a configuration using the LG_{33} mode, the radii of curvature of the mirrors must be changed if one wants to keep the clipping losses at a constant level. Similarly, beams from other families, such as multi-Gauss flat-top beams, can be created with different divergence (i.e. wave fronts) while retaining the same intensity profile on the cavity mirrors. The divergence should be optimized so that the setup is robust against small changes to the system, such as mirror alignment or static mirror surface deformations.

13.3 Beam shapes

In general, the beam shape describes the spatial properties of a beam along the transverse orthogonal x and y directions, which should be independent of the temporal properties of

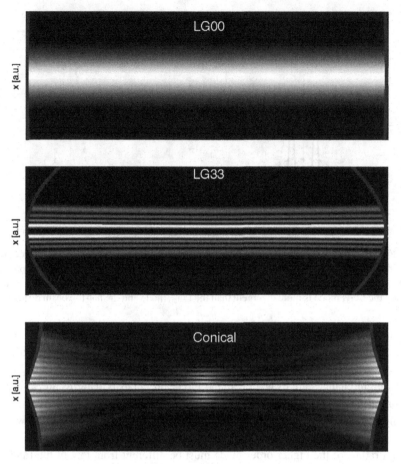

Figure 13.4 These plots show a cross section along the z–x-plane. The beam waists are located in the center and the vertical size is equal to 2.5 cm, the diameter of the cavity mirrors in this example. The beams are optimized for a 1 ppm clipping loss and a stable operation in this cavity of $\approx$ 22 cm length. At the left and right the required surface shapes of the cavity mirrors are indicated by an exaggerated trace of the beam wave fronts.

the beam. We can generally describe a beam as a sum of frequency and spatial components. For simplicity, we restrict the following description to a single frequency component at one moment in time ($t = 0$):

$$E(x, y, z) = \exp(-i\,kz) \sum_n a_n \, u_n(x, y, z), \tag{13.8}$$

with u_n as special functions describing the spatial properties of the beam, a_n as complex amplitude factors, and $k = 2\pi f / c$. Different types of spatial modes u_n can be used in this context. Of particular interest are the higher order *Gaussian modes*, which will be treated in Section 13.3.4. The most common beam can be described by the fundamental Gaussian mode; the light propagates along one axis, is well collimated around that axis, and the cross section of the intensity perpendicular to the optical axis shows a Gaussian distribution. The beam shape function of this mode is given as

$$u(x, y, z) = \sqrt{\frac{2}{\pi}} \, \frac{1}{w(z)} \, \exp\left(\mathrm{i}\,\theta(z)\right) \exp\left(-\mathrm{i}\,k\frac{x^2 + y^2}{2R_C(z)} - \frac{x^2 + y^2}{w^2(z)}\right), \tag{13.9}$$

with a corresponding radial intensity profile of:

$$I(r) = \frac{2P}{\pi\,w^2(z)} \exp\left(-2r^2/w(z)^2\right), \tag{13.10}$$

with P the light power and $w(z)$ the *spot size*, defined as the *radius* at which the intensity is $1/e^2$ times the maximum intensity $I(0)$. The other parameters in the above equations are defined as follows; w_0 is the minimum value of $w(z)$, $\theta(z) = \arctan\left((z - z_0)/z_R\right)$ is the Gouy phase, an extra phase factor associated with a Gaussian beam, z_0 is the position of the beam waist, $z_r = \pi w_0^2/\lambda$ is the Rayleigh-range of the beam, and $R_C(z) = z - z_0 + z_R^2/(z - z_0)$ is the radius of curvature of the spherical wave front of the beam. The intensity is a Gaussian distribution, see for example the trace "LG00" in Figure 13.1, hence the name *Gaussian beam*.[3]

One of the key features of a Gaussian beam is that it retains its shape over propagation. Equation 13.9 already contains the dependency of the shape as a function of propagation (along the z-axis). The size of the beam increases in the far field but the shape remains unchanged. Furthermore a whole framework exists to analytically describe the interaction of Gaussian beams with optical elements such as mirrors, lenses, optical fibres, etc., see, for example, Siegman (1986) and Freise and Strain (2010).

13.3.1 Multiple-Gauss flat-top beams

Bagini *et al.* (1996) showed that flat beams can be described well as a sum of LG modes and state

[...] although there are an infinite number of possible formulas representing such a type of field [flat], a desirable feature would be the ability of evaluating in an easy way the corresponding field distribution everywhere in space upon free propagation. From this standpoint some widely used profiles such as the super-Gaussian are not entirely satisfactory, because the corresponding propagation problem is to be treated numerically.

Flat beams can be described by sums of Laguerre–Gauss beams or Hermite–Gauss beams. Initially, this was of particular interest because the well understood propagation of Gaussian beams through optical systems can be utilized to describe flat beams. However, it turned out that this method also provides an intuitive handle on the design of new beam

---

[3] The fundamental mode in any family of Gaussian modes, such as the Hermite–Gauss or Laguerre–Gauss modes is often labelled HG00, LG00 or simply TEM00 (for Transverse Electro-Magnetic mode).

shapes which are optimized for certain properties, such as their potential thermal noise suppression.

13.3.2 Mesa beams and Mexican-hat mirrors

While multi-Gauss flat-top beams had been known and used for several years in the optical community, they were first considered as a possible means for reducing thermal noise by D'Ambrosio (2003), O'Shaughnessy *et al.* (2004) and D'Ambrosio *et al.* (2004b).

The original idea was to compose a flat beam by an integral of smaller beams. An ideal flat-top could be constructed mathematically as an integral over narrow Gauss functions:

$$u(x, y) = \iint_{x_0^2+y_0^2 \leq p} dx_0 dy_0 \sqrt{\frac{2}{w_0^2 \pi}} \exp\left(-\frac{1}{w_0^2}\left((x - x_0)^2 + (y - y_0)^2\right)\right), \quad (13.11)$$

with p and w_0 being the size of the total beam and of the narrow Gaussian modes. This becomes an ideal flat-top for $w_0/p \to 0$. However an ideal flat-top cannot retain a flat shape over propagation nor have a flat profile in the near and far field, see Figure 13.3. A compromise can be found for larger w_0 such that the beam profile remains rather flat at the location of both cavity mirrors.

As shown in Section 13.1.3, there exists a minimal waist size for Gaussian beams that can be supported by a cavity of length L. Using Equation 13.11 with such minimal Gaussians produces a flat beam that has been dubbed a *Mesa* beam. Mesa is the Spanish word for "table" and this term is used in the western United States to describe flat-top mountains similar in shape to this beam. The wave front of this beam in the far field is not spherical but resembles roughly the shape of a shallow Mexican hat, which needs to be matched in the mirror profile.

It was later recognized that the Mesa beam profile can be described in a compact and mathematically more useful way (see Vinet (2009) for the derivation of the following expression). The normalized Mesa beam profile can be written as

$$\Psi_{FM}(r, z) = \frac{2Z}{w_M \sqrt{\pi M}} \int_0^{w_M/w} \exp\left[-Z\left(\frac{r}{w(z) - x}\right)^2\right]$$
$$\times \exp\left(\frac{-2Zrx}{w(z)}\right) I_0\left(\frac{2Zrx}{w(z)}\right) x dx \quad (13.12)$$

with w_M the radius of the Mesa beam, w_0 and $w(z)$ the waist and beam size of the "small" Gaussians of which the Mesa beam is composed. Further we have $Z = 1 - iz/z_r$ and M as a normalization factor,

$$M = 1 - \exp\left(\frac{-w_M^2}{w_0^2}\right)\left[I_0\left(\frac{w_M^2}{w_0^2}\right) + I_1\left(\frac{w_M^2}{w_0^2}\right)\right], \quad (13.13)$$

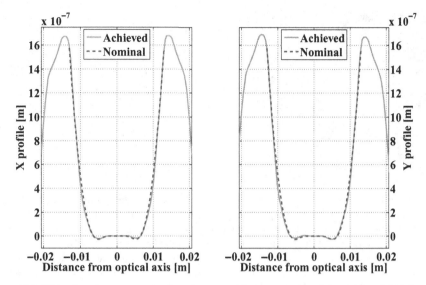

Figure 13.5 This plot shows a comparison between the theoretical and the achieved Mexican Hat profile from Miller (2010). (Courtesy of John Miller.)

where I_0, I_1 are the modified Bessel functions. The subscript "FM" in Equation 13.12 stands for "Flat Mesa" because the wave front of this beam on the cavity mirrors is almost flat with the Mexican hat shape superimposed.

The flat Mesa beams were the first alternative beam shape to be tested experimentally (Tarallo *et al.*, 2007). A prototype cavity has been set up to generate a Mesa beam with an optical design based on a folded cavity approximately 7 m in length, an end mirror with a Mexican hat profile and two flat mirrors. One of the flat mirrors is simply a folding mirror and the other, the input mirror, is positioned at the beam waist. This design is equivalent to a symmetric linear cavity of twice the length.

The "Mexican Hat" mirror was manufactured using a special coating technique invented to selectively deposit material on a mirror surface (Cimma *et al.*, 2006). See Section 2.7 for a discussion of creating a mirror for use with Mesa beams. The process was started with a flat cylindrical substrate, 50 mm in diameter and 30 mm thick. The entire Mexican Hat shape was created by depositing extra material, up to a thickness of 1 μm onto the polished front surface. The resulting shape compared well with the required shape, see Figure 13.5, with a roughly 5 nm random deviation. However the resulting coated mirror showed some anomalies, such as 150 ppm total scattering losses and a variation of 700 ppm in the transmission. Both of these values are in excess of what was expected and is typical for high-quality spherical mirrors.

Apart from the first successful creation of the Mesa beam, this prototype has been used to investigate the behavior of this beam shape in a suspended cavity. The research group reported the successful comparison between experimentally measured beam shapes and

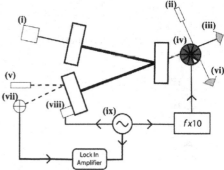

Figure 13.6 Photograph and schematic of the Mesa cavity setup. The left image shows the cavity structure during construction: A – Suspension system, B – INVAR rods, C – Cavity end plate housing input and Mexican Hat mirrors, D – Intracavity bracing plate. The right image shows the conceptual layout to measure the tilt of the Mexican hat mirror: (i) Mephisto laser, (ii) HeNe chopper monitor laser, (iii) mode profiler CCD camera, (iv) chopper wheel, (v) HeNe optical lever laser, (vi) chopper monitor photodiode, (vii) optical lever quadrant photodiode, (viii) mirror dithering PZT, and (ix) function generator. (Photograph on left courtesy of John Miller.)

numerical models using the FFT propagation method. They could successfully demonstrate a longitudinal control via the Pound–Drever–Hall technique (see Section 15.2.1) and developed a possible alignment control system. A detailed review of the experimental prototype for the Mesa beam cavity can be found in Miller (2010).

13.3.3 Hyperbolic and conical beams

While the original Mesa beams were studied in detail, both theoretically and experimentally, it was realized that their almost flat wave front would make the corresponding cavities very sensitive to mirror alignment and other surface distortions, similar to nearly flat-flat resonators for Gaussian beams. Bondarescu and Thorne (2006) then suggested a new way of creating the desired flat shape from multiple small Gauss beams. Instead of keeping the mathematical Gauss components all parallel they proposed to align them along hyperboloids with a characteristic twist angle α. The nearly flat configuration coincides with a twist angle $\alpha = 0$ while a near concentric beam can be created with a twist angle of $\alpha = \pi$. After further study they concluded that, similar to the current design, arm cavities for Gaussian beams with a near concentric design would provide the best results for cavity stability.

We will not recall the original mathematical derivation of the hyperbolic beams which is elegant and illustrative as it has since been replaced by a formalism which is more useful for practical optical design tasks. Galdi *et al.* (2006) analyzed the mathematical structure of the hyperbolic beams and realized that another parametrization can be realized by using a Gauss–Laguerre expansion, as first proposed by Sheppard and Saghafi (1996).

The hyperbolic beams can be described efficiently by a rapidly converging series,

$$U_\alpha(r, z) = \sum_{m=0}^{\infty} A_m^{(\alpha)} \Psi_m(r, z),$$ (13.14)

with $A_m^{(\alpha)}$ the α-dependent complex expansion coefficients and $\Psi_m(r, z)$ the Gauss–Laguerre propagator as base function

$$\Psi_m(r, z) = \frac{w_0}{w(z)} \psi_m \left(\frac{\sqrt{2}r}{w(z)} \right) \exp \left(i \frac{k_0 r^2}{2R_C(z)} \right) \exp \left(i \left(k_0 z - (2m + 1) \theta(z) \right) \right),$$ (13.15)

in which w_0, $w(z)$ are the Gaussian waist size and beam radius, $R_C(z)$ is the Gaussian radius of curvature of the wave front, and $\theta(z)$ the Gouy phase. The orthonormal Gauss–Laguerre basis functions $\psi_m(\xi)$ are given by

$$\psi_m(\xi) = \sqrt{2} \exp \left(-\frac{\xi^2}{2} \right) L_m(\xi^2),$$ (13.16)

with $L_m(\xi)$ being Laguerre polynomials. The family of hyperbolical Mesa beams can be well described using the expansion coefficients

$$A_m^{(\alpha)} = (-\cos \alpha)^m A_m^{(\pi)},$$ (13.17)

and

$$A_m^{(\pi)} = \frac{\sqrt{2}w_0^2}{R_0^2} P \left(m + 1, \frac{R_0^2}{2w_0^2} \right),$$ (13.18)

with $P(n, \xi)$ the incomplete Gamma function and R_0 the waist radius of the hyperbolical beam.

The Gauss–Laguerre formalism has been recognized as a powerful method to describe any multi-Gauss beam so far investigated. Pierro *et al.* (2007) derived a framework for optimizing the beam shape in an optical cavity with respect to thermal noise and then Bondarescu *et al.* (2008) showed that a maximal thermal noise suppression can be achieved by a beam they dubbed "Conical" because of its near conical wave front. The intensity pattern of this Conical beam is shown in Figure 13.1 and the intensity distribution within an example cavity has been illustrated in Figure 13.4. The Conical beam represents the state-of-the-art of research into multi-Gauss beams for thermal noise suppression. However, it should be noted that the optimization performed by Bondarescu *et al.* (2008) did not take into account several practical considerations such as cavity alignment control. Thus the Conical beam offers the maximal possible noise reduction but for practical use a slightly less aggressive noise optimization might prove beneficial.

13.3.4 Higher-order Gaussian modes

While the multi-Gauss flat-top beams promise a strong reduction of thermal noise, they have the disadvantage of requiring mirrors of special non-spherical shapes. Current laser interferometers make use of over 40 years of experience with Gaussian beams and spherical mirrors. The polishing and coating of mirrors to the highest precision has been optimized for spherical mirrors. Consequently Mours *et al.* (2006) proposed use of Laguerre–Gauss modes for thermal noise reduction. These beam shapes are part of the Gaussian family and thus offer all the advantages of Gaussian beams, such as a spherical wavefront. The higher orders of these modes feature a wider intensity profile and thus also provide thermal noise suppression, see Section 13.2.1.

The Laguerre–Gauss modes are a complete set of functions which solve the paraxial wave equation. Laguerre–Gauss modes are commonly given in their orthonormal form, see for example Freise and Strain (2010),

$$
\begin{aligned}
u_{p,l}(r, \phi, z) = {} & \tfrac{1}{w(z)} \sqrt{\tfrac{2p!}{\pi(|l|+p)!}} \exp(\mathrm{i}\,(2p + |l| + 1)\theta(z)) \\
& \times \left(\tfrac{\sqrt{2}r}{w(z)}\right)^{|l|} L_p^{(|l|)}\left(\tfrac{2r^2}{w(z)^2}\right) \exp\left(-\mathrm{i}\,k\tfrac{r^2}{2q(z)} + \mathrm{i}\,l\phi\right),
\end{aligned}
\tag{13.19}
$$

with r, ϕ and z as the cylindrical coordinates around the optical axis. The letter p is the radial mode index, l the azimuthal mode index and $L_p^{(l)}(x)$ are the associated Laguerre polynomials:

$$
L_p^{(l)}(x) = \frac{1}{p!} \sum_{j=0}^{p} \frac{p!}{j!} \binom{l+p}{p-j} (-x)^j.
\tag{13.20}
$$

The other parameters used in Equation 13.19 are the position of the beam waist along the z-axis, z_0, the Rayleigh-range, z_R, the beam radius, $w(z)$, the so-called *Gaussian beam parameter*, q, and the Guoy phase, θ. The dependence of the Laguerre modes on ϕ as given in Equation 13.19 results in a spiraling wave front, while the intensity pattern will always show unbroken concentric rings. These modes are called *helical* Laguerre–Gauss modes because of their special phase structure.

There exists a slightly different type of Laguerre–Gauss mode (compare the two plots in Figure 13.7) that features dark radial lines as well as dark concentric rings. Mathematically, these can be described simply by replacing the phase factor $\exp(\mathrm{i}\,l\phi)$ in Equation 13.19 by a sine or cosine function.

Initial investigations on using higher-order Laguerre-Gauss modes have been done with numerical simulations (Chelkowski *et al.*, 2009). Encouraged by the simulation results, Laguerre-Gauss modes are now being investigated further. A particular concern is the degeneracy of higher-order Gaussian modes in optical cavities. Typically, optical cavities are designed such that they are non-degenerate, i.e. the resonance for the fundamental Gauss mode does not overlap with a resonance of another low-order mode (inevitably some high-order mode resonance will overlap with that of the fundamental mode, but these modes have a much larger beam size and typically suffer larger optical losses, so that they can be ignored). This effect relies on the fact that the round trip Gouy phase, given by

$$
\theta_{\text{r.t.}} = 2(2p + |l|)(\theta(L) - \theta(0)),
\tag{13.21}
$$

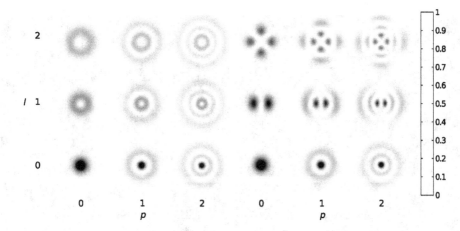

Figure 13.7 Intensity profiles for Laguerre–Gauss modes (dark areas are shown as white, high intensity zones are black): the left plot (three columns) shows the normalized intensity of helical modes u_{pl}. The u_{00} mode is identical to the Hermite–Gauss mode of order 0. Higher-order modes show a widening of the intensity and decreasing peak intensity. The number of concentric dark rings is given by the radial mode index p. The right plot (three columns) shows intensity profiles for sinusoidal Laguerre–Gauss modes u_{pl}^{alt}. The u_{p0} modes are identical to the helical modes. However, for azimuthal mode indices $l > 0$ the pattern shows l dark radial lines in addition to the p dark concentric rings.

is different between the fundamental mode and higher-order modes. However, for any non-zero order several modes share the same Gouy phase and are thus resonant simultaneously. The order of an LG_{33} mode is $(2p + |l|) = 9$ and there are in total 9 modes of order 9 (for each Gaussian family) which can be resonant together with the LG_{33} mode. This can be problematic, especially in a setup that uses optical cavities in the arms of a Michelson interferometer. Tiny distortions of the surface of the cavity mirrors could cause a significant coupling of degenerate modes which would then strongly reduce the contrast in the Michelson interferometer. Therefore, experimental demonstrations and further numerical investigations are required to confirm the suitability of these beam shapes for high-precision experiments.

The first experimental results have been produced by Fulda *et al.* (2010). In a table-top setup, higher-order Laguerre-Gauss modes have been created using a spatial light modulator. These modes were then used as the input light for standard mode cleaner cavities of linear and triangular design. Mode cleaners are middle- to high-finesse optical cavities which are used in the input optics setup of some high-precision interferometers. The laser light is transmitted through these cavities in order to reduce beam geometry fluctuations and in some cases laser frequency and amplitude noise (Drever *et al.*, 1983). The goal of the experimental demonstration was to validate the claims of Chelkowski *et al.* (2009) regarding the optical control signals and to test for possible degradation of the optical performance due

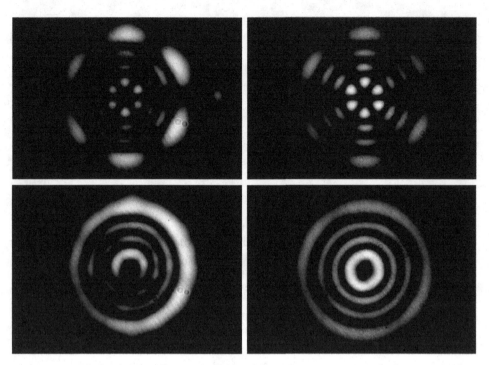

Figure 13.8 Intensity patterns of LG_{33} modes before and after a mode cleaner cavity. The top row shows sinusoidal modes, the bottom row contains helical modes. The pictures on the left show beams with about 60% in the desired LG_{33} mode. After passing the mode cleaner the purity is increased to approximately 99%, as shown on the right. Reprinted figure with permission from Fulda *et al.* (2010). Copyright 2010 by the American Physical Society.

to degeneracy. The results show that the Pound–Drever–Hall control (see Section 15.2.1) with higher-order modes works exactly as with the fundamental mode. The mode-cleaning performance was also confirmed; with ordinary off-the-shelf components the mode purity of an LG_{33} mode was increased approximately from 60% to 99% by transmitting it through a linear mode cleaner with a finesse of ≈ 170, see Figure 13.8. No problems due to the mode degeneracy were reported. However, these might become apparent only at a higher cavity finesse.

Fulda *et al.* (2010) further discussed and demonstrated practical problems related to LG modes: triangular cavities which are commonly used as mode cleaners cannot pass helical modes, as these modes lack the required symmetry around the vertical axis. This problem does not occur when the cavity features an even number of mirrors. However, in ordinary non-linear cavities at least one spherical mirror will be used under an angle (not normal incidence) and introduce astigmatism. This is a problem as LG modes cannot be eigenmodes of astigmatic cavities. Thus, linear cavities or special non-astigmatic ring cavities must be preferred over triangular cavities.

13.4 Conclusion

Most laser beams used in precision measurements today are fundamental Gaussian beams. Theoretical studies have shown that the effect of thermal noise from mirrors in such experiments can be reduced if beams with an alternative shape, i.e. a wider intensity distribution, are used. The reduction factor depends largely on the type of thermal noise and the specific experimental setup. However, noise reduction factors of 2 or larger seem possible.

It has further been shown theoretically and using computer simulations that such new beam shapes are, in principle, compatible with high-precision measurement setups. They can be used together with, for example, different optical materials, new coatings designs, cryogenic setups, and so on. This makes them a very interesting technology with a wide range of application and a potentially high impact.

However, alternative beam shapes set more stringent requirements on the optical parameters, especially the alignment and surface quality of the main mirrors and optical components. A number of mathematical frameworks have been developed to describe the propagation of beams with special shapes as well as their interaction with optical elements. Numerical simulations as well as first experimental prototypes have given promising results. More research, especially with table-top prototypes, is under way to test the alternative beams in realistic systems. While this technology has not been used for reducing thermal noise yet, the research and development has advanced so that implementations in actual devices such as optical clocks or gravitational wave detectors can be tested in the near future.

14

Gravitational wave detection

DAVID J. OTTAWAY AND STEVEN D. PENN

14.1 Introduction

Some of the most energetic and intriguing phenomena in our universe, including black hole coalescence and the first moments of the Big Bang, cannot be observed with traditional "light telescopes". These events can only be "seen" with gravitational wave observatories that detect minute ripples in the fabric of space-time. Gravitational wave detection has the potential to open a new window for humanity to view and understand the universe. However the experimental requirements for detection are extreme and meeting the sensitivity limits has spurred research in noise reduction. The study of thermal noise in optical coatings was driven by the need for mirror displacement sensitivities of 10^{-19} m in advanced gravitational wave interferometers (Levin, 1998; Harry and The LIGO Scientific Collaboration, 2010; Harry *et al.*, 2002).

Einstein's General Theory of Relativity states that an accelerating mass will produce a gravitational wave. The acceleration produces a distortion in space-time that propagates at the speed of light (see Einstein (1915), Einstein (1916), and Barish and Weiss (1999)). Gravitational waves interact extremely weakly with matter and as a result are very difficult to detect. Thus, an astrophysical event is required to produce gravitational waves that are strong enough to be detected on Earth. These weakly interacting waves are rarely scattered or absorbed, therefore the information they carry is virtually unaltered regardless of the time and distance they have traveled to us or the intervening objects and conditions they have encountered.

This chapter provides a general introduction to gravitational waves, their interferometric detectors, and the impact of coating thermal noise on this research. The detector overview (Section 14.4.2) includes sections for each of the major subsystems: optical, vacuum, seismic isolation, lasers and thermal noise. It concludes with a summary of all detectors either terrestrial (Section 14.5) or space-based (Section 14.6).

Optical Coatings and Thermal Noise in Precision Measurement, eds. Gregory M. Harry, Timothy Bodiya and Riccardo DeSalvo. Published by Cambridge University Press. © Cambridge University Press 2012.

14.2 Generation of gravitational waves

In 1915–1916, Einstein (Einstein 1915, 1916) developed the General Theory of Relativity (GR). GR describes now the mass–energy distribution defines the local curvature of space-time, and how this local curvature determines the trajectory followed by mass energy.[1] "Mass tells Space how to Curve, and Space tells Mass how to Move"[2] The relationship is given by the Einstein equation:

$$G_{\mu\nu} = 8\pi \, G \, T_{\mu\nu}, \tag{14.1}$$

where $G_{\mu\nu}$, the Einstein tensor, describes the space-time curvature and $T_{\mu\nu}$, the stress-energy tensor, describes the mass–energy and momentum flux density.

In GR, the local proper distance s between space-time points (events) is described by the metric tensor, $g_{\mu\nu}$, where

$$ds^2 = g_{\mu\nu} dx^\mu dx^\nu \tag{14.2}$$

and x^α, $\alpha = \{0, 1, 2, 3\}$, is the local coordinate system. $G_{\mu\nu}$ is calculated from $g_{\mu\nu}$ and its derivatives.[3] In empty space, $T_{\mu\nu} = 0$, and $g_{\mu\nu} = M_{\mu\nu}$ the flat metric of special relativity, where $g_{ii} = \{-1, 1, 1, 1\}$ and $g_{ij} = 0$ for $i \neq j$. When a small perturbation, h, is added to the metric $g_{\mu\nu} = M_{\mu\nu} + h_{\mu\nu}$, then, to first-order in h and utilizing the transverse-traceless gauge, the Einstein equation (14.1) becomes the wave equation

$$\left(\nabla^2 - \frac{1}{c^2} \frac{2}{2t^2} \right) h_{\mu\nu} = 0 \tag{14.3}$$

This gravitational wave is a quadrupolar strain that travels at the speed of light and oscillates in the plane transverse to the direction of propagation. Figure 14.1 illutrates the effect of the wave on a ring of unconstrained masses in this transverse plane. There are two polarizations, h_t and h_x, that have a relative angle of 45°.

The quadrupole nature of gravitational waves is supported by our understanding of oscillations in the mass-energy distribution. The conservation of energy prevents the mass (energy) creation required for a monopole wave. The conservation of momentum prohibits the mass dipole oscillation needed to form a dipole wave. Thus the leading order multipole is the oscillating quadrupole term, which is exemplified by two masses in orbit about their center of mass.

A major shortcoming of Newtonian gravity is that it allows for instantaneous action at a distance; a shift in the mass distribution alters the gravitational field and that change is known

---

[1] A clear and thorough introduction to General Relativity may be found in several texts, especially Misner *et al.* (1973), Hartle (2003), and Schutz (2009).

[2] A common adage used to describe General Relativity.

[3] The full equation for $G_{\mu\nu}$ can be found in most GR textbooks (Misner *et al.* (1973), Hartle (2003)).

D. J. Ottaway and S. D. Penn

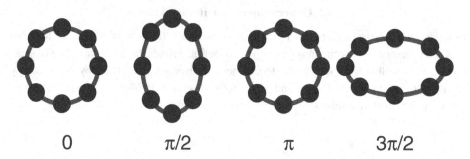

$$0 \qquad\qquad \pi/2 \qquad\qquad \pi \qquad\qquad 3\pi/2$$

Figure 14.1 The effect of a gravitational wave propagating into the page on a ring of free masses.

instantly everywhere. GR alleviates that problem. In GR, a shift in the mass distribution alters the space-time curvature and that change propagates out as a gravitational wave.

14.3 Gravitational wave sources

When Einstein theorized the existence of gravitational waves, he estimated that the strain waves too small to ever be measured. Any terrestrial source mass that is accelerated to its material strength limit will produce a wave that is many orders of magnitude smaller than the sensitivity limit of today's best detectors. For example, two tungsten spheres, each with a 1 m diameter and a mass of 10 metric tones, supported at the end of the strongest cabon nanotube composite rod, when spun about at 100 H_2 would produce a strain wave with an amplitude of only $h \approx 10^{-39}$, which is 10^{15} times smaller than the sensitivity of a second generation detector.

Even main-sequence, binary star system, with orbital frequencies in the ωH_2 range, are well outside the frequency range of terrestrial detectors (10 Hz to 10 kHz) and medium-term space-bosed detectors (1 mHz to 1 Hz). Instead, today's interferometric detectors search for signals from sources that were unknown in 1916, including neutron stars, black holes, and the Big Bang.

Only celestial sources with bodies of solar mass or more are expected to produce measurable gravitational waves. These celestial sources are generally divided into four categories: periodic sources, compact binary coalescence, burst events, and the stochastic background.[4]

14.3.1 Periodic sources

Mass systems that radiate gravitational waves at nearly constant frequency are called periodic sources. These sources include binary star systems and rapidly spinning compact

---

[4] Note that at the time of this writing there have been no direct detections of gravitational waves from any source.

masses. Main-sequence binaries revolve at frequencies at or below the µHz level. If one or both of the binary stars are compact masses (white dwarfs, neutron stars, or black holes), then the system can reach mHz frequencies, ($f < 1$ Hz). Compact mass binary systems comprised of black holes, neutron stars, and/or white dwarf stars can emit potentially detectable gravitational waves as they orbit. For the vast majority of the lifetimes of these systems, ranging from 10^7–10^9 years, the orbits are nearly constant. The orbital velocity is sufficiently low that the energy lost to gravitational radiation is proportionally small.

The most well-known periodic source of this type is PSR 1913+16, the Hulse–Taylor pulsar. Discovered in 1974, this pulsar is part of a binary system in which the orbit is slowly decaying and its orbital period is decreasing as a result of gravitational wave emissions (Weisberg *et al.*, 2010). The orbital period of 7.75 hours is diminishing by 76 µs per year and it is estimated that in 300 million years the system will have lost enough energy that it will rapidly inspiral and the stars will coalesce. Low frequency detectors, such as the planned space-based LISA detector (Jafry *et al.*, 1994) (see Section 14.6 below), are designed to be sensitive to radiation from these stable binary systems. However, advanced Earth-based interferometers have a low frequency limit around 10 Hz that prevents them from observing these binary systems until the final few minutes of their inspiral and coalescence.

For terrestrial based interferometers the only known potential sources of periodic gravitational radiation are spinning neutron stars. Neutron stars are formed when massive stars have exhausted enough nuclear fuel that they can no longer sustain themselves against gravitational collapse. The energy released during this collapse results in a brilliant explosion whose energy output can exceed that of a billion stars. The resulting neutron star typically has a mass of 1.4 $M_\odot$ and a radius of about 10 km. Conservation of angular momentum ensures that stars that are slowly rotating prior to a supernova can result in neutrons stars that are spinning with frequencies approaching 600 Hz.

Typically, these neutrons stars are extremely rotationally symmetric. A spinning neutron star that has perfect rotational symmetry emits no gravitational waves. However even tiny amounts of ellipticity can allow a spinning neutron star to generate significant amounts of gravitational radiation. At current sensitivity levels, a first generation detector is capable of detecting neutron stars with eccentricities of 1×10^{-7} at 100–300 Hz within 0.2 kpc (Abbott and The LIGO Scientific Collaboration, 2010).

14.3.2 Compact binary coalescence

Binary systems formed from combinations of black holes and neutrons stars can have relatively stable orbits for millions, or even billions, of years before they eventually lose enough energy that they begin to appreciably inspiral. As the orbit decays the gravitational wave frequency and amplitude increase in a characteristic waveform known as a "chirp" (Cutler *et al.*, 1993). During the final few minutes of a binary neutron star inspiral, the orbital period will increase from tens of milliHertz to 1 Hz. The rise from 1 Hz to hundreds of Hertz occurs during the last few seconds before the masses distort and merge. The predicted gravitational waveform just prior to merger is well understood using post Newtonian approximations to General Relativity (Blanchet *et al.*, 1996). However, the merger phase is extremely complex and numerical relativists have only recently succeeded in making predictions of the likely waveform of this phase (Baker *et al.*, 2007). Post merger, the resulting body is most likely a black hole whose quasi-normal modes decay by emitting further gravitational waves in the form of a damped sinusoid. Studies of short gamma ray bursts have suggested that these bursts are likely caused by the merging of coalescing compact bodies (Narayan *et al.*, 1992). Joint searches for gravitational waves and gamma rays from these events are underway.

14.3.3 Burst sources

Potential burst sources are thought possible from asymmetric supernova explosions and from cataclysmic events in neutron stars. During the last day in the life of a massive star, ($M \geq 8M_\odot$), the core rapidly fills with the iron produced by silicon fusion. When the mass exceeds the Chandresekhar limit of 1.4 $M_\odot$, the gravitational pressure exceeds the electron degeneracy pressure. The reaction $e + p \rightarrow n + \nu$ collapses the core to nuclear densities, thus forming a neutron star. The outer shells of the star then fall in towards the now-vacant core and are heated by the collapse and collisions with the extreme neutrino flux. Variations in the core collapse and the in-fall of the outer shell can lead to an asymmetry in the energy production and a non-spherical explosion. Any quadrupole component of the accelerating matter produces a gravitational wave burst. Given the enormous variability of supernovas, the exact form of the gravitational wave is difficult to predict beyond estimates of power and duration.

Rapid changes in the sphericity of neutron stars are also a potential source of burst gravitational waves. The intense gravity at the surface of a neutron star (10^{11} g) keeps the surface uniform to within a few millimeters. Starquakes are thought to form rifts or mountains of several millimeters that generate gravitational waves from a spinning neutron star. If the gravitational pressure quickly heals or reduces these deviations, then the wave is more burst-like.

When matter coalesces or is absorbed into a black hole, it may excite the black hole's quasi-normal modes. Quadrupole excitations will produce gravitational waves. However the interaction of the black hole's field with the wave's energy density quickly damps these oscillations leading to waveforms that look like damped sinusoids.

14.3.4 Stochastic background

The cosmic microwave background is the electromagnetic radiation emitted by the universe about 300 000 years after the Big Bang, when the universe transitioned from being radiation-dominated (opaque) to matter-dominated (transparent). There is expected to be an analogue of the cosmic microwave background in gravitational waves. In the very earliest moments of the Big Bang (at the Planck time of 10^{-43} s), the Universe was opaque to gravitational waves. The energy-matter density was so high that the waves were emitted and absorbed at near equal rates. However, as the universe expanded it became transparent to gravitational waves and thus this background carries information about the universe at the time of this transition. It is predicted that the universe became transparent to gravitational waves of a given wavelength when the diameter of the universe exceeded that wavelength. Thus, the number of gravitational waves of a given wavelength should be proportional to the energy density. Therefore, by measuring the gravitational wave background as a function of frequency, the energy density as a function of time during the first seconds of the universe can be determined.

The gravitational wave stochastic background is extremely weak and it will be a very long time, if at all, before a single detector will have sufficiently low noise to be able to observe it directly. However, by using coherence methods to combine the measurements of two detectors whose noise is uncorrelated it is possible to study the level of the stochastic background. Current upper limit measurements have already eliminated some theories for gravitational wave production in the early universe (Abbott and The LIGO and Virgo Collaborations, 2009).

14.4 Terrestrial gravitational wave detection

The first attempts to detect gravitational waves utilized resonant bar antennas (Weber, 1960; Mauceli *et al.*, 1996). However, it was recognized relatively early that interferometric detectors had significant advantages including potentially greater sensitivity and bandwidth. Given the quadrupolar form of gravitational waves, the most straightforward design for a detector is an L-shaped interferometer. These interferometers detect gravitational waves by sensing the relative displacement of the mirrors defining the arm length. These mirrors must be in "free fall", meaning free to move in the plane of the interferometer. These optics are independently suspended pendulums, which gives the detectors a wide bandwidth and the ability to gain sensitivity with increased mirror separation.

In 1972, Rainer Weiss published the first design proposal (Weiss, 1972) for a large-scale, gravitational wave interferometer. Weiss was quite thorough in describing the detection scheme, specifying the major system components, and estimating the primary noise sources, including seismic noise, mechanical thermal noise, laser noise, radiation pressure noise, and gravity gradient noise. From the kernel of this design grew the research project that eventually became the modern field of gravitational wave detection. Indeed, most major systems of current detectors can be traced back to concepts in that paper, including the vacuum system, seismic isolation, suspended optics, optical cavity arms, stabilized laser, and feedback control of the optics. In the 39 years since that proposal, Weiss' hard

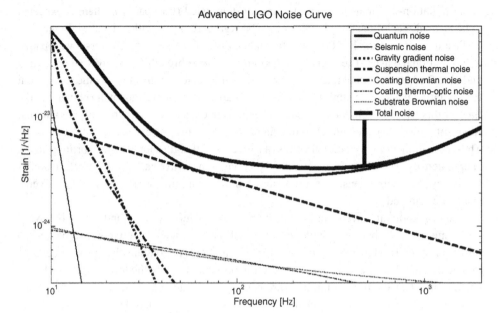

Figure 14.2 The Advanced LIGO sensitivity curve showing the relative importance of quantum noise, coating Brownian noise, seismic noise, gravity gradient noise, suspension thermal noise, coating thermo-optic noise, and silica substrate Brownian noise (Harry and The LIGO Scientific Collaboration, 2010). See also Figure 3.2 for noise curves of other thermal noises.

work and leadership have been a major driving force in the worldwide gravitational wave collaboration's success and growth.

14.4.1 Critical noise sources

The sensitivity of gravitational wave interferometers is limited by a number of noise sources. In this section we will summarize the main ones. The summaries will not be detailed, as many of the noise sources are covered in other parts of this book. For appreciation of the frequency bands that these noise sources dominate, see Figure 14.2 for a typical sensitivity curve of a second generation interferometer. For the first generation detectors most of these noise sources are higher and their relative importance somewhat alters.

Quantum noise

Quantum noise refers to the uncertainty that arises from the statistical variations of the laser light in the interferometer. For any period of observation, T, the expected number of photons produced by the laser is $N \pm \sqrt{N}$ where $N = TP\lambda/(hc)$, where P is the laser power and λ is the laser wavelength. Thus, the noise spectrum of the laser is given by

$$S_P = \frac{hcP}{\lambda}.$$

(14.4)

This uncertainty in photon flux gives rise to two noise mechanisms. At low frequencies, quantum noise is dominated by radiation pressure noise. At higher frequencies, shot noise is the dominant component of quantum noise and the major limitation on the high frequency performance of gravitational wave detectors.

Radiation pressure noise is mirror displacement noise caused by the varying pressure exerted on an optic by a changing photon flux. When light is reflected off a mirror of mass m the momentum change of the photons causes an impulse on the mirror that results in a force. The statistical fluctuations in the arrival times of the photons causes a fluctuating pressure and a fluctuating displacement of the mirror. The amplitude noise spectrum for an interferometer of length L is given by

$$\sqrt{S_{h,\text{r.p.}}(f)} = \frac{1}{Lmf^2}\sqrt{\frac{hP}{4\pi^4 c\lambda}}. \tag{14.5}$$

Shot noise is due to the reduction in precision of a phase measurement because of the statistical fluctuations in the light. Interferometers measure the phase shift imposed by a passing gravitational wave on light. Hence the statistical variations are a fundamental limit on the ability of light to sense the location of a free body. The amplitude noise spectrum of the shot noise obeys

$$\sqrt{S_{h,\text{ shot FP}}(f)} \propto 1/\sqrt{P}. \tag{14.6}$$

For a given measurement frequency there exists a power level that minimizes the combination of the radiation pressure noise, which scales with the square root of the power, and the shot noise, which scales as the inverse square root of the power. The displacement level that is achieved for this optimization is called the standard quantum limit (see Section 1.4). In general, this is a fundamental limit on our ability to detect gravitational waves with test masses of a given mass. However, the addition of a signal recycling mirror (see Section 14.4.2 below) or the injection of squeezed light into the interferometer (see Section 11.3) can reduce the noise below the standard quantum limit. Shot noise drives the need for having high powers reflecting off the test masses and is a reason why very low absorption coatings are needed on the test masses (see Chapter 10).

Thermal noise

Thermal noise is the label given to measurement uncertainties that result from fluctuations driven by thermal energy. An experimental system that is in thermal equilibrium at a temperature, T, will have thermal energy per degree of freedom of $k_B T/2$, where k_B is Boltzmann's constant. Microscopically, the thermal energy is expressed as the average kinetic energy of the atoms.

A thermal bath connected to the experimental system by a dissipation mechanism creates thermal noise in the system. In a perfectly elastic solid (i.e., no dissipation), the thermal energy would be completely expressed as a superposition of the normal modes of the system. The motion of this system would not be noise. If at any time the amplitude and phase of the modes was measured, then the future motion could be calculated.

Dissipation is a non resonant response that couples energy is or out of a mode of the system. Although random in time, the effect of dissipation on the power spectral density is described by the Fluctuation-Dissipation Theorem (Callen and Greene (1952); Callen and Welton (1952)).

In gravitational wave detectors, thermal noise is important in the test mass suspensions, substrates, and mirror coatings, all of which can cause displacements of the front face of the mirror. The primary strategy for reducing this noise is to reduce the mechanical loss, thus concentrating the thermal excitations in narrow bands around the resonant frequencies. The thermal noise due to the coatings on the high reflectance surfaces of the mirrors is expected to be the dominant noise mechanism in second generation gravitational wave detectors. See Chapter 4 for a detailed discussion of (Brownian) coating thermal noise, Chapters 3 and 9 for other coating thermal noises, and Chapter 7 for a detailed discussion of substrate thermal noise.

Seismic noise

Earth-based detectors are subject to strong seismic noise from the ground, with micron-level displacements below about 1 Hz and a sharp drop in noise above 10 Hz. The exact distribution and amplitude of this noise is location and condition dependent. A series of masses and springs can isolate the test masses from this seismic noise, but in general, it is hard to build a low-noise, seismic isolation system below about 10 Hz. The goal for seismic noise in first generation detectors was to reach 10^{-19} m/$\sqrt{\text{Hz}}$ at 40 Hz. Heavy traffic, trains, construction and logging, and other anthropogenic activity all add to the seismic background.

Gravity gradient noise

When seismic waves travel through the Earth they cause density fluctuations that alter the local gravitational field. These variations exert an oscillating force on the test masses. This noise source will primarily affect third generation detectors. Unfortunately, it is not possible to isolate against this noise source. Plans to combat this noise include siting the detectors deep underground, beneath the dominant surface seismic waves, and using local seismometer data to predict the noise and correct for it.

Laser frequency and intensity noise

Interferometer asymmetries between the two arms make the experiment sensitive to classical laser frequency and intensity noise. Frequency noise directly mimics the signal from the motion of a test mass, but would be interferometrically canceled except for asymmetries. Intensity noise comes from motion of the test masses from radiation pressure. If the circulating power in the arms is identical, then both interferometer arms experience the same displacement and no change in light intensity is observed. However, slight differences in the finesse of the two arm cavities can result in intensity fluctuations being transferred to the signal.

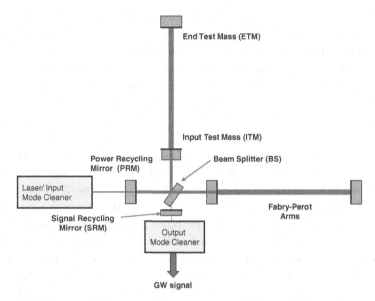

Figure 14.3 Schematic of the optical layout of a generalized gravitational wave detector.

14.4.2 Critical components of Earth-based interferometers

In this section we will discuss the main components that make up an interferometric gravitational wave detector. Detector designs vary so this section is a general description which serves as a basis to which the specific detectors can be discussed in Section 14.5.

Interferometer topology

The farther the test masses of a gravitational wave interferometer are apart then the larger the effective displacement that the gravitational wave produces. The finite speed of light means that the accumulated phase shift imposed on the light is maximized when the round trip travel time is equal to half a period of the gravitational wave. Longer round trip travel times mean that the additional phase accumulated starts subtracting from the gravitational wave interferometer signal. The optimum length for a gravitational wave detector that has a maximum sensitivity at 150 Hz (a canonical frequency for a binary neutron star inspiral) is therefore 500 km long. It is clearly not practical to build terrestrial gravitational wave detectors with such large test mass separations. However, the use of long Fabry–Perot cavities or delay lines in the interferometer arms can achieve the same accumulated phase shift on considerably shorter baselines, hundreds or thousands of meters. The only disadvantage is that the light bounces off many more mirror surfaces and hence the effects of mirror displacement noise and losses are enhanced.

All current and planned large scale interferometric gravitational wave detectors utilize variations of the Michelson interferometer. Figure 14.3 shows a schematic of a Michelson based gravitational wave detector made up of the two test masses (labeled ETM, for End Test Mass), the beamsplitter (BS) and the laser. All the other components are used to increase the sensitivity of the interferometer in an economical way. To simulate the optimum length of 500 km for the arms, over-coupled Fabry–Perot cavities can be installed by adding input test masses (ITM) near the beamsplitter. However, the Fabry–Perot cavities do not improve the signal-to-noise ratio for noise sources arising from the test masses (i.e. thermal noise) because the cavity reflections increase the phase shifts from these physical displacements as well.

One of the dominant noise sources is shot noise on the light, as discussed above in Subsection 14.4.1. This has two implications for the design of a gravitational wave detector. First, the power incident on the beamsplitter must be high to increase the number of photons used and thereby increase the signal-to-noise ratio. The limit due to shot noise improves as the square root of the power incident on the beamsplitter. The second implication is that the optimum noise performance is achieved when the port opposite to that in which the laser enters is made intentionally dark from destructive interference of the light. This port is often referred to as the dark port or the antisymmetric port. Conservation of energy means that all of the input laser light is then reflected back towards the laser. The purpose of the power recycling mirror (PRM) is to reflect this potentially wasted light back towards the beamsplitter. The addition of this power recycling mirror can build up the light incident on the beamsplitter by up to a factor of 50. Imperfections in the interferometer mean that light leaks out of the antisymmetric port which increases the background shot noise. The output mode cleaner is an optical cavity that is designed to spatially filter this unneeded light and help ensure quantum noise limited performance.

Optics

The optics that have the highest demands placed on them are the test masses that make up the Fabry–Perot arm cavities. If Fabry–Perot arm cavities are not used, increased demands are placed on the beamsplitter. The substrates of the optics must be made of a material that has very low mechanical loss, such as fused silica or sapphire. The high reflectance coatings that are used on the test masses must have extremely low absorption, excellent, surface figure, very low scatter, and very low mechanical loss. Low optical losses will maximize the interferometer's power recycling gain and reduce shot noise. In addition, light that is multiply scattered out and back into the interferometer will introduce phase noise (see Chapter 11). Excellent surface figure is needed to ensure the greatest overlap of the wavefronts that recombine at the beamsplitter. This condition minimizes the shot noise and the power leaking out of the dark port. Low absorption minimizes the impact of thermal lensing on the performance of the interferometer (see Chapter 10). Low mechanical loss in the coatings and substrates minimizes the motion of the reflective surfaces of the Fabry–Perot arm cavities due to thermal noise (see Chapters 4 and 7). For more on optics for gravitational wave detectors see Camp *et al.* (2002).

Light sources

Gravitational wave detectors require lasers whose output is extremely spatially pure and which have very low frequency and intensity noise. The input mode cleaner is a suspended cavity around 10 m long that filters higher order spatial modes from the input light and provides a frequency discriminator for reducing the frequency noise. All of the first and second generation interferometers use single frequency Nd:YAG lasers operating at 1.064 μm. The first generation detectors use around 10 Watts of power whereas the second generation detectors are designed to use around 150 W of laser power. The slight imbalances between the two arms mean that the frequency noise on the lasers must be extremely low. Typically, the lasers use the long arms of the interferometer as a frequency reference, meaning that relative frequency noise is often less than $1 \ \mu Hz/\sqrt{Hz}$ at 100 Hz. The intensity noise requirements are also extremely challenging. For second generation detectors the requirements are $3 \times 10^{-9}/\sqrt{Hz}$ at 10 Hz. For more on lasers for gravitational wave detectors see Willke *et al.* (2008).

Control systems

To achieve the build up of power in first the recycling cavity and then the two arms it is necessary to stabilize the lengths of the arms to less than 100 pm for first generation instruments and 10 pm for second generation. The control system must achieve this without injecting significant noise into the gravitational wave band between 10 Hz and 10 kHz. For second generation detectors to operate at full sensitivity requires in excess of 100 control loops to operate simultaneously. This global control of the interferometer is done using variations of the Pound–Drever–Hall (PDH) method (see Section 15.2.1). The input laser beam passes through an electro-optic modulator where the beam is phase modulated creating frequency sidebands on the light. The intensity modulations that these sidebands cause when they become unbalanced in either amplitude or phase is detected at various parts of the interferometer to generate error signals. These are then fed back to a variety of actuators including electromagnets, electrostatic drives, and even quiet hydraulics to keep the test masses in their desired locations. For more details on control systems of gravitational wave interferometers see Acernese and The Virgo Collaboration (2010).

Seismic isolation and suspensions

In order for an interferometer to function as a gravitational wave detector, the mirrors that define the arms must be free to move in the plane of the interferometer, i.e. in free fall. Free fall is usually established by having the optics suspended as pendulums, making them essentially free along the direction of the beam. If this were not the case, then the detector signal would be convolved with the frequency response of the mirror support structure. The optics must appear free to the gravitational wave for frequencies in the detection band, while at the same time being isolated from other external forces and held in an optically stable "lock" position.

Seismic isolation systems decouple the test masses from motions of the ground. These systems use either passive or active techniques or a combination of both. Passive isolation systems achieve isolation by creating mechanical resonances which decouple the test mass motion from ground motion for frequencies significantly above their resonant frequencies. An example of such a mechanical resonance is a simple pendulum. Earth-based gravitational wave detectors require significantly more isolation than that which is possible with a single pendulum system. In second generation detectors, typically nine orders of magnitude of isolation are required at 10 Hz. Often compound pendulums or spring-mass stacks are used to increase the isolation. Very low frequency mechanical resonators can be made by utilizing soft springs made from inverted pendulums or nonlinear springs. This system allows mechanical resonances as low as 30 mHz. Passive isolation techniques often use mechanical feedback to damp resonances that might cause excess motion in the control band (generally DC – 10 Hz). The alternative to passive isolation is completely active isolation. In these systems an optical platform is stabilized by measuring its motion using inertial sensors such as accelerometers or seismometers. Actuators then move the table to minimize its motion. These actuators can be electromagnetic or even quiet hydraulic systems.

The connections between the seismic isolation systems and the test masses must minimize thermal noise. This goal is achieved by making the last attachment of very pliant, low mechanical loss material such as steel piano wire for first generation detectors and silica fibers for second generation. See Abbott *et al.* (2002) and Section 5.3.1 for more on seismic isolation and suspension thermal noise.

Vacuum envelope

Small scale pressure fluctuations in the air lead to refractive index variations. These refractive index changes lead to optical path length fluctuations which would completely mask a gravitational wave signal. This noise source is called residual gas noise. For this reason, all Earth-based detectors need to operate within a vacuum envelope. Very low pressures are required and gravitational wave detectors are housed in the largest volume ultra high vacuum systems ever built. Photographs of vacuum beam tubes and optics chambers are shown in Figure 14.4. The vacuum in these vessels is sufficiently high that residual gas noise will not be a limiting noise source until third generation detectors. See Section 5.3.4 for more on residual gas noise.

The vacuum systems are formed by roughly 1 m diameter beam tubes connecting larger vacuum chambers that house the optics. The chambers contain seismic isolation platforms that are supported independent of the chambers. These isolation platforms hold all of the optics and suspension systems. Connecting the corner station to the end station is a kilometer-scale long beam tube that is formed from low oxygen stainless steel connected with a single spiral weld. The beam tubes contain a series of baffles with irregularly shaped edges to block any coherently scattered light. After installation, the vacuum system is

Figure 14.4 The vacuum beam tube covered by a concrete vault (left) and optics chambers in the corner station (right). These large chambers each house a single optic (beamsplitter or test mass) and its associated seismic isolation and suspension. Photos courtesy of the LIGO Laboratory.

pumped down and baked. The system is maintained near 10^{-8} Torr using only cryopumps and ion pumps during operation.

14.5 Terrestrial gravitational wave detectors

14.5.1 First generation

The construction of the first generation of large scale interferometers began in the last part of the twentieth century and operation at the design sensitivity was achieved in the first few years of this century. The main task of these first generation detectors was to develop and demonstrate the technology needed to operate high performance interferometers over kilometer length scales. Astrophysical models predicted that detection of a gravitational wave was possible but unlikely with these interferometers. Therefore, the goal was to pave the way for gravitational wave astronomy commencing with the second generation. We here briefly describe the primary first generation gravitational wave detectors.

TAMA 300

TAMA 300 is a gravitational wave detector located on the Mitaka campus of the National Astronomical Observatory in Tokyo, Japan. In 2001 it was the first of the large scale, first generation detectors to report sensitivity sufficient to detect a binary neutron-star inspiral within our galaxy (Ando and The TAMA Collaboration, 2001). TAMA 300 makes use of arms that are 300 m long; this short arm length compared to LIGO and Virgo (see below) means that the optimum sensitivity can be achieved using high finesse arms without power recycling (although power recycling was installed later; see Arai and The TAMA Collaboration (2008)). The mirrors of the main interferometer are mounted on three stage seismic isolation platforms that provide over 165 dB isolation at 150 Hz. See Arai and The TAMA Collaboration (2008) for more on TAMA 300.

Figure 14.5 Aerial views of the LIGO Hanford Observatory (Left) and the LIGO Livingston Observatory (Right). Photos courtesy of the LIGO Laboratory.

Initial LIGO

The Laser Interferometer Gravitational-wave Observatory (LIGO) consists of three interferometers located at two facilities: a 4-km interferometer in Livingston, Louisiana, and a 4 km and 2 km interferometer nested within a single vacuum system in Hanford, Washington. Figure 14.5 is an aerial view of the LIGO Livingston and Hanford observatories. With the exception of their length and location, the interferometers are essentially identical. The two detectors at the Hanford site provide a diagnostic for distinguishing site-specific noise from possible gravitational wave signals. The two sites are separated by 3000 km, which minimizes background noise coherence, and results in a 10 ms timing separation. The timing resolution of the joint LIGO-Virgo detectors allows for a theoretical angular resolution of <0.5°, but typical noise levels and signal uncertainties yield a practical limit of 5–7° (Abbott and The LIGO Scientific Collaboration, 2009; Wen and Chen, 2010). The Initial LIGO detectors had a design sensitivity of $h = 2.5 \times 10^{-22}$ ($\Delta L = 10^{-18}$ m) in the frequency band of 40–4000 Hz. That design goal was achieved during 2006–2007, the detector's fifth science run (Abbott and The LIGO Scientific Collaboration, 2009).

The Initial LIGO interferometers were power recycled interferometers with Fabry–Perot arm cavities. The optics were supported using single piano wire suspensions mounted on passive isolation stacks. Increased anthropogenic noise at the Livingston Observatory meant that further seismic isolation was required to achieve continuous operation there. This was achieved by utilizing an active platform that was actuated using quiet hydraulics. One of the best sensitivity curves from the Hanford 4 km is shown in Figure 14.6.

Initial Virgo

Virgo is a French–Italian collaboration that operates a 3 km-long gravitational wave interferometer in Cascina, Italy (Accadia and The Virgo Collaboration, 2010a). The Virgo and LIGO detectors are of similar design with the greatest difference being the seismic isolation systems. Virgo uses superattenuator towers (Braccini and The Virgo Collaboration, 2005) to isolate its optics, while Initial LIGO used stacks of masses and high loss springs.

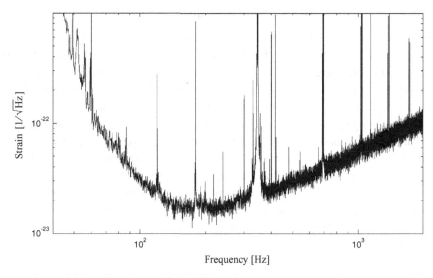

Figure 14.6 Sensitivity of the 4 km Initial LIGO interferometer at Hanford, Washington in 2010. The large spikes are due to 60 Hz lines, mechanical resonances of the suspension, and other narrowband artifacts.

The superattenuators allow superior seismic isolation to be achieved at lower frequencies than the passive LIGO stacks. The Initial Virgo detector has achieved a sensitivity of $h = 10^{-21}/\sqrt{Hz}$ at 20 Hz (Accadia and The Virgo Collaboration, 2010b).

GEO 600

GEO 600 (Lück and The GEO 600 Collaboration, 2006) is a gravitational wave interferometer operated by a British–German collaboration and located outside Hannover, Germany. The interferometer uses power recycling and an output cavity technique known as signal recycling (Meers, 1988) but no Fabry–Perot cavities in the arms. This requires an additional mirror at the output port known as the Signal Recycling (SR) mirror, see Figure 14.3. Instead of Fabry–Perot arms, GEO 600 uses folded arms that allow a 1.2 km-long optical path to fit within its 600 m-long vacuum envelope. This shorter interferometer makes GEO 600 inherently less sensitive than the LIGO or Virgo detectors, particularly at low frequencies. However, the GEO 600 detector also acts as a test-bed for new technologies. The smaller-sized facility allows the GEO scientists to more rapidly test and deploy the technologies that will be used in second generation detectors, including monolithic silica suspensions (Plissi *et al.*, 2000) and signal recycling. GEO 600 also uses an innovative corrugated vacuum tube design that is produced with less stainless steel. The combination of these technologies has allowed GEO 600 to achieve a sensitivity within a factor of four at 1 kHz of the large scale detectors.

14.5.2 Second generation detectors

There are currently four second generation gravitational wave interferometer projects in various states of production. The LIGO and Virgo instruments are being upgraded so that their sensitivity increases by a factor of ten. A third large scale second generation interferometer project, the Large-scale Cryogenic Gravitational-wave Telescope (LCGT), has recently received initial funding and will be built in Japan. In addition, the GEO 600 instrument is being upgraded to a high sensitivity, high frequency instrument.

The distance gravitational wave detectors can detect a given event scales directly with the sensitivity. So an increase by a factor of ten will result in roughly a thousand-fold increase in the volume of space sensed, which is roughly proportional to the rate of gravitational wave events. Accounting for the astrophysical uncertainties, the predicted rate of neutron star binary inspirals that these advanced interferometers can detect ranges from a few per year to a few thousand per year (O'Shaughnessy *et al.*, 2010).

Advanced LIGO

Advanced LIGO (Harry and The LIGO Scientific Collaboration, 2010) is the successor to the LIGO detectors with upgrades to all of the major detector systems, including seismic isolation, suspensions, optics, and the laser. The original upgrade plan included extending the 2 km interferometer to 4 km. However, a recent initiative has proposed siting the new 4 km detector in Australia, if that country will agree to fund the necessary facilities and operations. An Australian detector will significantly improve the angular resolution of the worldwide array because it is located out of the near-plane formed by the US, European, and Japanese detectors (Blair and Munch, 2009). This increased resolution is particularly important when an event requires a follow-up observation by an electromagnetic wave telescope.

Advanced LIGO will include a new seismic isolation system in each optic chamber which consists of an optical table with three stages of active isolation. The optical table supports a four-stage pendulum that ultimately supports the interferometer test masses. The last stage of the pendulum system consists of silica fibers to minimize thermal noise. This combination of a four-stage pendulum, active seismic isolation, and silica fiber suspensions means the sensitivity of Advanced LIGO will not be limited by either thermal noise from the suspensions or seismic noise in its normal operating mode.

The laser in Advanced LIGO is designed for 180 W of power at 1064 nm. This is a significant increase over the 10 W achieved in Initial LIGO. Since the signal-to-noise ratio in the shot noise regime scales as the square root of power, this large of an increase is necessary to gain sensitivity at higher frequencies. The laser must also satisfy stringent requirements in intensity and frequency noise.

The test masses in Advanced LIGO have been increased in mass from 10 kg to 40 kg to minimize the impact of radiation pressure noise (see Section 14.4.1). The coating on the test masses will be a test of a coating specifically designed to minimize coating thermal noise, using titania doped tantala and silica layers (see Chapter 4) and layer thickness optimization (see Chapter 12).

Advanced Virgo

Advanced Virgo (Acernese and The Virgo Collaboration, 2006), like Advanced LIGO, is designed to gain a factor of ten in sensitivity over its first generation instrument. For Advanced Virgo, this sensitivity improvement is brought about by increased laser power, larger optics suspended from silica fibers, and lower loss mirror coatings.

Large-scale Cryogenic Gravitational-wave Telescope

The Large-scale Cryogenic Gravitational-wave Telescope (LCGT) (Kuroda and The LCGT Collaboration, 2010) is a 3 km-long interferometer proposed by a Japanese collaboration and to be built in the Kamioka mine. Building off the work of the Cryogenic Laser Interferometer Observatory (CLIO) (Yamamoto *et al.*, 2008) test facility, LCGT will be the first kilometer-sized detector to be built underground and the first large detector to combat thermal noise by operating some of its mirrors at cryogenic temperatures (see Chapter 8). The test masses are planned to be made from sapphire to take advantage of the low temperature (see Chapter 7 for more on sapphire substrates). This technology is also being explored for third generation detectors, see Section 14.5.3 below. A part of the LCGT budget has recently been approved, with additional funding anticipated in the future.

GEO-HF

GEO-HF (high frequency) (Willke and The GEO-HF Collaboration, 2006) is an upgrade to GEO 600 that will include higher laser power and light squeezing (see Section 11.3). These improvements should result in a factor of ten increase in sensitivity at frequencies above 1 kHz.

14.5.3 Third generation detectors

In excess of 15 years of research was required to bring about the technical developments needed to design and build the second generation detectors. As the construction of the second generation detectors is starting, research into the design of third generation detectors is already commencing. To improve on the performance of second generation detectors a number of noise sources will need to be overcome. The limiting noise sources for second generation detectors are coating thermal noise, gravity gradient noise, and quantum noise. Of the innovations that are being investigated, some are potential add-ons to second generation detectors while others will require a complete redesign of the interferometer and its potential relocation. The Einstein Telescope (Punturo and The Einstein Telescope Collaboration, 2010), a third generation detector being considered in Europe, is the furthest along in planning. It is still just at the conceptual stage and many years of research and planning will be needed before a practical detector will arise.

Coating thermal noise

Nearly all the methods of improving coating thermal noise discussed in this book are being considered for third generation gravitational wave detectors. These include improved materials (Chapter 4); changing laser wavelength, Khalili cavities, waveguide gratings, and improved coating design (Chapter 6); cryogenics (Chapter 8); and beam shaping (Chapter 13).

Substrate thermal noise

Improvements in substrate thermal noise, as discussed in Chapter 7, are possibilities for third generation detectors. Both silicon and sapphire are possible substrate materials at cryogenic temperatures and have thermal noise advantages. At room temperatures, sapphire has better Brownian thermal noise than silica and its disadvantage with thermoelastic noise is not limiting at higher frequencies. Silica is still the preferred substrate material at room temperature and low frequencies, although research into other materials continues.

Quantum noise

The quantum noise that limits a significant fraction of the frequency band of gravitational wave detectors is caused by vacuum fluctuations entering the interferometer through the dark or anti-symmetric port. These vacuum fluctuations can be suppressed by injecting squeezed light into the interferometer through the dark port, as discussed in Section 11.3. To achieve the full potential benefit of the squeezing it is necessary to minimize optical losses in the interferometer. This means keeping the scattering losses of coatings to a minimum, as discussed in Chapter 11. The use of squeezed light can in principle be retrofitted with minimal changes to improve the performance of any of the second generation detectors.

Gravity gradient noise

At the lowest frequencies, gravity gradient noise will begin to be an issue for second generation detectors. Various mitigation schemes have been proposed that utilize an array of seismometers to detect ground motion then subtract the estimated noise as coupled into the detector. Another alternative being actively pursued would place the interferometers underground where seismic motion is considerably reduced.

14.6 Space-based detectors

To achieve sensitivity below 1 Hz it is likely to be necessary to put detectors in space. Gravitational wave detectors in space have two main advantages; seismic noise and gravity gradient noise are eliminated and the constraint on length is significantly relaxed. The increase in arm lengths means the change in test mass separation for a given gravitational wave strain increases. There are currently three space-based missions in various stages of maturity: LISA, BBO and DECIGO, which are discussed below.

14.6.1 LISA

The Laser Interferometer Space Antenna (LISA) (Jafry *et al.*, 1994) is a collaborative mission between the European Space Agency (ESA) and the US National Aeronautical and Space Administration (NASA) with a current launch date of 2020.[5] The LISA mission will consist of three spacecraft that are separated by a distance of 5 million kilometers and will orbit around the Sun trailing the Earth by 20°. The LISA spacecraft are drag free which means that each one consists of two satellites one inside the other. The inner satellite is a test mass which is protected from collisions with space particles by the outer satellite. Micro thrusters orientate the outer satellite such that it maintains its orientation relative to the test mass. Each of the three LISA spacecrafts transmits a laser beam in the direction of the other two. This enables the distance between the outer satellites to be measured. An interferometric measurement between the test mass and the outer satellite then enables the distance between the test masses to be determined.

The sensitivity band of LISA ranges from 0.1 mHz to 0.1 Hz, limited below 3 mHz by residual acceleration noise (Jennrich, 2009). The sensitivity above 30 mHz is limited by a combination of shot noise and nodes in the gravitational wave signal.

14.6.2 Big Bang Observer (BBO)

The design of LISA is such that there still exists a significant hole in the observable gravitational wave spectrum centered around 1 Hz. Preliminary research has been done on a mission named the Big Bang Observer (BBO) that will fill in the gap. It is designed to have enough sensitivity to detect the stochastic background of gravitational waves from the Big Bang (see Section 14.3.4), hence the name.

An arm length of 50 000 km, a 300 W, 355 nm wavelength laser, and locked arms (unlike LISA) are all being considered for BBO (Harry *et al.*, 2006b). Of particular interest in this book is that coating thermal noise will possibly be a limiting noise source if coatings have not improved by the time of this mission. Many of the technologies proposed for BBO do not currently exist and significant technical progress will be required before this mission becomes feasible. However, this mission is not expected to be launched for at least 20 years so there is time for the technical advances to be made.

---

[5] NASA's financial support of USA was removed in the proposed Fy 2012 budget. In response, the project is being redefined as an ESA mission.

14.6.3 DECIGO

DECIGO is an advanced space based gravitational wave detector which, like BBO, is designed to fill in the sensitivity gap (around 0.1 Hz) between ground based interferometers and the LISA mission. DECIGO stands for the DECI-hertz Interferometer Gravitational wave Observatory and it is being proposed in Japan. The DECIGO mission is planned to have three spacecraft, like LISA and BBO, which will be separated by 1000 km. DECIGO's launch is planned to be around 2027 with a couple of technology demonstrators missions to fly earlier (Ando and The DECIGO Collaboration, 2010).

15

High-precision laser stabilization via optical cavities

MICHAEL J. MARTIN AND JUN YE

15.1 Introduction

Optical cavities are extremely useful devices in laser-based research. Within the context of precision measurement, they enable tests of the laws that govern the macroscopic structure of the universe, embodied in the search for gravitational waves (Abbott and The LIGO Scientific Collaboration, 2009) (see also Chapter 14). At the other end of the length scale, cavity-stabilized lasers are powerful tools for precision spectroscopy that probe nature at the quantum mechanical level, through tests of quantum electrodynamics (QED) (Kolachevsky *et al.*, 2009). Furthermore, cavities enable high-sensitivity broadband spectroscopy (Thorpe *et al.*, 2006), which has practical applications in trace gas sensing; exploration of new light–matter interaction regimes in cavity QED (Miller *et al.*, 2005) (see also Chapter 17); tests of fundamental physical principles including relativity (Brillet and Hall, 1979; Hils and Hall, 1990; Eisele *et al.*, 2009), local position invariance (Blatt *et al.*, 2008), and the time invariance of the fundamental constants of nature (Fortier *et al.*, 2007b); and nonlinear optics, including coherent light build-up for studies of extremely nonlinear effects (Gohle *et al.*, 2005; Yost *et al.*, 2009). In general, optical cavities have become indispensable tools at the heart of many modern experiments.

In conjunction with optical frequency combs, cavity-stabilized laser systems have enabled the development of highly accurate frequency standards based on neutral atoms (Sterr *et al.*, 2004; Takamoto *et al.*, 2005; Ludlow *et al.*, 2006; Le Targat *et al.*, 2006; Ludlow *et al.*, 2008; Lemke *et al.*, 2009) and trapped ions (Diddams *et al.*, 2001; Madej *et al.*, 2004; Margolis *et al.*, 2004; Rosenband *et al.*, 2008). In recent years, two ion-based standards (Oskay *et al.*, 2006; Chou *et al.*, 2010), and also a neutral atom-based standard (Ludlow *et al.*, 2008) have surpassed the fractional frequency uncertainty of the primary cesium frequency standards that define the SI second (Heavner *et al.*, 2005; Bize *et al.*, 2005).

In addition to better accuracy, the real power of optical frequency standards is precision and stability (Hollberg *et al.*, 2005b). Ultrastable lasers paired with ultranarrow atomic

Optical Coatings and Thermal Noise in Precision Measurement, eds. Gregory M. Harry, Timothy Bodiya and Riccardo DeSalvo. Published by Cambridge University Press. © Cambridge University Press 2012.

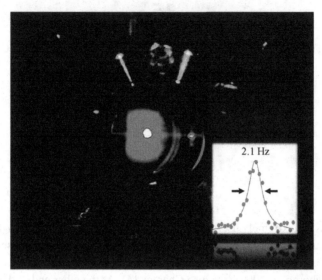

Figure 15.1 Strontium atoms trapped in a magneto-optical trap at JILA. Florescence from laser cooling at 461 nm, the first stage in a cooling process that creates optical lattice-trapped samples at μK temperatures, is visible from the atomic cloud (center). This same fluorescence serves as the clock readout. Inset: precision spectroscopy of the ultranarrow $^1S_0 \rightarrow {}^3P_0$ clock transition in ^{87}Sr at 429 THz. This is one of the highest Q spectroscopic features ever observed (Ludlow *et al.*, 2007). (Figure from Zelevinsky *et al.* (2008), Copyright Wiley-VCH Verlag GmbH & Co. KGaA. Reproduced with permission.)

transitions in the optical domain (Figure 15.1) have allowed the realization of optical clocks that are orders of magnitude more stable than current microwave-based frequency references. With increased stability, highly precise measurements of intricate physical effects can be made in short periods of time. For example, collisional effects between ultracold atoms that cause frequency shifts at the 10^{-16} level can be resolved within a few hours time (Campbell *et al.*, 2009). This measurement precision is only possible because the lasers at the heart of the best optical atomic clocks now operate at or below the 1×10^{-15} fractional frequency instability level with a mere 1 s averaging time.

The desire to further improve the stability of optical clocks continues to drive advances in cavity-stabilized laser systems. However, the cavity mirror coating and substrate thermal noise limits the stability of these optical systems in a fundamental way. Presently, this noise is a limiting factor in some of the best optical standards.

15.2 Review of optical cavities

A basic optical cavity is formed by an array of two opposite-facing mirrors (Figure 15.2). For high-precision frequency stabilization, these mirrors are typically held apart by a rigid spacer and are kept under vacuum to eliminate a varying intra-cavity index of refraction due to air. Although more complicated cavity geometries exist, including ring-type cavities

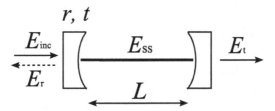

Figure 15.2 Schematic of an optical cavity in the standing wave configuration. The mirror amplitude reflectivity and transmission coefficients are given by r and t, respectively. E_{SS} is the steady-state electric field for a given incident field, E_{inc}. Some of E_{inc} is transmitted as E_t, while some is reflected as E_r.

(e.g. a cavity formed from a triangular mirror configuration) where the optical field is a running wave, we consider only this basic configuration as it is most common for precision frequency stabilization. For an incident laser of electric field amplitude E_{inc}, the steady-state electric field inside such a cavity, E_{SS}, is obtained by enforcing the condition

$$E_{SS} = E_{SS}e^{i\varphi}r^2 + E_{inc}t. \tag{15.1}$$

Here, φ is the round-trip phase accumulated by the light, t is the mirror field amplitude transmission coefficient and r is the corresponding amplitude reflectivity. In the absence of mirror absorption and scatter, it is possible to relate the magnitude of the r and t coefficients by $|r|^2 + |t|^2 = 1$, however we choose to allow for the real-world situation, where mirror losses are influential, by leaving these distinct. The cavity phase shift, φ, can be re-written in terms of the cavity length, L, and the laser's optical frequency, $\omega = 2\pi\nu$, as

$$\varphi = \frac{2L\omega}{c} + \tilde{\varphi}. \tag{15.2}$$

The term $\tilde{\varphi}$ is due to an additional mode-dependent diffraction phase term. While its consideration is necessary for finding the longitudinal mode-dependent frequency structure of an optical cavity, we omit this term for the remainder of this chapter because we will consider only a single optical mode. We also note that the cavity length, L, includes the effects of optical field penetration into the mirror coating, which typically requires a correction to the physical length on the order of an optical wavelength. For macroscopic cavities, this effect is negligible, but it becomes important for cavities whose size is of the order of an optical wavelength (Hood *et al.*, 2001).

By solving Equation 15.1 for the steady-state field, we find that the transmitted field amplitude, given by $E_t = tE_{SS}$, is

$$\frac{E_t(\nu)}{E_{inc}(\nu)} = \frac{e^{i\varphi/2}t^2}{1 - e^{i\varphi}r^2}. \tag{15.3}$$

Similarly, the reflected field, E_r, is given by

$$\frac{E_r(\nu)}{E_{inc}(\nu)} \equiv \mathcal{R} = r\left[\frac{1 - e^{i\varphi}\left(r^2 + t^2\right)}{1 - e^{i\varphi}r^2}\right]. \tag{15.4}$$

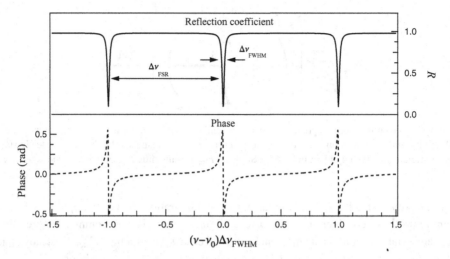

Figure 15.3 Reflection coefficient and corresponding phase shift of the reflected light incident upon an optical cavity. If there are no mirror losses, there is a discontinuity in phase as the reflected light drops to zero. Here, a small mirror loss term has been included, causing the reflection dip to not reach zero.

The cavity reflection transfer function, $\mathcal{R}$, is plotted in Figure 15.3. As can be seen from Equations 15.3 and 15.4, the transmission (reflection) is maximized (minimized) when the round-trip phase is a multiple of 2π. When this condition is met, the cavity is said to be on resonance. This results in the resonance condition

$$\nu_n = n\frac{c}{2L},\tag{15.5}$$

where L is the distance separating the mirrors and c is the speed of light. This condition is simply the requirement for a standing wave resonance within the cavity. Thus, the modes are spaced in frequency by $c/2L$, a quantity known as the free spectral range ($\Delta\nu_{FSR}$). By analyzing the denominator of Equation 15.4, the width of the cavity resonance (i.e. the width of a dip in $|\mathcal{R}|^2$), denoted as $\Delta\nu_{FWHM}$, is related to the free spectral range by

$$\mathcal{F} = \frac{\Delta\nu_{FSR}}{\Delta\nu_{FWHM}} = \frac{\pi\,|r|}{1 - |r|^2} = \frac{\pi\sqrt{R}}{1 - R}.\tag{15.6}$$

Here $R \equiv |r|^2$, and is the intensity reflection coefficient. This ratio, $\mathcal{F}$, is known as the cavity finesse and, as Equation 15.6 shows, depends only on the mirror properties.

15.2.1 Pound–Drever–Hall locking

As seen in the previous section, an optical cavity defines a series of narrow resonances. A common way to stabilize a laser to such a resonance is through a frequency modulation locking technique. The most commonly used and successful frequency modulation technique for laser stabilization is the Pound–Drever–Hall (PDH) stabilization scheme

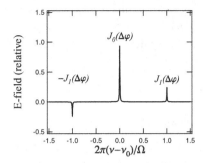

(a) Electric field amplitude in Fourier space in the presence of phase modulation of depth $\Delta\varphi$.

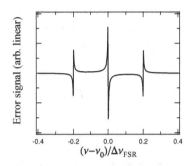

(b) Pound-Drever-Hall error signal as a function of laser detuning from the cavity resonance.

Figure 15.4 Pound–Drever–Hall sidebands and error signal.

(Drever *et al.*, 1983), where the frequency modulation is performed at a much higher frequency than the cavity linewidth.

There are several reasons for the widespread adoption of PDH locking. First, there are no restrictions upon the phase modulation frequency, as long as it is larger than the cavity linewidth. A higher modulation frequency gives the lock immunity to common laser amplitude noise offsets and also permits the use of resonant electro-optic modulators (EOMs). Additionally, in the PDH scheme, the lock bandwidth is not restricted by the cavity linewidth, allowing extremely narrow cavity resonance features to provide high-bandwidth stabilization.

Several reviews of PDH locking and laser feedback control theory exist (Day *et al.*, 1992; Mor and Arie, 1997; Black, 2001; Bava *et al.*, 2006). Here, we briefly discuss the important results of the PDH locking technique.

In order to measure the PDH error signal from an optical cavity, the laser must first be phase modulated. A phase modulated signal can be decomposed into a carrier and sidebands using the Jacobi–Anger expansion (Hils and Hall, 1987):

$$E_0 e^{-i2\pi f_0 t - i\Delta\varphi \sin(\Omega t)} = E_0 J_0\left(\Delta\varphi\right) e^{-i2\pi f_0 t}$$

$$+ E_0 \sum_{n=1}^{\infty} J_n(\Delta\varphi) \left[e^{-i(2\pi f_0 + n\Omega)t} + (-1)^n e^{-i(2\pi f_0 - n\Omega)t} \right]. \quad (15.7)$$

Here, the term $\Delta\varphi$ is the phase modulation depth and Ω is the phase modulation frequency. From this expansion, it is clear that the first order sidebands are 180 degrees out of phase (as are all odd order sidebands), as pictured in Figure 15.4. When the phase modulation frequency is well outside the cavity bandwidth, these sidebands are sufficiently detuned from the cavity resonance such that they are promptly reflected from the cavity unaffected. The carrier, which is near the cavity resonance, is affected by the complex response of the cavity (as shown in Figure 15.3 and given in Equation 15.4) and interferes with the phase modulation sidebands upon reflection.

This interference term can be explored by assuming that the phase modulation sidebands are completely reflected and finding the time-dependent reflected optical power, $P_{\text{ref}}(t)$, when the carrier is near resonance. Making use of only the first order sidebands of Equation 15.7,

$$P_{\text{ref}}(t) = \frac{1}{2} \left| E_{\text{c, ref}} + E_s e^{-i\Omega t} - E_s e^{i\Omega t} \right|^2$$

$$= P_{\text{c, ref}} + 2P_s - 2\Im\left\{ E_{\text{c, ref}} E_s^* \right\} \sin(\Omega t) + 2\Omega \text{ terms.} \tag{15.8}$$

Here, $E_{\text{c, ref}}$ ($P_{\text{c, ref}}$) is the reflected electric field amplitude (power) in the carrier and E_s (P_s) is the reflected electric field amplitude (power) in the sidebands. Keeping everything that oscillates at Ω or below, and making use of Equations 15.7 and 15.4,

$$P_{\text{ref}}(t) = P_0 \left[J_0^2(\Delta\varphi) |\mathcal{R}|^2 + 2J_1^2(\Delta\varphi) \right]$$

$$+ 4P_0 J_0(\Delta\varphi) J_1(\Delta\varphi) \Im\{\mathcal{R}\} \sin(\Omega t). \tag{15.9}$$

When the carrier is less than a cavity linewidth from resonance,

$$\Im\{\mathcal{R}\} \simeq \frac{-2|t|^2 \mathcal{F}}{(1-|r|^2)\Delta\nu_{\text{FSR}}} \delta\nu. \tag{15.10}$$

Here, $\delta\nu$ is the laser detuning from cavity resonance given by $\delta\nu = \nu_{\text{laser}} - \nu_{\text{cavity}}$. The term in Equation 15.9 that oscillates as $\sin(\Omega t)$ is thus given by

$$\mathcal{D}\delta\nu \sin(\Omega t), \tag{15.11}$$

where we have used the definition

$$\mathcal{D} \equiv -\frac{8P_0 J_0(\Delta\varphi) J_1(\Delta\varphi)}{\Delta\nu_{\text{FWHM}}} \left(\frac{|t|^2}{1-|r|^2} \right), \tag{15.12}$$

along with the relationship $\Delta\nu_{\text{FSR}}/\mathcal{F} = \Delta\nu_{\text{FWHM}}$ to derive Equation 15.12.

Equation 15.11 gives the component of optical power that oscillates at the phase modulation frequency. For small detunings, the amplitude is linear in $\delta\nu$, and can thus be used to lock the laser to the optical cavity after the optical power has been detected on a photodiode and demodulated. The degree to which the amplitude changes for a given detuning is characterized by the parameter $\mathcal{D}$, which, as should be expected, varies inversely with cavity linewidth and is proportional to the product of the zero and first-order Bessel functions. In passing, we note that this can be used to define an optimal modulation depth, given by $\Delta\varphi = 1.08$. By measuring this oscillating RF signal, and demodulating by mixing in the proper quadrature at the frequency Ω, a linear error signal can be obtained with which to feed back upon the laser frequency.

While Equation 15.11 describes the behavior of the error signal near the cavity resonance, one may go a step further and include the frequency dependence of $\Im\{\mathcal{R}\}$ to calculate the shape of the error signal over a broader range, as shown in Figure 15.4 (Mor and Arie, 1997). When this detail is included, it can be shown that the error signal of Equation 15.11 needs to be multiplied by a single-pole low-pass filter function with a corner frequency equal to the cavity half-width. This effect can be compensated for by appropriate servo design, such that the lock bandwidth need not be limited by the cavity linewidth. Physically, this low-passing effect represents a transition from a regime in which the cavity is sensitive to frequency fluctuations of the laser to one in which the error signal is proportional to phase fluctuations of the laser. As expected, this transition occurs at the cavity half-width.

15.2.2 Sources of lock error

There are two important considerations when frequency stabilizing a laser by locking it to an optical cavity. The first is that the reference cavity optical length must be as stable as possible and will be discussed in Section 15.3. However, an equally important question is whether there are any issues that can prevent a laser from precisely tracking the reference cavity resonance, due to either technical or quantum effects.

The most fundamental but least important source of error in cavity locking systems is quantum noise (see, e.g., Salomon *et al.* (1988); Day *et al.* (1992)). The optical power spectrum of shot noise on the light at the detector is given by the single-sided power spectral density

$$G_P = 2h\nu P_{\text{opt}} \quad [\text{W}^2/\text{Hz}]. \tag{15.13}$$

Thus, assuming $|r|^2 + |t|^2 = 1$, the expected frequency noise due to shot noise in the most ideal case (and with an ideal demodulator) is

$$G_\nu = \frac{G_P}{\eta \mathcal{D}^2} = \frac{h\nu \Delta \nu_{\text{FWHM}}^2}{16\eta P_0 J_0^2} \quad [\text{Hz}^2/\text{Hz}], \tag{15.14}$$

where η takes into account the detector quantum efficiency and $\nu = c/\lambda$ is the optical frequency. Substituting in the very modest parameters $\lambda = 1\,\mu\text{m}$, $P_0 = 10\,\mu\text{W}$, $\eta = 0.5$, and $\Delta \nu_{\text{FWHM}} = 10\,\text{kHz}$, the very low shot noise floor of $G_\nu = 4 \times 10^{-7}\,\text{Hz}^2/\text{Hz}$ can be achieved. This can be related to the locked laser linewidth by

$$\Delta \nu_{\text{locked}} = \pi G_\nu, \tag{15.15}$$

where it is assumed that G_ν is white noise. For the parameters given above, this results in a locked linewidth of $1\,\mu\text{Hz}$. Thus, for high-finesse cavities, the shot noise locking limit is far below any cavity locking result, even in experiments designed to exclude other technical and thermal effects (Salomon *et al.*, 1988). This indicates that in practical situations, "technical" effects are most important.

Residual amplitude modulation (RAM) is a term that collectively describes a variety of effects that induce amplitude modulation at the phase modulation frequency. For example, temperature-dependent parasitic etalons within the optical system can induce RAM which is then demodulated along with the cavity error signal. This causes an offset to be introduced in the locking system. Other effects that can cause RAM are typically related to the phase modulation device, most often an electro-optic crystal-based (e.g. LiNbO_3, ADP, KDP) modulator. For instance, stress-induced birefringence in crystal devices can rotate the principal crystal axis, creating not only an electric field-dependent phase shift, but also a corresponding polarization rotation, which can be distributed across the optical wave front.

While in many cases a temperature-controlled electro-optic (EO) crystal has low enough RAM for acceptable performance, a certain degree of success has been achieved by using a DC electric field on crystal-based EO modulators to actively servo the RAM (Wong and Hall, 1985). However, beam pointing deviations can cause slightly different regions of the crystal to be sampled, causing the phase of the RAM to shift, and limiting the effectiveness of the active system.

It is important to note that the effect of RAM on laser stability is reduced for cavities with narrower resonances due to the fact that a given fractional change in RAM results in a smaller change in frequency for a narrower resonance. Thus, for a given cavity length, higher cavity finesse is always desirable to help mitigate RAM-induced line pulling.

15.3 Mechanical design of optical reference cavities

Despite the considerable challenges present in building a high-quality cavity locking system that is free from residual frequency offsets, these effects are not what limit most high-finesse optical cavity frequency references. Instead, perturbations to the length of optical reference cavity are ultimately what limit the frequency stability. These perturbing effects fall into two categories: mechanical and thermal perturbations that are not fundamental, i.e. that are non-statistical in origin; and fundamental statistical fluctuations in the cavity spacer, substrate, and coatings that arise from their contact with a thermal reservoir at room (or cryogenic) temperature.

In this section, we discuss non-statistical perturbing effects and describe methods for their mitigation. These mechanical effects can be divided into two categories: those caused by vibrations (accelerations) that structurally deform the cavity, and those that couple through the coefficient of thermal expansion (CTE) of the cavity materials. Use of finite element analysis to optimize cavity geometries and choice of materials has drastically reduced and elucidated these effects.

15.3.1 Vibration sensitivity

Although optical cavity mirrors and spacers are typically made out of a rigid substance, such as ultra low expansion glass (ULE), the length stability requirements are extremely stringent for sub-Hz lasers. As can be seen from Equation 15.5, the fractional frequency change of a cavity resonance is directly related to the fractional length change by

$$\frac{\Delta v}{v} = -\frac{\Delta L}{L}. \tag{15.16}$$

The current world record for cavity stabilization was set in 1999 with 1 s stability at 3×10^{-16} and employed a very impressive vibration isolation scheme in order to keep the cavity length constant – effectively suspending the entire optical table on giant rubber

(a) FEA model of a vertical cavity mounted at the midplane, verifying insensitivity (Chen et al., 2006a).

(b) Picture of completed cavity. The spacer material is ultra low expansion glass.

Figure 15.5 Vertical cavity mounting geometry using the symmetric supporting scheme. Finite element analysis (FEA), figure (a), can be used to fine-tune the mounting geometry.

bands (Young *et al.*, 1999). Although the cavity length we discuss is the effective length sensed by the optical field, which is averaged over the mirror surface, it is still astounding that the length stability needed for a sub-Hz laser is sub-fm (10^{-15} m) – the length scale of the proton radius! It should come as no surprise that length stability at or below one part in 10^{15} takes significant engineering effort.

One approach that significantly reduces the dependence on vibration-isolating structures is to design the cavity spacer such that the mirror–spacer system is insensitive to vibrations. An intuitive way to achieve this is to mount the cavity at its midplane in the vertical direction (Notcutt *et al.*, 2005). In this way, the top and bottom mirrors move equal amounts when subject to vibrations along the vertical axis. However, one can only get so far exploiting intuitive geometry for the simple reason that the support structure breaks perfect vertical symmetry. Thus, to finalize any cavity design, finite element analysis (FEA) must be employed (Chen *et al.*, 2006a). This technique can be applied to a variety of cavity geometries and tailored to a specific design goal, such as insensitivity in a specific direction.

One system that exploits the benefits of vertical symmetry (Ludlow *et al.*, 2007), shown in Figure 15.5(b), has a measured vibrational sensitivity in the vertical direction of $30 \text{ kHz}/\left(\text{m/s}^2\right)$ representing a fractional sensitivity of $7 \times 10^{-11}/\left(\text{m/s}^2\right)$. A more recent example of a stable cavity based on a vertical geometry has shown even better vibrational insensitivity by the FEA technique at the level of $10^{-11}/\left(\text{m/s}^2\right)$ (Millo *et al.*, 2009). These results are quite good given that the highest grade commercial isolation platforms can give isolation performance at the 50 ng/$\sqrt{\text{Hz}}$ level, resulting in a vibration-limited frequency noise performance of order 10 mHz/$\sqrt{\text{Hz}}$ for the sensitivities exhibited by modern cavities in the visible spectrum.

In principle, one is not restrained to vertical configurations. In fact, there may be good reason to choose a horizontal configuration, especially if it is expected that the majority of

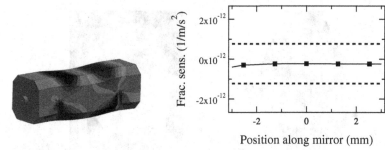

(a) FEA model for a horizontal cavity geometry. In this Figure, the model has been optimized to limit length change and mirror tilt for vertical accelerations.

(b) Mirror displacements per m/s² of vertical acceleration as a function of vertical distance along the mirror surface. The support positions have been optimized to reduce length sensitivity for vertical accelerations. The dashed horizontal lines indicate the estimated uncertainty due to finite mesh size effects.

Figure 15.6 Finite element analysis (FEA) of a 40 cm cavity in development at JILA.

vibrations will be in the vertical direction. Through FEA, results comparable to and even better than those obtained with vertical mounting schemes have been obtained (Webster *et al.*, 2007; Millo *et al.*, 2009). Other motivating factors include structural stability, especially for larger cavities, and the experimental ease of access for horizontal geometries. Most importantly, in the horizontal configuration, the coupling of vertical accelerations to deviations along the optical axis is reduced by the Poisson's ratio, representing roughly an 80% reduction in sensitivity. However, horizontal accelerations can still couple into mirror displacement, although the inherent symmetry in the two horizontal axes limits this effect. Figure 15.6(a) shows an FEA model of a 40 cm ULE cavity under development at JILA. Sensitivity to vertical accelerations has been eliminated in the FEA model by choice of support points. (See Figure 15.6(b).)

15.3.2 Thermomechanical perturbations

The most common cavity spacer and mirror substrate materials for ultrastable reference cavities are ULE, fused silica (FS), sapphire, and silicon. The coefficients of thermal expansion (CTEs) of these materials are shown in Figure 15.7. Owing to large room-temperature CTEs, the latter two materials have primarily been used at cryogenic temperatures (Richard and Hamilton, 1991; Seel *et al.*, 1997; Müller *et al.*, 2003a). See Chapter 8 for more on cryogenics.

In general, a cavity made of a uniform-CTE material will experience a fractional length change given by

$$\Delta L/L = \alpha(T)\delta T + \frac{1}{2}\alpha'(T)\delta T^2 + \mathcal{O}\left(\delta T^3\right), \qquad (15.17)$$

where $\alpha(T)$ is the CTE at the operating temperature, T, $\alpha'(T)$ is its first derivative, and δT are small temperature variations. Ideally, one operates a reference cavity around a

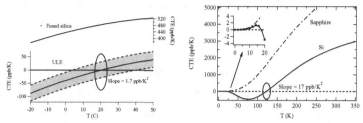

(a) CTEs of fused silica and ultra low expansion glass (ULE) (Fox, 2009) near room temperature. The gray band on the ULE curve represents the manufacturer's stated uncertainty in the zero-crossing temperature.

(b) CTEs of sapphire (White, 1993; Taylor et al., 1997) and silicon (Swenson, 1983). The inset shows that silicon has a second zero-crossing 17 K in addition to the zero-crossing at 123 K.

Figure 15.7 Temperature dependence of coefficients of thermal expansion (CTEs) of common cavity spacer and substrate materials. See also Section 7.3.

zero-crossing temperature of a material's CTE, T_0, defined as

$$\alpha(T_0) = 0. \tag{15.18}$$

In the case of sapphire, which has no zero crossing, optimal operation is at cryogenic temperatures which reduce the CTE to an acceptable level. Taking Equation 15.17 in the vicinity of a zero crossing and substituting in $\alpha'(T_0) = 1.7 \times 10^{-9}/\text{K}^2$ for ULE, it can be seen that the temperature needs to be stable below 1 mK for relative length stability at the 10^{-16} level. While this is in principle quite difficult, the large thermal mass of the ULE spacer tends to limit temperature effects to longer time scales. Mechanical coupling from the vacuum chamber itself, not the CTE of the spacer material, can introduce the biggest temperature-dependent frequency shift. Thus, care should also be taken to mechanically decouple the cavity support structure from the chamber. As seen in Figure 15.5(b), this can be accomplished with a separate ULE piece (in this case a ring).

In the case of a cavity with a mirror substrate made out of a different material than the spacer, extra complications arise. The two materials are typically optically contacted very firmly, such that the mismatch of CTEs causes an auxiliary mechanical effect, effectively causing the mirror substrate to bend. This modifies the effective CTE, α_{net}, such that (Notcutt *et al.*, 1995)

$$\alpha_{\text{net}} = \alpha_{\text{spacer}} + 2\delta \frac{R}{L} \left(\alpha_{\text{mirror}} - \alpha_{\text{spacer}} \right) + \Gamma. \tag{15.19}$$

Here, R is the mirror radius and L is the cavity length. The term δ describes thermomechanical stresses coupling into length change and Γ accounts for deviations from the ideal model (Fox, 2009). Thermo-mechanical finite element analysis can be used to find δ for a given cavity geometry, such that α_{net} can be found and minimized by operating at the new zero-crossing temperature (Fox, 2009; Legero *et al.*, 2010).

Finally, it is worth noting that there has been good success tuning the zero cross point by contacting an additional piece of ring-shaped ULE glass to the back of the mirror substrate, effectively "sandwiching" the substrate between two equivalent CTE materials. By varying the parameters of the ring-shaped piece of glass, one can engineer a more favorable T_0 (Legero *et al.*, 2010).

15.4 Statistical thermal noise

The second class of thermally driven length fluctuations in optical cavities are more fundamental in origin. These fluctuations have been discussed in Chapters 1, 3, 4, 7, and 9. Here we apply the developed formalism to the case of optical cavities.

15.4.1 Brownian motion

The first, and typically most important, type of thermal noise for precision optical measurements is the Brownian motion of the constituents of the optical system. In the case of a cavity, the components of concern are the spacer, mirror substrate, and mirror coatings. At temperature T, after appropriately weighting the displacement by the beam profile (see Chapter 1), the one-sided power spectral density of position fluctuations (in units of m^2/Hz) for each of these components at Fourier frequency f are (Harry *et al.*, 2002; Numata *et al.*, 2004)

$$G_x^{\text{substrate}}(f) = \frac{2k_B T}{\sqrt{\pi^3} f} \frac{1 - \sigma^2}{w_m Y_s} \phi_s, \tag{15.20}$$

$$
\begin{aligned}
G_x^{\text{coating}}(f) = \frac{2k_B T}{\sqrt{\pi^3} f} \frac{1 - \sigma^2}{w_m Y_s} & \left\{ \frac{1}{\sqrt{\pi}} \frac{d}{w_m} \frac{1}{Y_s Y_c \left(1 - \sigma_c^2\right) \left(1 - \sigma_s^2\right)} \right. \\
& \times \left[Y_c^2 \left(1 + \sigma_s\right)^2 \left(1 - 2\sigma_s\right)^2 \phi_{\parallel} + Y_s Y_c \sigma_c \left(1 + \sigma_s\right) \left(1 + \sigma_c\right) \left(\phi_{\parallel} - \phi_{\perp}\right) \right. \\
& \left. \left. + Y_s^2 \left(1 + \sigma_c\right)^2 \left(1 - 2\sigma_c\right) \phi_{\perp} \right] \right\},
\end{aligned}
\tag{15.21}
$$

$$G_x^{\text{spacer}} = \frac{2k_B T}{f} \frac{L}{3\pi^2 R_{\text{spacer}}^2} \frac{\phi_{\text{spacer}}}{Y_{\text{spacer}}}. \tag{15.22}$$

The parameters are the same as defined previously in Chapters 4 and 7, and are presented again in Table 15.1 for completeness.

Two qualitative remarks can be made at this point. First, both the substrate and coating displacement noise power spectral densities, given by Equations 15.20 and 15.21, respectively, do not depend on the length of the cavity. This is due to the fact that the fluctuations are localized to the mirror surface and this property can be exploited in order to reduce frequency noise. By increasing the cavity length, the fractional length fluctuations decrease, resulting in a substrate and coating thermal noise-induced frequency noise spectral density that is proportional to $1/L^2$. Secondly, while the spacer thermal noise contribution scales with length, and inversely with spacer radius, R_{spacer}, its contribution to the total fractional length change of the cavity in fact decreases with length. This is due to the conversion from power spectral densities to fractional frequency fluctuations involving division by

Table 15.1 *Summary of the parameters used in the text.*

Definition of parameters	
w_m	Beam $1/e^2$ intensity radius
d	Coating thickness
ϕ_s	Substrate loss angle
$\phi_{\perp(\parallel)}$	Coating loss angle perpendicular (parallel) to substrate
$Y_{s(c)}$	Substrate (coating) Young's modulus
$\sigma_{s(c)}$	Substrate (coating) Poisson's ratio
$\alpha_{s(c)}$	Substrate (coating) coefficient of thermal expansion
$\kappa_{s(c)}$	Substrate (coating) thermal conductivity
$C_{s(c)}$	Substrate (coating) heat capacity
$f_c^{\text{sub(coat)}}$	Substrate (coating) cutoff frequency

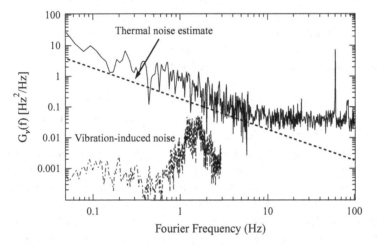

Figure 15.8 Measured frequency noise of the cavity shown in Figure 15.5. The noise frequency spectrum is thermal noise limited to 10 Hz, where the photon shot noise and detector electronic noise begin to dominate.

L^2. Additionally, longer optical cavities typically have larger radii in order to maintain favorable mounting geometry, meaning that as L increases, so too does R_{spacer}.

Current experiments are at or near the expected thermal noise limitations set by the substrate and coating Brownian thermal noise alone. This is because the spacer contribution of Equation 15.22 is generally an order of magnitude below mirror thermal noise for typical cavity aspect ratios. Figure 15.8 shows an example of an experimentally obtained frequency spectrum taken from a comparison between two ultra-stable lasers (of the type shown in Figure 15.5), which approximately agrees with the Brownian thermal noise limit (Ludlow *et al.*, 2007). See also Chapter 5 for direct observations of coating thermal noise in optical cavities. Also shown in Figure 15.8 is the vibration-limited frequency noise spectrum, which is well below the thermal noise floor, indicating that vibration noise does not contribute to the observed spectrum.

One clear avenue to decrease this noise limit is to decrease the material losses (see Chapter 4), lower the temperature (see Chapter 8), or do both. However, the situation is not so simple. For example, the loss angle of fused silica begins to increase sharply at temperatures below ∼250 K, ultimately suffering an almost four orders of magnitude increase before it levels off at 50 K, completely eliminating the benefit of operating at these temperatures (Braginsky *et al.*, 1985; Schnabel *et al.*, 2010). However, crystalline materials such as sapphire (Braginsky *et al.*, 1985; Uchiyama *et al.*, 1999), calcium fluoride (Nawrodt *et al.*, 2007a), and silicon (McGuigan *et al.*, 1978; Rowan *et al.*, 2003; Schnabel *et al.*, 2010) offer the benefits of low thermal expansion and low loss angle at cryogenic temperatures. Unfortunately, typical coating loss angles exhibit an approximate factor of 3 increase at cryogenic temperatures (see Chapters 4 and 8), which offsets some of the gains of operating at low temperatures. We note in passing that there is an active search for low-loss coating materials or dopants to reduce the mechanical loss of the existing coating materials (Harry *et al.*, 2006a). It also becomes increasingly difficult to shield cryostat vibrations at very low temperatures, due to the large cooling powers required. Chapter 8 contains a more complete discussion of the advantages and disadvantages of cryogenics for thermal noise.

A powerful alternative approach to reduce Brownian coating and substrate thermal noise relies not on reducing the temperature, but instead on using specially shaped beams, such as Mesa, conical, or higher order Laguerre–Gauss beams (Bondarescu *et al.*, 2008). This has the effect of better averaging the position fluctuations of the mirror surface due to thermal noise. While one could simply envision working near the stability edge of an optical cavity with typical spherical mirrors to create larger mode-areas on the mirrors, the input pointing stability requirements become more stringent in these regimes. Specially shaped beams have their own challenges, however, especially controlling the manufacturing process to create satisfactory mirror profiles in the small-scale optics used in optical cavities (Tarallo *et al.*, 2007) (see also Section 2.7). See Chapter 13 for a detailed discussion of beam shaping for thermal noise reduction.

15.4.2 *Thermo-optic noise*

Substrate thermoelastic and coating thermo-optic noise have been studied as a noise source for gravitational wave detectors (Cerdonio *et al.*, 2001; Braginsky and Vyatchanin, 2003a; Evans *et al.*, 2008) where it potentially has important sensitivity implications. Substrate thermoelastic noise is discussed in Chapter 7, coating thermo-optic noise is covered in Chapter 9, and gravitational wave detectors in general are the topic of Chapter 14. In contrast to Brownian motion of the mirror substrate and coatings, thermo-optic noise arises from fundamental temperature fluctuations in the bulk material coupling to the coefficient of thermal expansion. These temperature fluctuations can be described by the well-known expression (Braginsky *et al.*, 1999)

$$\langle \delta T^2 \rangle = \frac{k_B T^2}{\rho C V}. \tag{15.23}$$

Here ρ is the material density, C is the heat capacity per unit mass, and V is the volume over which the temperature fluctuations are considered.

There are two Fourier frequency regimes in the analysis of thermo-optic noise. The first is where the thermal diffusion length scale is smaller than the laser spot size, allowing an averaging effect to take place. This regime is known as the adiabatic limit and only applies to time-domain Fourier frequencies f that satisfy $f \gg f_c$ where the cutoff frequency, f_c, is given by

$$f_c = \frac{\kappa}{\pi w_m^2 \rho C}. \tag{15.24}$$

Owing to the small beam sizes (~ 100 μm) and interest in the frequency noise spectrum all the way to DC in cavity-stabilized laser systems, one must be aware that f_c is typically in the 1 Hz range. Thus, consideration of thermo-optic noise in the second regime, at Fourier frequencies $f < f_c$, is necessary for a complete picture of the various contributions to the frequency noise of cavity stabilized lasers. See also Section 8.2.5.

Substrate thermoelastic noise

To date, many optical cavities have employed mirrors made from ULE substrates (Ludlow *et al.*, 2007; Alnis *et al.*, 2008b). As a result, consideration of substrate thermoelastic noise is not necessary for these systems, as the material CTE is close to zero. (See Chapter 7 for the relationship between CTE and thermoelastic noise.) This approximation has also been made in the case of fused silica substrates (Numata *et al.*, 2004). In fact, while alarming predictions for substrate thermoelastic noise can be obtained by extrapolating the high-frequency behavior of fused silica to DC, using the appropriate expression for the low-frequency behavior verifies that the substrate thermoelastic noise is at least an order of magnitude below the Brownian noise of the substrates and coatings.

It has been shown (Braginsky *et al.*, 1999; Cerdonio *et al.*, 2001) that the one-sided power spectral density of mirror length fluctuations due to the substrate is

$$G_x^{\text{TE,sub}}(f) = \frac{4}{\sqrt{\pi}} \alpha_s^2 (1 + \sigma_s)^2 \frac{k_B T^2 w_m}{\kappa_s} J[\Omega(f)]. \tag{15.25}$$

(See Chapter 7 for a full discussion of substrate thermoelastic noise.) Here, $\Omega(f) = f/f_c^{\text{sub}}$, and $J[\Omega]$ is given by

$$J[\Omega] = \sqrt{\frac{2}{\pi^3}} \int_0^\infty du \int_{-\infty}^\infty dv \frac{u^3 e^{-u^2/2}}{(u^2 + v^2) \left[(u^2 + v^2)^2 + \Omega^3 \right]}. \tag{15.26}$$

While the integral can be evaluated numerically, it is more instructive to calculate thermal noise in the low and high frequency limits. Specifically,[1]

$$G_x^{\text{TE,sub}} \rightarrow \frac{8\sqrt{2}}{3\pi}\alpha_s^2\,(1+\sigma_s)^2\,\frac{k_B T^2}{\sqrt{2\pi f \kappa_s \rho_s C_s}}, \quad \Omega \ll 1 \tag{15.27}$$

$$G_x^{\text{TE,sub}} \rightarrow \frac{16}{\sqrt{\pi}}\alpha_s^2\,(1+\sigma_s)^2\,\frac{k_B T^2 \kappa_s}{(2\pi f \rho_s C_s)^2\,w_m^3}, \quad \Omega \gg 1. \tag{15.28}$$

These equations indicate that at low frequencies, thermoelastic noise rises less rapidly than extrapolated from the high-frequency behavior. Qualitatively, this effect can be explained as a crossover from the regime where the thermal diffusion length is smaller than the spot size to one where it is larger. Thus, this change in behavior can be thought of as an averaging effect that is no longer valid at low frequencies (Cerdonio *et al.*, 2001; Braginsky and Vyatchanin, 2003a).

Coating thermo-optic noise

A second way that optical cavities are sensitive to thermodynamic temperature fluctuations is through a pair of correlated mechanisms present in the mirror coatings: thermorefractive and thermoelastic effects, collectively called thermo-optic noise. Thermo-optic noise is discussed in detail in Chapter 9. It has been shown that the typically opposite signs of these coherent mechanisms reduces their impact (Evans *et al.*, 2008) and the total effect can be written as

$$G_x^{\text{TO}}(f) = G_{\Delta T}(f)\left(\bar{\alpha}_c - \bar{\beta}_c - \bar{\alpha}_s\frac{C_c}{C_s}\right)^2. \tag{15.29}$$

The term in parentheses is the coherent sum of thermoelastic and thermorefractive effects; the thermo-optic noise. The parameter $\bar{\alpha}_c$ ($\bar{\alpha}_s$) is the effective coating (substrate) coefficient of thermal expansion, and $\bar{\beta}_c$ is the effective coating thermorefractive coefficient. The term $G_{\Delta T}(f)$ is the power spectral density of temperature fluctuations given by

$$G_{\Delta T}(f) \rightarrow \frac{2k_B T^2}{\pi w_m^2 \sqrt{\pi f \kappa \rho_c C_c}}, \quad \Omega \gg 1. \tag{15.30}$$

15.4.3 Total thermal noise contribution to cavity frequency stability

The total thermal noise is given by

$$G_x^{\text{tot}} = \sum_{\text{L,R}} G_x^{\text{TO}} + \sum_{\text{L,R}} G_x^{\text{TE}} + \sum_{\text{L,R}} G_x^{\text{substrate}} + \sum_{\text{L,R}} G_x^{\text{coating}} + 2G_x^{\text{spacer}}, \tag{15.31}$$

where the sum over left and right (L, R) takes into account that the beam waist is potentially different at the left and right mirrors. The factor of two in front of G_x^{spacer} is because this term is always equivalent at each mirror.

---

[1] Equation 15.27 differs from Equation 8.4 in the numerical prefactor, but only by less than 10%. The reason for this discrepancy is under investigation, but the numerical difference should not be significant in most applications.

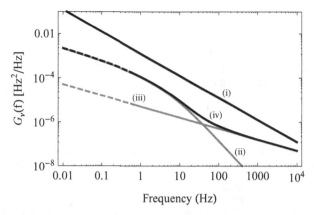

Figure 15.9 Thermal noise for a 40 cm-long (approximate radius 7 cm) cavity at 698 nm with fused silica mirror substrates and 1 m radius of curvature/planar mirror geometry (see also Figure 15.6(a)). The temperature is 300 K. (i) Sum of mirror substrate and coating, and cavity spacer Brownian noise. (ii) Substrate thermoelastic noise. (iii) Coating thermo-optic noise. (iv) Sum of coating thermo-optic and substrate thermoelastic noise contributions. The dotted regions indicate frequency regimes where $f \leq f_c$ (i.e. frequencies below the adiabatic limit).

Converting the total length fluctuation power spectral density, G_x^{tot}, into optical frequency deviations can be accomplished by use of Equation 15.16, which directly relates fractional length change to frequency fluctuations. We obtain

$$G_\nu^{tot} = \nu_0^2 G_x^{tot}/L^2, \tag{15.32}$$

where L is the length of the cavity and ν_0 is the laser's optical frequency. Current state-of-the-art systems have a thermal noise floor that is approximately an order of magnitude above the vibration-limited noise floor (Ludlow *et al.*, 2007). The results of Equation 15.32 are shown in Figure 15.9, detailing the different contributions of the thermal noise for the specific case of the 40 cm-long JILA ULE cavity with FS mirrors investigated in Figure 15.6.

15.5 Atomic clock applications of frequency-stabilized lasers

A confluence of two key technologies – femtosecond laser frequency combs and ultrastable lasers – has enabled a new class of atomic clocks based not upon microwave frequency transitions but instead upon extremely narrow optical transitions in neutral atoms and ions. This revolution in precision science continues to progress as the physics behind minute effects continues to be unraveled and quantum technologies are increasingly being employed to gain signal size and robustness, further increasing clock accuracy. Additionally, neutral atom clocks stand to gain an order of magnitude in stability with the advent of next-generation ultrastable laser systems with lower thermal noise.

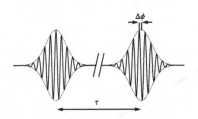

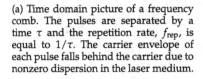

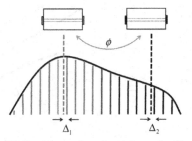

(a) Time domain picture of a frequency comb. The pulses are separated by a time τ and the repetition rate, f_{rep}, is equal to $1/\tau$. The carrier envelope of each pulse falls behind the carrier due to nonzero dispersion in the laser medium.

(b) Frequency domain picture of a frequency comb. The comb-like structure arises from the periodic and phase-coherent nature of each pulse. The relative beat frequencies between the comb and clock lasers can be used to compare two optical frequency references.

Figure 15.10 Time and frequency representations of a frequency comb.

15.5.1 Frequency combs

No discussion of ultrastable lasers and atomic clocks would be complete without introducing optical frequency combs. Femtosecond laser-based optical frequency combs have revolutionized the field of optical frequency metrology (Udem *et al.*, 2002; Cundiff and Ye, 2003). With laser media ranging from bulk Ti:Sapphire and optical fibers to microtoroidal resonators, the frequency comb revolution shows no signs of slowing down. The spectral coverage of frequency combs has been demonstrated to span the mid-IR to the vacuum ultraviolet (Gohle *et al.*, 2005; Yost *et al.*, 2009; Adler *et al.*, 2010).

At the heart of a comb's utility is the equation that describes the optical frequency of a given mode, v_n, as

$$v_n = nf_{\text{rep}} + f_0. \tag{15.33}$$

Here, $f_{\text{rep}} = 1/\tau$ is the comb pulse repetition rate, where τ is the time between successive pulses. The carrier envelope offset frequency, f_0, arises from the group and phase velocities inside the laser cavity being different. It is related to the pulse to pulse carrier envelope phase slippage ($\Delta\phi$ in Figure 15.10(a)) by

$$f_0 = \Delta\phi f_{\text{rep}}/(2\pi). \tag{15.34}$$

In principle, f_{rep} and f_0 are the comb's only degrees of freedom when describing the frequency of a given "tooth" in the frequency domain.

By locking a frequency comb to an optical source and stabilizing f_0 by the self-referencing technique (Jones *et al.*, 2000; Cundiff and Ye, 2003), the comb degrees of freedom are completely constrained and directly related to the optical phase of the reference laser. By making a heterodyne beat with a second laser, the phase of the two optical sources can be directly compared (Figure 15.10(b)), often across > 100 THz of spectral bandwidth (Foreman *et al.*, 2007). This technique can be used to compare optical atomic clocks based upon different atomic species to constrain the drift of fundamental constants (Rosenband *et al.*, 2008), and also allows optical frequencies to be measured against primary frequency standards with sub-Hz accuracy (Oskay *et al.*, 2006; Campbell *et al.*, 2008; Lemke *et al.*, 2009).

15.5.2 Precision spectroscopy and optical standards

Optical atomic clocks have now reached unprecedented levels of stability and accuracy. At the forefront of accuracy, a clock located at the National Institute of Standards and Technology (NIST), based on a single aluminum ion and probed via quantum logic spectroscopy, now has a fractional frequency uncertainty of 8.6×10^{-18} (Chou *et al.*, 2010). A second clock based on a mercury ion, also at NIST, is at the 2×10^{-17} fractional frequency uncertainty level (Rosenband *et al.*, 2008). These remarkable advances in ion-based clocks are followed closely by a new class of optical clocks based on lattice-trapped ensembles of ultracold neutral atoms, with the most accurate at the level of 10^{-16} fractional uncertainty (Ludlow *et al.*, 2008; Lemke *et al.*, 2009).

Ions and atoms make good frequency references because they are quantum systems whose transition frequencies depend very directly on the fundamental laws of physics. An atom or ion that is considered a good candidate for a clock also exhibits strong immunity to external perturbations, such as magnetic fields. Additionally, atoms or ions serving as optical standards are typically trapped sufficiently tightly that problems such as Doppler broadening and recoil shifts can be mitigated by the tight trap. These traps consist of RF Paul traps for ions (Jefferts *et al.*, 1995) and magic-wavelength optical lattices for neutral atom optical clocks (Katori *et al.*, 2003; Ye *et al.*, 2008). In addition to these general considerations, the transition used for the clock must be sufficiently narrow to provide a useful correction signal. This last condition is met by using multiply forbidden transitions in the clock atoms or ions, resulting in extremely long excited state lifetimes (in some cases >100 s) and correspondingly very narrow resonance linewidths.

Atomic clocks of all types (microwave, optical trapped ion, and optical neutral atom) all rely on the same principle of operation. An oscillator with good short-term stability, the local oscillator, is used to interrogate a transition in the ion or atomic ensemble as shown in Figure 15.11(a). The local oscillator very precisely probes the energy difference of a given transition, ΔE. The difference in energy is related to optical frequency by the well known formula

$$\Delta E = h\nu, \tag{15.35}$$

with the frequency ν being the useful clock signal. Since ν is an optical frequency, a frequency comb is needed if phase-coherent dissemination of a useful microwave signal is desired.

While the accuracy of single ion-based clocks is extraordinary, there is a compelling reason to pursue in parallel standards based upon ensembles of atoms: signal-to-noise.

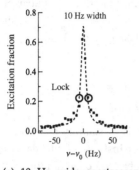

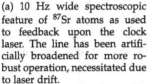

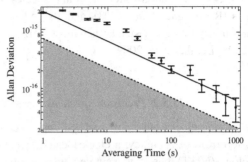

(a) 10 Hz wide spectroscopic feature of ^{87}Sr atoms as used to feedback upon the clock laser. The line has been artificially broadened for more robust operation, necessitated due to laser drift.

(b) Allan deviation of a comparison between Sr and Yb neutral atom optical standards. The dotted line indicates the quantum projection noise (QPN)-limited stability of the current system, while the shaded region shows the region of QPN accessible by use of a narrower resonance, and by increasing atom number.

Figure 15.11 Spectroscopic feature used to discipline the local oscillator laser to Sr (a) and Allan deviation of two neutral atom clocks (b).

Roughly speaking, making N parallel measurements versus one single measurement should yield a $\sqrt{N}$ enhancement of the signal-to-noise ratio (SNR). This enhancement of the SNR for ensembles of atoms is known as quantum projection noise (Itano *et al.*, 1993). The quantum projection noise-limited stability of an optical atomic clock based on a quantity, N, of quantum references (neutral atoms or ions) is given by (Lemonde *et al.*, 2001; Sterr *et al.*, 2009)

$$\sigma(\tau) = \frac{\chi}{Q\sqrt{N}}\sqrt{\frac{T_c}{\tau}}. \tag{15.36}$$

Here, χ is a constant of order unity that accounts for the details of spectroscopy and fraction of atoms excited, T_c is the clock operation cycle time, and Q is the fractional line quality factor, which for optical standards can be $>10^{14}$. Current neutral atom systems have a quantum projection noise-limited frequency stability at the sub $10^{-15}/\sqrt{\tau}$ level (Ludlow *et al.*, 2008), and near term advances in both spectroscopic resolution and atom number make reducing this effect to below $10^{-17}/\sqrt{\tau}$ a realistic possibility (Lodewyck *et al.*, 2009).

One roadblock to benefiting from the SNR afforded by thousands of atoms is broadband laser noise, which ends up contaminating the error signal through the Dick effect. The Dick effect is a process through which a periodic clock interrogation with spectroscopic dead time writes noise onto the correction signal, degrading long term stability (Santarelli *et al.*, 1998; Quessada *et al.*, 2003). For example, for every 1 s of time per cycle a neutral atom system might spend cooling and trapping atoms in an optical lattice, only 100 ms might be time during which spectroscopy is being performed. Thus, there is an inevitable dead time between spectroscopy sequences, resulting in a periodic sampling of the laser phase noise, and leading to aliasing of higher-frequency laser noise, deteriorating the stability.

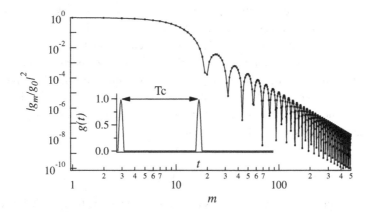

Figure 15.12 Fourier and time-domain (inset) representation of the Dick sensitivity function. With low duty-cycle clock operation, the sensitivity function covers the entire spectral region over which the clock laser is thermal noise-limited.

Specifically, it can be shown that the Dick effect-limited Allan deviation due to the aliasing mechanism is given by (Audoin *et al.*, 1998)

$$\sigma^2(\tau) = \frac{1}{\tau} \sum_{m=1}^{\infty} \frac{|g_m|^2}{g_0^2} \frac{G_\nu(m/T_c)}{\nu_0^2}. \tag{15.37}$$

Here, g_m and g_0 are given by

$$g_m = \frac{1}{T_c} \int_0^{T_c} g(t) e^{i2\pi mt/T_c} dt, \qquad g_0 = \frac{1}{T_c} \int_0^{T_c} g(t) dt \tag{15.38}$$

and $G_\nu(f)$ is the laser frequency noise power spectral density. The function $g(t)$ describes the spectroscopic sensitivity to a phase shift in the local oscillator laser. It is zero during the dead time, during which no spectroscopy takes place, and the details of its shape during spectroscopy are beyond the scope of this chapter. However, the important point is that the smaller fraction of the total experimental cycle the spectroscopic probe occupies, the more harmonics extend into Fourier m-space as $g(t)$ becomes more "comb-like". A typical sensitivity function for an optical clock with >90% dead time is shown in the inset to Figure 15.12. The normalized harmonic content of $g(t)$, $|g_m/g_0|^2$, is shown in Figure 15.12, and only begins to roll off after the 10th harmonic in this specific case. Comparing this result to Figure 15.8, we see that for the experimentally accurate cycle time of 1 s, and the corresponding Fourier frequency range of 1–10 Hz, the most heavily weighted values of $G_\nu(m/T_c)$ are squarely in the thermal noise-dominated portion of the frequency noise spectrum.

Thus, the stability of current state-of-the-art neutral atom clocks is directly tied to the thermal noise present in ultrastable cavity mirror coatings and substrates. Advances in ultrastable laser technology will permit narrower atomic resonance features to be obtained due to longer laser coherence time, while reducing fractional dead time. This represents a triple-sided attack on the effects that currently limit neutral atom clocks: longer coherence times allow longer probe times, which increase spectroscopic line Q linearly with probe time, lowering the quantum projection noise limit; decreasing fractional dead time decreases the harmonic content of g_m, sampling less of the thermal noise spectrum; and the thermal noise level itself is lowered, making it less of a contributing factor. While reducing the experimental dead time alone is also a critical and very promising step (Lodewyck *et al.*, 2009), reducing the thermal noise-induced laser frequency noise will have immediate impact on neutral atom clocks via these three mechanisms.

15.6 Conclusion

Precision laser locking via optical cavities is a valuable scientific tool. With this technique, extraordinary fractional frequency stabilities as low as 3×10^{-16} have been reached (Young *et al.*, 1999), limited by fundamental mirror substrate and coating thermal noise. Thus, cavity-stabilized laser systems are a key enabling technology for optical frequency standards and precision measurement.

While there is no clear path to easily reducing thermal noise without making other sacrifices (e.g. larger cavities or cryogenic systems), the thermal noise-limited performance of ultrastable lasers seemingly has room for one or two more orders of magnitude improvement before the limitations of laboratory-scale technology are reached. However, state-of-the-art optical clocks based on neutral atoms are poised to benefit from even modest reductions in thermal noise, allowing quantum-limited operation to be realized.

Furthermore, exploration of effects that can degrade clock accuracy, such as density-dependent shifts (Campbell *et al.*, 2009), can be realized with the short term stability of lasers alone. By making a series of differential measurements, many residual effects that limit clock accuracy can be better understood, leading in turn to better clock accuracy.

Thus, as we look towards the next generation of optical references, we can expect advances in atomic clock technology to go hand in hand. While neutral atom clocks have the largest stake in the future of stable lasers, all frequency references in the optical domain stand to benefit from better lasers. As accuracy and precision continue to increase, so too does our ability to test fundamental physical principles.

16

Quantum optomechanics

GARRETT D. COLE AND MARKUS ASPELMEYER

16.1 Introduction

High-quality optical cavities – and hence optical coatings – have emerged as an indispensable tool in physics. One of their key uses is in precision sensing and spectroscopy, where high-performance coatings bridge a broad spectrum of applications from large-scale gravitational wave interferometers (see Chapter 14) to atomic optical clocks (see Chapter 15). Radiation pressure forces inside of such cavities, i.e. the momentum transfer of photons onto the cavity boundaries, impose a fundamental limit on their performance, the so-called standard quantum limit (see Section 1.4), when combined with unavoidable quantum fluctuations in a laser beam. Another limitation arises from random phase fluctuations that are induced by the thermal noise of the coating's mechanical motion (see Chapter 3). With the improved performance of optical coatings, optical cavities are gradually becoming sensitive enough to observe these optomechanical effects. Curiously, what is being considered a nuisance by some fields of research has led within only a few years to the development of a completely new field of optical physics – cavity optomechanics – that exploits optomechanical interactions in high-finesse optical cavities as a new interface between light and matter. A common goal in this field is to exploit the quantum regime of mechanical devices through quantum optical control. This work has immediate implications with respect to fundamental questions, such as the quantum-classical transition of massive objects, and new applications including quantum limited mechanical sensors or mechanical interfaces for quantum information processing. A prominent example from this nascent field is the demonstration of laser cooling of mechanical modes close to and even into the quantum ground state of motion (Gröblacher *et al.*, 2009a; Schliesser *et al.*, 2006; Park and Wang, 2009; Rocheleau *et al.*, 2009; Teufel *et al.*, 2011; Chan *et al.*, 2011).

In this chapter we will provide an introduction into the basic working principles of cavity quantum optomechanics. We will discuss the fascinating perspectives of the field and present proof-of-concept experiments that have been performed with optical coatings both at room temperature and at cryogenic temperatures. Finally, we will show how the

Optical Coatings and Thermal Noise in Precision Measurement, eds. Gregory M. Harry, Timothy Bodiya and Riccardo DeSalvo. Published by Cambridge University Press. © Cambridge University Press 2012.

required mechanical coating properties for cavity optomechanics experiments have resulted in new strategies for the development of low-noise optical coatings.

16.1.1 Basic physics of cavity optomechanics

The conjecture that light may show mechanical effects can be traced back at least to Johannes Kepler, who in his work *De Cometis Libelli Tres* argued that mechanical forces from the radiation of the sun are responsible for the inclination of a comet's tail. The first theoretical backing for this idea was provided 150 years later by Maxwell who predicted the effects of radiation pressure within his electrodynamic theory of light. By the turn of the century, the independent experiments by Lebedev (1901) and by Nichols and Hull (1901) confirmed this effect in their famous light-mill demonstrations (for more on early history, see Nichols and Hull (1901)). Of the resulting developments, the most important application making use of the mechanical effects of light is optical trapping and laser cooling of atoms and ions.

The idea that the properties of a mechanical object can be modified by radiation pressure forces inside of an optical cavity goes back to the pioneering works of Braginsky (Braginsky *et al.*, 1970; Braginsky *et al.*, 2001; Braginsky and Vyatchanin, 2002). In essence, the optical cavity response to the motion of a mechanical object can lead to forces that depend on both the position and the momentum of the mechanics (due to the finite cavity lifetime). For sufficiently strong forces, this results in a modification of the mechanical susceptibility. This is the underlying mechanism for optical control over mechanical degrees of freedom and vice versa. Such coupling between the cavity response and the mechanical motion may be implemented in a variety of ways, either via direct changes in the cavity length (as will be outlined in the following for a Fabry–Pérot configuration), by dispersion (Thompson *et al.*, 2008), or by optical near-field effects (Li *et al.*, 2008; Eichenfield *et al.*, 2009).

To introduce the basic concepts we consider a single-sided Fabry–Pérot cavity of length L, frequency ω_c, and linewidth κ with a suspended end mirror of mass m and mechanical resonance frequency ω_m. We will specify frequencies with the angular frequency ω in this discussion instead of f as in the rest of the book to simplify the presentation. They are related by $\omega = 2\pi f$. The general configuration shown in Figure 16.1 covers all essential features of cavity optomechanics and can be transferred easily to all other optomechanical geometries. For more detailed descriptions of the variety of cavity optomechanics experiments see Kippenberg and Vahala (2008); Aspelmeyer and Schwab (2008); Favero and Karrai (2009); Marquardt and Girvin (2009); Genes *et al.* (2009); Aspelmeyer *et al.* (2010).

Let us first look at the most fundamental process, the reflection of a single photon. Within the cavity round-trip time $\tau = \frac{2L}{c}$ (with c the vacuum speed of light) the photon will transfer momentum $\Delta p = 2\hbar k = \frac{2h}{\lambda}$ (with λ the photon wavelength and h Planck's constant) and hence exert a radiation pressure force $F_{rp} = \frac{\Delta p}{\tau} = \frac{\hbar \omega_c}{L}$ on the cavity boundaries.[1] This results in a relative mechanical displacement $\frac{\Delta x}{x_{zp}} = \frac{F_{rp}}{x_{zp} m \omega_m^2} = \frac{g_0}{\omega_m}$ and a shift

---

[1] We assume that the photon wavelength is resonant with the cavity and hence $\frac{2\pi c}{\lambda} = \omega_c$.

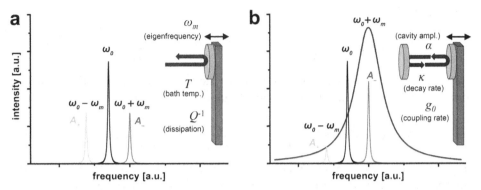

Figure 16.1 Radiation-pressure-based optomechanical interactions. (a) Laser light reflected from a mechanical resonator (in this case a simple cantilever) generates Doppler shifted sidebands at ± the eigenfrequency of the mechanical resonator, ω_m (focusing on a single mechanical resonance for simplicity). (b) With a second mirror introduced, the filtering response of the optical cavity modulates the relative strengths of the sidebands. In the example configuration shown, the incident laser is red-detuned by ω_m with respect to the optical cavity resonance, suppressing the "heating" Stokes sideband at $-\omega_m$ while enhancing the higher frequency Anti-Stokes sideband at $+\omega_m$. The higher frequency photons scattered into resonance with the optical cavity remove energy (in the form of phonons) from the mechanical system, thus decreasing the effective temperature of the eigenmode.

of the cavity frequency by $\Delta\omega_c = \frac{d\omega_c}{dx}\Delta x = \frac{g_0^2}{\omega_m}$. Here we have defined the fundamental optomechanical coupling rate $g_0 = \frac{d\omega_c}{dx}x_{zp}$ as the shift in cavity frequency when the mechanical system is displaced by its ground state size, i.e. $\Delta x = x_{zp}$. For a Fabry–Pérot cavity with a mechanically moving end mirror, $g_0 = \frac{x_{zp}}{L}\omega_c$. For other optomechanical architectures, e.g. a toroidal microcavity with radial mechanical breathing modes (Armani *et al.*, 2003), a dispersive mechanical element inside a rigid optical cavity (Thompson *et al.*, 2008), a free-standing waveguide with near-field interaction to a close-by surface (Li *et al.*, 2008) or localized mechanical motion inside a photonic crystal cavity (Eichenfield *et al.*, 2009), g_0 has to be computed individually. In general, g_0 characterizes the fundamental optomechanical interaction between photons in a cavity and the displacement x of the mechanical element, which is described by the interaction Hamiltonian $H_{int} = -\hbar g_0 n x$ (where n is the intra-cavity photon number).

One can now distinguish the following scenarios. For the case of a resonantly pumped cavity, i.e. $\omega_L = \omega_c$, where ω_L is the optical frequency, the radiation pressure results in a static displacement Δx of the mechanical system. Consequently, an optical bistability can occur for strong driving when the associated cavity shift exceeds the cavity linewidth κ (Dorsel *et al.*, 1983). For off-resonant driving of the cavity, i.e. $\omega_L - \omega_c = \Delta \neq 0$, two effects occur. First, the intra-cavity photon number now depends on the position of the mechanical system which, as a result, experiences a position-dependent radiation-pressure force. Second, due to the finite cavity lifetime $1/\kappa$, this force is retarded in time and hence allows work to be performed on the mechanical system. In other words, the radiation-pressure forces inside of a detuned optical cavity modify the mechanical susceptibility according to

$$\chi_{\text{eff}}(\omega) = \frac{1}{m}\left(\omega_m^2 - \omega^2 - i\omega\gamma - \frac{g_0^2\alpha^2\Delta\omega_m}{(\kappa - i\omega)^2 + \Delta^2}\right)^{-1}. \qquad (16.1)$$

The modification of the real component of the susceptibility, $\chi_{\text{eff}}(\omega)$, arises from the position-dependent force and leads to a modification of the mechanical spring constant to

$$\omega_{\text{eff}}(\omega) = \left(\omega_m^2 - \frac{g_0^2\alpha^2\Delta\omega_m(\kappa^2 - \omega^2 + \Delta^2)}{[\kappa^2 + (\omega - \Delta)^2][\kappa^2 + (\omega + \Delta)^2]}\right)^{1/2}. \qquad (16.2)$$

This effect is also known as the "optical spring" (Sheard *et al.*, 2004; Corbitt *et al.*, 2007; Cuthbertson *et al.*, 1996). In turn, the modification of the imaginary part is due to the retardation and leads to a modification of the mechanical damping γ to

$$\gamma_{\text{eff}}(\omega) = \gamma + \frac{g_0^2\alpha^2\Delta\omega_m\kappa}{8[\kappa^2 + (\omega - \Delta)^2][\kappa^2 + (\omega + \Delta)^2]}. \qquad (16.3)$$

Here α is proportional to the square root of photon number.

For blue detuning, i.e. $\Delta > 0$, mechanical damping is reduced and additional optical energy is pumped into the system, which leads to parametric amplification of the mechanical motion (Cuthbertson *et al.*, 1996; Carmon *et al.*, 2005). For red detuning, i.e. $\Delta < 0$, mechanical damping increases while work is being extracted through the radiation field, which leads to optical cooling of the mechanical motion. This effect of cavity enhanced radiation-pressure cooling has been discussed theoretically in the context of gravitational wave antennas (Khalili, 2001) and was first experimentally realized in 2006 (Gigan *et al.*, 2006; Arcizet *et al.*, 2006; Schliesser *et al.*, 2006).

Besides controlling mechanical degrees of freedom, the optomechanical interaction resembles an effective Kerr medium and can hence generate strong optical nonlinearities. These nonlinearities can be exploited for a number of interesting applications, including quantum nondemolition measurements of light amplitudes (Milburn *et al.*, 1994; Pinard *et al.*, 1995) and the generation of optical squeezing (Fabre *et al.*, 1994; Mancini and Tombesi, 1994). It is also very instructive to discuss these effects in a full quantum framework (Marquardt *et al.*, 2007; Wilson-Rae *et al.*, 2007; Genes *et al.*, 2008b). There, the interaction between cavity photons and the mechanical system leads to scattering of photons into optical sidebands of higher and lower frequency, analogous to Anti-Stokes and Stokes scattering processes in atomic physics. By choosing the proper detuning, the cavity can resonantly enhance either the Stokes ($\Delta = +\omega_m$) or the Anti-Stokes ($\Delta = -\omega_m$) scattering process while suppressing the other. In the first case, photons can only enter (leave) the cavity when they generate (annihilate) a mechanical excitation, which results in an effective interaction $H_{\text{int}}^{\text{eff}} = a^+b^+ + ab$. Here a and $b(a^+ + b^+)$ are the optical and mechanical annihilation operators, respectively. These interactions can be used to generate strong optomechanical correlations via joint excitations. This interplay resembles the well-known 2-mode squeezing interaction from quantum optics and forms the basis for the generation of quantum entanglement between light and mechanics (Vitali *et al.*,

2007; Paternostro *et al.*, 2007; Aspelmeyer *et al.*, 2010). In the second case, the effective interaction $H_{int}^{eff} = a^+ b + ab^+$ resembles the quantum optical beamsplitter and describes the coherent energy exchange between photons and mechanical excitations. In this case, photons can only enter the cavity by annihilating mechanical excitations and hence cooling of the mechanics takes place. This cooling mechanism is fully analogous to resolved side-band cooling of ions (Leibfried *et al.*, 2003; Wilson-Rae *et al.*, 2007) and as in that case, the cavity-assisted optomechanical cooling method enables the attainment of the quantum regime of mechanical motion. The ultimate temperature limit for mechanical laser cooling is $n_{final} = \frac{\kappa^2}{16\omega_m^2}$, which is well below 1 in the regime of sideband resolution, i.e. $\kappa < \omega_m$.

One can see from the above discussion that cavity optomechanics allows for full optical control over mechanical degrees of freedom. In combination with quantum optics, such interactions enable full *quantum* control of mechanical resonators allowing for a whole range of new perspectives. Extending the established technology of mechanical sensing to the quantum regime of mechanical resonators may allow, for example, new sensing capabilities at or even beyond the quantum limit by utilizing squeezed mechanical states (Mari and Eisert, 2009; Jähne *et al.*, 2009; Vanner *et al.*, 2011). Another remarkable feature of micro- and nanomechanical resonators is their versatility in coupling to many different physical systems, which can be achieved by functionalizing the mechanics. In combination with their on-chip integrability, such devices are unique candidates for transducers in quantum information processing, enabling large scale coupling between otherwise incompatible quantum systems (Rabl *et al.*, 2010; Cleland and Geller, 2004; Stannigel *et al.*, 2010). From a completely different point of view, preparing quantum superposition states of massive mechanical objects (containing up to 10^{20} atoms) opens up a new avenue for novel fundamental tests of macroscopic quantum physics (Leggett, 2002; Arndt *et al.*, 2009; Romero-Isart *et al.*, 2011; Marshall *et al.*, 2003), where spatially macroscopic distinct states of mechanical quantum systems could serve as a paradigm example for Schrödinger's cat.

16.1.2 Summary of requirements

Approaching and eventually entering the quantum regime of mechanical resonators through optomechanical interactions essentially requires the following three conditions to be fulfilled: (1) sideband-resolved operation in which the cavity amplitude decay rate κ is small with respect to the mechanical frequency ω_m; (2) both an ultra-low noise cooling laser input and low absorption of the optical cavity field (phase noise at the mechanical frequency can act as a finite-temperature thermal reservoir (Rabl *et al.*, 2009) and absorption can increase the mode temperature and even diminish the cavity performance); and (3) sufficiently small coupling of the mechanical resonator to the thermal environment entailing a low environment temperature T and large mechanical quality factor Q. The thermal coupling rate is given by $k_B T/\hbar Q$, where k_B is the Boltzmann constant, $\hbar$ is the reduced Planck constant, and Q is the modal quality factor which depends on the mechanical loss ($Q = 1/\phi$ for a homogeneous resonator). Ultimately, the individual dissipation rates of the mechanical and optical system (represented by γ and κ respectively) should be exceeded by the optomechanical coupling coefficient g_0. With micromechanical devices fabricated directly from optical coatings, as discussed below in Section 16.2, the properties of the multilayer coating drive the ultimate performance of the mechanical resonator. In this light, requirements (2) and (3) will eventually be limited by the intrinsic materials properties (with respect to both the loss angle and absorption) of the coating. Recent experimental thrusts in quantum-optomechanics simultaneously require structures with both high optical reflectivity and low mechanical dissipation. Thus, there is an intimate link between advances in the performance of optical coatings applied to the widely disparate fields of macroscopic interferometric applications and micro- and nanoscale optomechanical systems.

16.2 Sculpted mirrors: micromechanical resonators from optical coatings

Microstructures of high mechanical and optical quality have become a leading candidate to achieve quantum optical control of mechanical systems. Each of these schemes relies crucially on mechanical structures that combine high optical reflectivity and low mechanical dissipation of the mechanical mode of interest. In pursuing research in optomechanical systems, micrometer-scale mechanical resonators are routinely fabricated directly from multilayer mirrors, thus generating high reflectivity, suspended structures. The multilayer elements from which these resonators are fabricated consist of alternating high- and low-index quarter wave (in terms of optical thickness) films that produce a high-reflectivity stop band over a specific frequency range. At the so called "Bragg wavelength" the individual interface reflections add constructively to produce a spectral region of high reflectivity, and in accordance with the underlying mechanism, these structures are commonly referred to as distributed Bragg reflectors or simply Bragg mirrors. See Chapters 2 and 12 for more on coating design.

Fabrication of the micro-resonators relies upon the selective removal of an underlying sacrificial material in order to create free-standing devices (Böhm et al., 2006; Cole et al., 2008; Gröblacher et al., 2008; Gröblacher et al., 2009a,b; Cole et al., 2010b). Sculpting the mechanical resonator from the coating itself allows for the development of low mass devices (with typical mechanical eigenfrequencies in the MHz range) that simultaneously achieve the desired optical characteristics – including high reflectivity and low optical absorption. Additionally, such structures represent an ideal tool for probing the intrinsic mechanical damping of the constituent materials of the Bragg mirror, assuming all other extrinsic dissipation mechanisms have been accounted for. Micromechanical resonators may be limited by unique loss mechanisms not experienced by their macroscopic counterparts (see Chapter 4); important damping mechanisms in these structures will be discussed in Section 16.2.3. In the following, we briefly summarize historical results for micromechanical resonators fabricated from three classes of optical coatings; (1) dielectric coatings based on silica/titania (SiO_2/TiO_2) as utilized for the initial demonstration of cavity-assisted laser cooling, (2) high-performance ion-beam sputtered silica/tantala (SiO_2/Ta_2O_5) multilayers which have recently been implemented for the demonstration of ultra-low temperature mechanics as well as optomechanical strong coupling, and most recently (3) monocrystalline $Al_xGa_{1-x}As$ heterostructures that simultaneously achieve excellent reflectivity and a high mechanical quality factor.

16.2.1 Silica/titania

In 2006 a milestone was achieved in the first demonstration of laser-cooling of micromechanical resonators via passive radiation-pressure interactions in a high-finesse optical cavity (Gigan *et al.*, 2006; Arcizet *et al.*, 2006). The implementation detailed here, from Gigan *et al.* (2006), relied on an optical cavity using a high reflectivity micro-resonator as one of the end mirrors. The high-quality resonator was fabricated using a combination of direct laser writing and gas phase lateral etching. In this process, doubly clamped beams with fundamental frequencies near 300 kHz were developed (approximate dimensions of 520 μm × 120 μm × 2.4 μm) (Böhm *et al.*, 2006). These devices were constructed from SiO_2/TiO_2 multilayers deposited via ion beam sputtering (see Chapter 2) on a crystalline silicon substrate. Through a process of ultraviolet excimer-laser-based ablation, the dielectric coating was patterned, generating the sculpted mechanical resonator and exposing the underlying silicon substrate. In order to free the resonator, a gas-phase pulsed XeF_2 etch (Chu *et al.*, 1997) was utilized in order to remove the silicon from the exposed areas. Given the isotropic nature of this etch (i.e. an equal etch rate in all directions), the beam was eventually left free-standing once the underlying silicon had been removed. The mechanical resonators generated in the fabrication procedure consist solely of the Bragg-mirror materials (TiO_2 and SiO_2). As can be seen in Figure 16.2, these structures demonstrated mechanical quality factors approaching 1×10^4 at room temperature (Gigan *et al.*, 2006). However, transmission losses and finite optical absorption at 1064 nm (potentially from non-optimized TiO_2 films or residual materials from the ablation procedure) limited the mirror reflectivity, while the latter resulted in additional photothermal (i.e. bolometric) effects (Metzger and Karrai, 2004) (see also Section 3.10 for a similar effect). Regardless of the initial limitations, these devices were successful in demonstrating successful laser cooling, with a final mode temperature of 8 K realized with 2 mW of detuned laser power.

16.2.2 Silica/tantala

In order to improve the optical performance of the resonators, a second generation of devices was developed. These structures utilized high performance SiO_2/Ta_2O_5 optical coatings as pioneered for the precision measurement community (see Chapters 2 and 4). Given the significant effort towards the optimization of these coatings, this material system exhibits a number of advantages including ultra-low optical loss resulting from sub-ppm ($<10^{-6}$) absorption (see Chapter 10) and Ångstrom-level surface roughness (see Chapter 11) in the

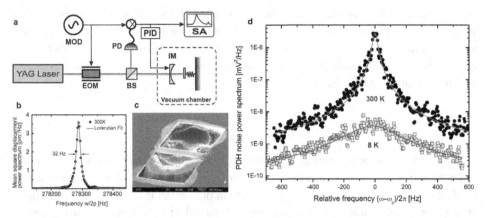

Figure 16.2 Initial demonstration of cavity assisted laser cooling of a mechanical resonator. (a) Schematic of the experimental setup: a cavity is formed between the microfabricated cantilever, as shown in panel (c), with a resonator reflectivity $\approx 99.6\%$ at 1064 nm and a macroscopic concave mirror of 25 mm focal length and 99.3% reflectivity. With a spot size of ≈ 20 μm this system attains a measured finesse of 500. To minimize losses due to residual gas damping, the mechanical resonator was placed in a vacuum chamber maintained at 10^{-5} Torr. (b) Cooling is realized for a mechanical mode near 280 kHz with a natural width of 32 Hz, corresponding to $Q \approx 9000$ at 300 K. (d) Power spectrum of the mechanical mode at two detuning levels Δ of the cavity for an input power of 2 mW. Experimental points at two different detuning levels (filled points, $\Delta = 0$; open points, $\Delta = 0.44\kappa$) are displayed along with Lorentzian fits to the data. The areas obtained from the fits correspond to an initial mode temperature of 300 K, followed by a radiation pressure cooled final temperature of 8 K.

constituent films. Whereas the initial resonators based on SiO_2/TiO_2 films were limited to reflectivities $\approx 99.6\%$, optimized SiO_2/Ta_2O_5 coatings are capable of R values in excess of 99.99% (see Chapter 12 for more on coating design). The initial devices fabricated from this materials system were again simple cantilever and fixed-fixed beam geometries, with resonant frequencies in the 10^2 kHz range. The reduced absorption in these coatings greatly improved the optical performance of the resonators, with the most important achievement being the complete elimination of photothermal effects (as indicated by the linear cooling curves shown in Figure 16.3).

Experiments carried out with these devices utilized cryogenic operation of the resonator (with an optical cavity constructed in a continuous flow ^{4}He cryostat; see Chapter 8 for more details on cryogenic optical cavity systems. As with the previous experiment, the input coupler of the Fabry–Pérot cavity is a concave massive mirror with a radius of curvature of 25 mm (reflectivity at 1064 nm: 99.93%) that is, in this case, attached to a ring piezo (PZT) in order to actively modify the optical cavity resonance for locking purposes. Here the total cavity length is slightly shorter than for the semi-concentric case ($L = 25$ mm) in order to have a stable cavity and a small cavity-mode waist w_m on the micromirror ($w_m \approx 10$ μm). Within the cryostat, a 3-axis translation stage allows precise positioning of the micromirror with respect to the cavity beam. Using an external imaging system

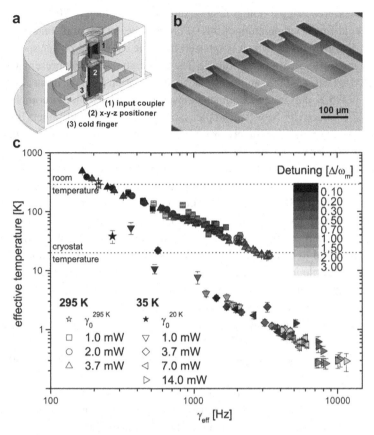

Figure 16.3 Radiation-pressure cooling in a cryogenic high-finesse cavity. (a) Cross-sectional schematic of the cryogenic cavity. Here the optical cavity is constructed inside a continuous-flow ^{4}He cryostat. In this configuration the sample chip rests on a small copper mount thermalized with the cold finger and atop a 3-axis attocube positioning stack. The input coupler is also contained within the cryostat but is thermally insulated from the cold finger (see details in text). (b) Scanning electron micrograph of a subset of the Ta_2O_5/SiO_2 resonators, in this case consisting of cantilevers 50 μm wide with lengths varying between 50 μm and 500 μm. (c) Plot of the effective temperature T_{eff} and the effective damping γ in a detuned cavity for various laser powers (corresponding to different symbols), while values of detuning (in units of ω_{m}) are encoded in the gradient scale shown in the upper right. The use of cryogenic pre-cooling results in a final effective mode temperature of 290 mK (or $\langle n \rangle = 1 \times 10^4$), limited by the attainable finesse and mechanical quality factor.

it is possible to monitor both the position and size of the cavity mode. In operation, the cryostat is first evacuated to 10^{-6} Torr. Cryogenic cooling is achieved by a continuous flow of helium in direct contact with the cold finger. The additional cryogenic freeze-out reduces the pressure to below 2×10^{-7} Torr. On cooling the cryostat from room temperature to approximately 6 K (measured temperature at the cold finger), the thermal contraction of the cavity can be compensated by the translation stage. This implementation results in a measured sample holder temperature of approximately 20 K (compared with a measured cold-finger temperature of 6 K). Furthermore, an initial resonator temperature (bath) of 35 K was realized, which is inferred from the calibrated power spectrum of the micromirror motion at zero optical detuning. The temperature gradient present in the system was attributed to heating of the sample by blackbody radiation from the input coupler, which was kept at $\approx$295 K only a few centimeters away from the sample, in combination with finite thermal conductivity between sample, sample holder and cold finger. Both at room temperature and at cryogenic temperatures stable locking of the cavity was maintained for a finesse of up to 8000.

Although these microresonators were found to exhibit excellent optical characteristics (i.e. high reflectivity and low absorption), it was found that the mechanical quality factor of the free-standing Ta_2O_5/SiO_2 resonators is limited to an order of magnitude of $\approx 10^3$. Note that the modal quality factor of a microresonator comprised solely of a coating is related to the coating mechanical loss by $Q = 1/\phi_{coat}$ in the absence of external losses. Particular Q values were found to be between 3000 and 6000 (Gröblacher et al., 2008; Brodoceanu et al., 2010), with little improvement seen at cryogenic temperatures. A lack of geometric dependence on quality factor points towards a materials limited phenomenon dominating the overall damping in these structures (in contrast with anchor losses or thermoelastic damping processes that would display unique frequency, temperature, and geometric dependencies as discussed in detail in Section 16.2.3). The low Q-value recorded in these resonators is consistent with the coating loss angles observed in Harry et al. (2002, 2006a) and discussed in Chapter 4.

This damping is most likely dominated by the Ta_2O_5 layers, which unfortunately display severe acoustic attenuation. As investigated previously, dielectric multilayer coatings based on this materials system exhibit particularly large acoustic damping as a consequence of their intrinsic amorphous materials structure (Harry et al., 2002, 2006a). The lack of a periodic ordering of the constituent elements (as in a crystalline material) allows for the presence of 2-level fluctuators, atoms may be mechanically excited between two equilibrium positions in an asymmetric double-well potential (Anderson et al., 1972). Tunneling transitions between these two stable configurations may be driven by thermal phonons, thus attenuating mechanical vibrations in the coating, resulting in an enhancement in the dissipated mechanical energy in the system. As discussed in Chapters 1 and 4, mechanical losses are directly related to thermal fluctuations via the Fluctuation–Dissipation Theorem, and thus the limiting thermal noise in an interferometer (Callen and Welton, 1951; Callen and Greene, 1952).

In order to minimize mechanical damping introduced by the Ta_2O_5 layers, a specifically engineered mechanical resonator was devised that separated the mechanical and optical functionality of the device. As can be seen in Figure 16.4, these resonators effectively decouple the optical and mechanical requirements of the structure; in this case employing a mechanical resonator based on chemical vapor deposited (CVD) low-stress silicon nitride (Si_xN_y) (see Chapter 2), carrying a high reflectivity, low-absorption SiO_2/Ta_2O_5 mirror

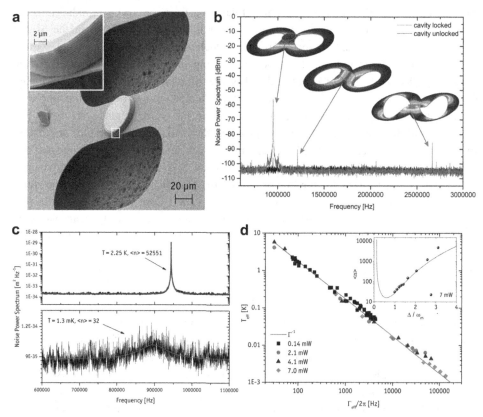

Figure 16.4 High-performance dielectric micro-optomechanical resonator. (a) Scanning electron micrograph of the hybrid mechanical system, which is formed by a doubly clamped Si_xN_y beam supporting a circular, high-reflectivity Bragg mirror. The Bragg mirror consists of a low-absorption dielectric stack based on alternating Ta_2O_5/SiO_2. (b) Displacement spectra shown as noise power spectra of the readout-beam phase quadrature for a locked and an unlocked cavity. The fundamental mode at $f_m \approx 945$ kHz and all higher mechanical modes are identified by finite element modeling (FEM). For the cases that involve large Bragg mirror displacements, the simulated mode profile is provided. (c) Calibrated noise power spectra for the fundamental mechanical mode at 5.3 K environmental temperature with small cavity cooling (top) and at maximum cooling (bottom). The thermal energy is reduced from $\approx 53\,000$ quanta at 7 µW laser power to 32 ± 4 quanta at 7 mW. The vertical axes in both plots are logarithmic. (d) Plot of the calibrated effective temperature T_{eff} versus the observed damping γ_{eff} for various power and detuning values of the cooling beam. No deviations from the theoretically expected power-law dependence (solid line, generated using expressions taken from Genes *et al.* (2008a)) can be observed. The inset shows the mean thermal occupation as a function of detuning for maximal laser power. Cavity instability prevents detunings arbitrarily close to resonance. The solid curve is a simulation that uses only experimentally obtained parameters.

pad (Gröblacher *et al.*, 2009a,b). Mechanical damping from the Ta_2O_5 is reduced as the strain generated via displacement of the resonator is primarily confined in the Si_xN_y. This device exploits standard microfabrication techniques and thus does not require serial assembly of the individual components. Using this separately optimized hybrid design, a reflectivity $>99.99\%$ was realized, while simultaneously boosting the quality factor of the resonator to $\approx 3 \times 10^4$ at cryogenic temperatures. Beyond the improvement to the mechanical design of the microresonator, the experimental system was further improved by including the use of a properly thermalized input coupler and extensive radiation shielding, allowing for base temperatures near 5 K. With this updated test setup, the fundamental mode of the doubly clamped Si_xN_y resonator (eigenfrequency near 1 MHz) was utilized for the demonstration of a mechanical system with an ultra-low final phonon occupation of ≈ 30 (Gröblacher *et al.*, 2009a), as well as the first demonstration of strong coupling between an optical field and a mechanical resonance (Gröblacher *et al.*, 2009b). Similar results have been obtained in different optomechanical architectures (Teufel *et al.*, 2011).

16.2.3 Monocrystalline optomechanical resonators

Attacking this problem from an entirely different angle, two generations of mechanical resonators have additionally been realized in a novel materials system from the perspective of the precision measurement community. For these devices the resonators are fabricated directly from a single-crystal Bragg reflector, utilizing ternary alloys of $Al_xGa_{1-x}As$ (where aluminum is a substitutional alloying element on a Ga lattice site in zinc-blende GaAs, or vice versa for high Al content structures) as the mirror material. For simplicity, from this point on the compound will simply be abbreviated as AlGaAs. This materials system allows for a relatively high index contrast (with refractive indices of 3.480 41 for GaAs and 2.938 33 for AlAs at 1064 nm at 300 K) and thus high reflectivity for a reasonable number of layer pairs. Furthermore, AlGaAs offers transparency for wavelengths greater than the intrinsic band edge of 1.424 eV (871 nm) for binary GaAs at 300 K (shifting to 1.519 eV (817 nm) at 4 K), enabling the development of low absorption Bragg mirrors at 1064 nm, with the potential for transparency out to ≈ 4 μm. Here, the long wavelength cutoff is limited by significant free-carrier absorption for photon energies less than 0.3 eV (Spitzer and Whelan, 1959). Using epitaxial growth techniques such as molecular beam epitaxy (MBE) (Parker, 1985; Herman and Sitter, 1989; Tsao, 1993) or metalorganic chemical vapor deposition (MOCVD) (Ludowise, 1985; Stringfellow, 1989) (see also Chapter 2), almost arbitrarily thick multilayers of these materials may be grown while still retaining a single-crystalline structure. A concise review of the properties, applications, and growth techniques for AlGaAs may be found in Davies (1998). The initial impetus for the development of AlGaAs-based devices was to investigate the limits of mechanical dissipation in crystalline resonators, specifically via the removal of the loss originating from two-level fluctuators thought to dominate the damping in high-performance dielectric coatings (see Section 4.2.6).

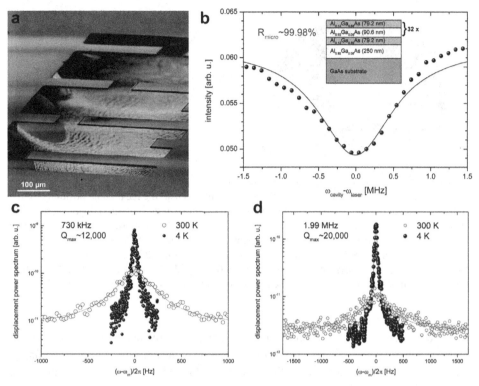

Figure 16.5 Monocrystalline optomechanical resonators fabricated from an epitaxial AlGaAs multi-layer. (a) Scanning electron micrograph of a set of completed cantilevers fabricated from a 32 period $Al_{0.12}Ga_{0.88}As$ / $Al_{0.92}Ga_{0.08}As$ Bragg stack. The beams shown have a width of 50 μm and vary in length between 50 and 500 μm. (b) Measured cavity resonance using the microresonator as a cavity end mirror. The observed linewidth ($\sim$ 1.1 MHz) corresponds to a finesse of 5500. (c) Fundamental mechanical mode of a 150 μm long doubly clamped resonator at 300 and 4 K, central frequencies are 731 and 697 kHz, respectively. The corresponding quality factors are 2200 and 12 000. (d) First harmonic of the same resonator resulting in quality factors of 5000 and 20 000 for frequencies of 1.997 and 1.971 MHz at 300 and 4 K, respectively.

The use of an epitaxially grown AlGaAs Bragg mirror has resulted in significant improvements in the achievable mechanical quality factor – corresponding to a significant reduction in the coating loss angle – for the optomechanical microresonators. The initial AlGaAs devices consisted of 32.5 periods of alternating $Al_{0.12}Ga_{0.88}As$ (high index) and $Al_{0.92}Ga_{0.08}As$ (low index) with nominal layer thicknesses of 79 nm and 90 nm, respectively, grown via MBE (Cole *et al.*, 2008). Resonators fabricated from this material system utilized an identical geometry to the early dielectric devices (again employing simple doubly and singly clamped beams) and exhibited nearly an order of magnitude improvement in the mechanical dissipation as compared with Ta_2O_5-based devices of similar geometry, as shown in Figure 16.5. These resonators displayed Q-values greater than 2×10^4 at

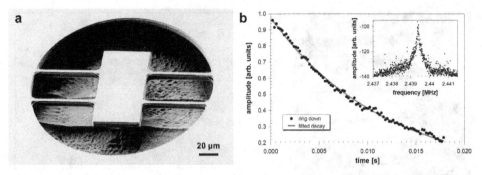

Figure 16.6 Free-free monocrystalline resonators based on MOCVD-grown AlGaAs. (a) Close-up view of a single completed resonator with the auxiliary beams near the ideal nodal position (for the fundamental flexural mode of the central resonator). (b) Envelope of the measured ringdown response and accompanying exponential fit for the free-free resonance of a $130 \times 40 \ \mu m^2$ resonator fabricated from a 40.5 period mirror stack. This measurement was performed at ≈ 20 K and 2.0×10^{-7} Torr in a custom cryogenic fiber interferometer described in Cole *et al.* (2010c). For the eigenmode at 2.44 MHz, the extracted exponential decay time (t = 0.012 s) yields a Q value exceeding 90 000. The inset displays the frequency response of the resonator when driven with a white noise signal.

eigenfrequencies up to 2 MHz at cryogenic temperatures for the first harmonic of a doubly clamped beam (Cole *et al.*, 2008). Further investigation of the limiting damping mechanism in these devices indicated that the quality factors were limited by support-induced losses for the geometries studied. In addition to the significant improvement in the achievable quality factors, the AlGaAs Bragg mirrors simultaneously exhibited promising optical properties. Building an optical cavity using a curved macroscopic input coupler (similar to that described in the cooling experiments discussed in Section 16.2.2) a reflectivity (intensity) of 99.975% was extracted for the 32.5 period AlGaAs Bragg mirror. Atomic force microscopy (AFM) characterization of the surface quality yields RMS roughness values below 2 Å, verifying the excellent quality of the MBE-grown films and in this case the maximum reflectivity was found to be limited solely by transmission loss through the mirror.

Moving to a novel free-free resonator design in order to minimize phonon tunneling losses to the surrounding substrate (Cole *et al.*, 2010a), the total damping of the resonators is now beginning to approach the intrinsic mechanical loss of the multilayer. Recently recorded cryogenic quality factors approach 10^5 (9.5×10^4 at 2.4 MHz) (Cole *et al.*, 2010c), as shown in Figure 16.6. These devices utilize an updated materials system with a total of 40.5 mirror periods in order to reduce transmission losses through the mirror. The epitaxial multilayer used for the fabrication of these devices employs a nominally identical composition and layer thicknesses as described previously, although in this case the films are deposited via MOCVD. This growth method was utilized in order to take advantage of the excellent *in situ* layer-thickness control afforded by modern chemical vapor deposition systems, as pioneered in the manufacturing of vertical-cavity surface-emitting lasers (VCSELs) (Zorn *et al.*, 2002).

Cryogenic operation is required for the micromechanical resonators to minimize thermoelastic damping, which limits the room-temperature quality factor to below 10^4 for the current geometry and operating frequencies. For more details on thermoelastic damping in coatings see Chapter 9 and in substrates see Chapter 7. Cavity-based measurements of the optical performance of the 40.5 period mirror yield a reflectivity greater than 99.99% (Cole *et al.*, 2010c). This reflectivity essentially matches that achieved in the SiO_2/Ta_2O_5 microdevices (e.g., using an input coupler with an amplitude reflectance of 99.995%, a finesse of 3.0×10^4 is realized for the dielectric Bragg mirrors compared with 2.7×10^4 for the microfabricated AlGaAs mirror).

Recently, AlGaAs-based resonators employing the free-free mechanical design have been employed for an in-depth experimental effort aimed at quantifying the effects of support-induced damping, a key dissipation mechanism in high-quality-factor micro- and nanomechanical resonators. Such studies are vitally important not only for optomechanical experiments, but additionally for pushing the limits of mechanical loss, as such resonators have emerged as ubiquitous devices for use in advanced technological applications. Unfortunately, the performance of these devices is in many cases limited by the negative effects of mechanical damping, which is further confounded by the lack of quantitative understanding of many damping mechanisms. As can be seen in Figure 16.7 the free-free design provides an ideal platform to isolate and measure phonon tunneling dissipation. Altering the attachment position of the auxiliary beams allows for a significant variation in geometry, while approximately preserving the frequencies and effective surface-to-volume ratios of the resonators. As these characteristics are kept constant, one can rule out the influence of additional damping mechanisms on the variation in Q and hence isolate support-induced losses. To further remove extraneous damping mechanisms, characterization of the resonator quality factor is carried out at cryogenic temperatures (20 K) and at high vacuum (10^{-7} Torr) in order to remove thermoelastic damping and residual gas damping, respectively.

With the AlGaAs resonators, thermoelastic damping is found to be the limiting loss mechanism at room temperature. As described in Chapters 7 and 9, this mechanism is a coupled thermomechanical process involving the scattering of thermal phonons with the acoustic phonons responsible for the vibrations of the resonator. From a continuum point of view, flexural vibrations produce alternating regions of tensile and compressive strain on opposite sides of the neutral axis of the resonator. These strains in turn generate a thermal gradient in the structure, and energy is dissipated through irreversible heat flow. For the case of a simple flexural mode, this heat flow occurs through the thickness of the multilayer. Damping is especially problematic when the frequencies of the thermal and vibrational modes overlap. The center frequency of the thermal mode is controlled by the thickness and thermal properties of the structure. For the free-free resonators fabricated from the AlGaAs Bragg mirror (with a nominal central resonator size of 130 μm × 40 μm × 7 μm) the maximum Q at 300 K ($\approx$4500 at an average frequency of 1.9 MHz up to $\approx$8000 near 4 MHz) is accurately described using analytical expressions derived for flexural vibrations in Zener (1937). For more complex geometries, a numerical solver employing the

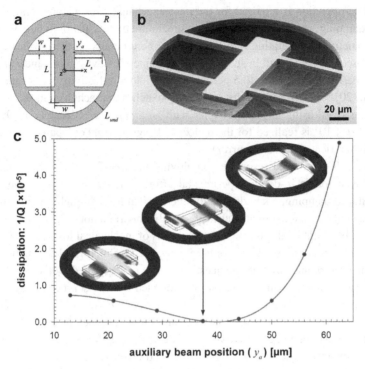

Figure 16.7 Mapping out phonon-tunneling dissipation in a free-free resonator. (a) Schematic diagram of the resonator geometry. (b) Scanning electron micrograph highlighting a single suspended structure. (c) Simulated dissipation as a function of the attachment point (y_a) of the auxiliary beam. Values corresponding to 8 discrete geometries were calculated here with the resonator thickness $t = 6.67$ μm, auxiliary beam width $w_s = 7$ μm, central resonator width $w = 42$ μm, central resonator length $L = 132$ μm, outer radius $R = 116$ μm, and undercut distance $L_{und} = 27$ μm – the line is simply a guide for the eye. The FEM calculated mode shapes correspond to the three extreme examples of the resonator design. The theoretical clamping loss limit, $1/Q_{th}$, for nodal positioning is always finite with the geometry closest to this position (indicated by the arrow) yielding $1/Q_{th} \approx 2 \times 10^{-7}$. (From Cole *et al.* (2011).)

finite element method has been developed (Duwel *et al.*, 2006). The thermoelastic limited quality factor has an inverse temperature dependence and thus thermoelastic damping can be minimized through cryogenic operation of the optomechanical system. Fortunately, this is the typical operating environment for the majority of cavity optomechanics experiments.

To numerically predict support-induced losses in micro- and nanoscale mechanical resonators an efficient FEM-enabled numerical solver has recently been developed (Cole *et al.*, 2010a). Employing the recently introduced "phonon-tunneling" approach (Wilson-Rae, 2008), this solver represents a substantial simplification over previous approaches. A key feature of this method is to combine a standard FEM calculation of the resonator mode, allowing for the investigation of complex geometries, as well as taking proper account of interference effects between the radiated waves. The results highlighted here establish the first systematic test of phonon tunneling dissipation in mechanical resonators. In combination with existing models for other relevant damping channels (e.g. residual gas and thermoelastic damping), the phonon-tunneling solver makes further strides towards the accurate prediction of Q in micro- and nanoscale mechanical resonators.

16.3 AlGaAs as a macroscopic mirror material

Through investigations of micro-structured optical coatings, it appears that monocrystalline AlGaAs Bragg mirrors exhibit extremely promising characteristics for use not only in micro and nanomechanical resonators (particularly in optomechanical experiments), but also for the precision measurement community as a whole. This materials system is especially beneficial with regards to the significant enhancement in the achievable mechanical quality factor (or equivalently, a low coating loss angle), while maintaining excellent reflectivity. Given these exceptional results, AlGaAs is an interesting candidate for minimizing phase noise in ultra-high sensitivity interferometric applications. By employing monocrystalline materials as the reflective elements of the cavity end mirrors, the deleterious effects of mechanical dissipation – arising from limitations in the amorphous constituents of currently employed materials systems – may be minimized. Simultaneously, sufficient reflectivity in a standard configuration can be maintained with no significant changes to the overall interferometer being required, unlike the use of gratings (see Sections 6.5.2 and 11.4) or total internal reflection techniques (see Section 6.5.1).

The development of high-performance monocrystalline Bragg mirrors builds upon the vast knowledge-base of epitaxially grown mirror materials previously generated for the development of active optoelectronic devices, including vertical-cavity surface-emitting lasers (VCSELs) (Iga, 2008; Wilmsen *et al.*, 1999), vertical-cavity semiconductor optical amplifiers (Bjorlin *et al.*, 2004; Cole *et al.*, 2005), semiconductor disc lasers (SDLs) (Kuznetsov *et al.*, 1999), and semiconductor saturable absorber mirrors (SESAMs) (Keller *et al.*, 1996). Over the previous 25 years a wealth of experience has been gained in the construction of high performance single-crystal Bragg mirrors for these applications. The most mature materials system for producing high-quality single-crystal Bragg stacks consists of lattice-matched ternary alloys based on AlGaAs. These materials may be epitaxially grown as crystalline heterostructures via MOCVD or MBE on a suitable lattice matched substrate, enabling the production of arbitrary stacks of high index-contrast materials that maintain nearly perfect crystalline order.

Recent measurements have focused on the absorption of AlGaAs Bragg mirrors, both the initial 32.5 period MBE-grown material (Cole *et al.*, 2008) and the 40.5 period Bragg mirrors deposited via MOCVD (Cole *et al.*, 2010c). These experiments have yielded absorption values of <10 ppm for the MBE-grown material while the absorption is currently greater than 50 ppm for the MOCVD-derived material. These measurements were performed using a photothermal common-path interferometry system (Alexandrovski *et al.*, 2009) (see also Section 10.1.1) employing a continuous wave pump operating at power levels

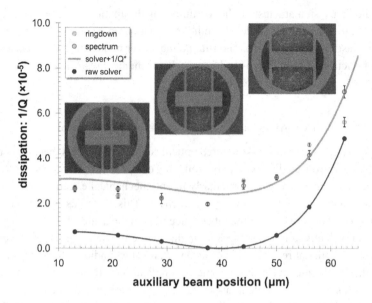

Figure 16.8 Comparison of experimental measurements performed at $T = 20$ K with theoretical dissipation values for the free-free mode of resonators with measured central dimensions of 132 μm × 42 μm and radius $R = 116$ μm. Both ringdown and spectrally derived data are included, with values averaged over two nominally identical chips. The plot includes both raw simulated data as well as fitted data (continuous lines are simply a guide to the eye) incorporating a constant offset representing the background material's damping and denoted as $1/Q_*$. Electron micrographs with overlaid CAD models are shown for the three extreme geometries as depicted in Figure 16.7. It is important to note that this model only captures support-induced losses, while this background dissipation channel still contributes to the overall damping $1/Q_{tot}$. However, the fact that the experiment utilizes sets of resonators for which the frequencies and effective surface-to-volume ratios are kept approximately constant implies that any additional damping mechanism that is relevant at low temperatures and high vacuum, but is insensitive to the variation in geometry, should contribute a constant offset. In this heteroepitaxial materials system (Cole *et al.*, 2010b) the extracted offset is found to be $1/Q_* = 2.41 \times 10^{-5}$. Although the exact nature of the corresponding dissipation mechanism is currently unknown, it appears to be caused by materials losses in the resonator epitaxial structure. (From Cole *et al.* (2011).)

up to 2 W (optical spot size of 70 μm), with no polarization or power dependence seen in the AlGaAs mirrors. Combined with the calculated optical penetration depth of 196 nm this leads to a maximum absorbance, α, of 1.8 cm^{-1}, comparable to previous measurements on GaAs-based optical microcavities (extracted value of 2.6 cm^{-1} at 980 nm) (Michael *et al.*, 2007). Note that the absorbance, given in units of cm^{-1}, is simply found by dividing the absorption value by the total interaction length, in this case twice the penetration depth. The source of the excess absorption in the MOCVD-grown mirrors is currently unknown, but most likely originates from impurities incorporated in the structure during the growth process, with the most likely candidate being carbon. In MOCVD, carbon is incorporated as a decomposition product of the metalorganic reactants. MBE typically exhibits lower unintentional dopant concentrations as the process operates at ultra-high vacuum (total pressure $<10^{-10}$ Torr) and with elemental sources (Wilmsen *et al.*, 1999).

In contrast, the MBE grown samples yield an absorbance of ≈ 0.2 cm^{-1}, nearly an order of magnitude less than that found in the MOCVD material with the same composition. This level of attenuation approaches the best values found in the literature, with the lowest AlGaAs Bragg mirror absorbance reported to date of 0.15 cm^{-1} (Reitzenstein *et al.*, 2007) for MBE-grown AlGaAs microcavities. For comparison, state-of-the-art dielectric coatings are capable of sub-ppm absorption values at 1064 nm (see Chapter 10). Note that all optical measurements have been performed at 300 K while there is interest in measuring absorption at cryogenic temperatures. Even if the background absorption can be fully eliminated, at sufficient optical intensities AlGaAs will be susceptible to two-photon absorption effects. Characterization of GaAs single crystals show two photon coefficients of approximately 0.03 cm/MW (Kleinman *et al.*, 1973; Boggess *et al.*, 1985). Fortunately, the small penetration depth into these mirrors helps to reduce the total absorption in the system (Obeidat *et al.*, 1997). Nonetheless, care must be taken in the system design in order to maintain the probe laser intensity below the threshold for significant levels of two-photon absorption.

In terms of the substrate selection, in order to generate single-crystal multilayers via epitaxial growth, two stringent conditions must be met: (i) the deposited films must have the same crystal symmetry as the underlying substrate, and (ii) the materials should have a nearly identical lattice parameter. Requirement (i) is typically satisfied for most relevant III–V semiconductor compounds (and also the common elemental semiconductors). As luck would have it, in the AlGaAs materials system, requirement (ii) is fulfilled over the entire composition range with a total lattice constant variation of less than 0.15% for all values of the aluminum content of the alloy. This minuscule variation in lattice constant is readily apparent in Figure 16.9. This allows for the growth of high index-contrast multilayers lattice-matched to crystalline GaAs substrates. The current maximum commercially available substrate diameter is 200 mm, with typical thicknesses between 500 μm and 1 mm. It is also possible to realize high quality AlGaAs structures on germanium wafers (Ting and Fitzgerald, 2000), although this hetero-epitaxial system is still in nascent research stages when compared with homo-epitaxial growth of AlGaAs on GaAs. Unfortunately, the choice of materials with this epitaxial materials system is quite limited, as opposed to the plethora of choices with dielectric mirrors. The ion beam sputtering process (see Chapter 2) enables deposition of high quality dielectric multilayers onto nearly any substrate, including sapphire, fused silica, and silicon. This enables the use of these mirrors in a wide variety of experimental architectures.

Rather than developing complex crystal growth techniques to realize epitaxial AlGaAs multilayers on foreign substrates, it may be possible to transfer these coatings to an arbitrary, ideally transparent, mirror substrate. This would significantly enhance flexibility in the application of monocrystalline Bragg mirrors. The envisioned fabrication procedure would entail separating the epitaxial layers from the original growth substrate via well-developed chemo-mechanical substrate removal or epitaxial lift-off processes (Konagai *et al.*, 1978; Yablonovitch *et al.*, 1987; Schermer *et al.*, 2006; Yoon *et al.*, 2010; Cole *et al.*, 2010b). This

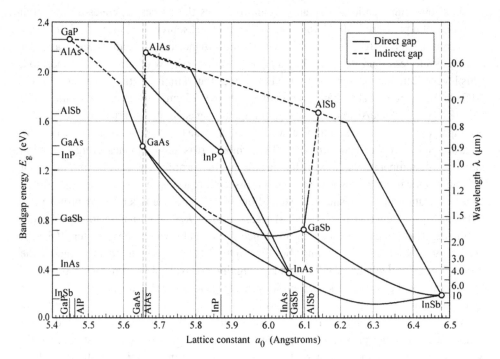

Figure 16.9 Bandgap versus lattice constant for common III–V semiconductor alloys. The *x*-axis highlights the typical substrate compositions (binary wafer materials including GaAs, InP, etc.), while the *y*-axis displays the bandgap energy and corresponding bandgap wavelength of the desired compound. Solid lines between the binary compounds denote ternary alloys, while quaternary compounds fall within the enclosed areas. Reproduced from Schubert (2003).

would be followed by direct wafer bonding of the Bragg mirror to the desired component. One interesting aspect of AlGaAs Bragg mirrors is the nearly identical coefficient of thermal expansion with *c*-axis oriented single-crystal sapphire. The absence of thermal strain in this materials combination has enabled successful large area direct bonding (over a 3 inch wafer) of GaAs to sapphire (Kopperschmidt *et al.*, 1997), with thermal cycling between room temperature and 77 K. However, these experiments were performed with flat structures, radius of curvature $= \infty$. Further investigation will be necessary to determine the limitations on radius of curvature attainable with this procedure.

An additional, and potentially interesting, advantage of epitaxial AlGaAs Bragg mirrors is the ability to modulate the conductivity through doping as demonstrated in a variety of electrically injected active optoelectronic devices. Complex doping schemes previously investigated for the constituent epilayers of electrically injected VCSELs, can be used to produce conducting films (even the intrinsic, or unintentionally doped, semiconducting state may be useful). For historical reference, all-epitaxial current injected microcavity lasers were first demonstrated in 1989, see Jewell *et al.* (1989). Although electrical conductivity is typically realized at the expense of enhanced absorption due to free carrier effects, tremendous advancements have since been made in the development of low optical loss and simultaneously low series resistance (i.e. vertically conducting) mirrors. A review of such structures can be found in Wilmsen *et al.* (1999). Design rules developed for conducting mirrors for electrically injected VCSELs should be readily applicable for efforts related to charge mitigation in high performance optical interferometer applications. For more on charge and precision measurements see Sections 3.5 and 10.1.2.

Currently, the fundamental reflectivity limits of AlGaAs materials system are unexplored. Although there have been innumerable devices developed, the limiting absorption has yet to be seriously investigated. Future research efforts should be aimed at further reducing the optical absorption and also at exploring the intrinsic limits of mechanical damping in this materials system.

16.4 Summary and conclusions

Quantum optomechanics is a rapidly growing interdisciplinary field focused on the ultimate goal of achieving quantum optical control of massive mechanical resonators. Employing a Fabry–Pérot configuration as described in the preceding chapter, the ultimate performance of such systems is dependent upon the development of high quality factor, high reflectivity micromechanical resonators, leading to very similar requirements to the optical coatings utilized by the precision measurement community. The requirements include low absorption, low mechanical loss, low surface roughness, and low transmission loss. Here we have reviewed the basic underlying physics as well as a number of experimental examples of resonators that have been implemented in this nascent field. In the course of this work, a promising alternative materials system has been explored – AlGaAs monocrystalline multilayers – that demonstrate very promising characteristics for overcoming the current challenges facing existing dielectric coating formulations.

17

Cavity quantum electrodynamics

TRACY E. NORTHUP

Cavity quantum electrodynamics (cavity QED) describes the interactions of single atoms and single photons in a resonator. In the laboratory, state-of-the-art experiments are able to observe the dynamics of these simple yet fundamental quantum systems. The recent development of quantum information science has also introduced a new role for cavity QED systems as quantum interfaces, in which information can be coherently mapped between photonic and atomic quantum bits.

Groundbreaking optical cavity QED experiments in the past twenty years would not have been possible without the technological development of mirror coatings with extremely low scatter and absorption losses. Thermal noise has typically played a smaller role in cavity QED experiments but in certain instances is nevertheless important to consider.

In this chapter, we begin with an overview of optical cavity QED experiments and some of the central questions that they address. It is hoped that this description, while by no means exhaustive, will provide a window for the nonspecialist into basic techniques and motivations in the field. We then turn to focus on the roles played by scattering and absorption loss and mechanical loss. We discuss how these losses have set limits on feasible experiments and the extent to which improvements may be possible.

17.1 Experimental realizations of cavity QED systems

An idealized cavity QED system consists of one atom placed between two highly reflective mirrors (Figure 17.1). In this model, a single atomic transition interacts with photons in a single cavity mode; the atom is often pictured at an antinode of the cavity standing wave, where it has a maximal dipole interaction with the cavity field. For the purpose of this book, we limit ourselves to experimental systems whose cavities conform quite closely to this model, i.e., resonators constructed from high-finesse mirrors with multilayer dielectric coatings in the optical range. However, it is worth noting that the earliest cavity QED research was carried out with Rydberg atoms in spherical microwave cavities (Raimond

Optical Coatings and Thermal Noise in Precision Measurement, eds. Gregory M. Harry, Timothy Bodiya and Riccardo DeSalvo. Published by Cambridge University Press. © Cambridge University Press 2012.

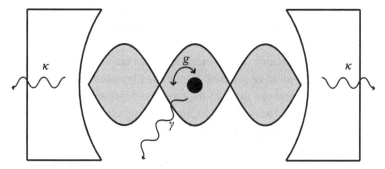

Figure 17.1 A model cavity QED system, in which a single atom is centered at the point of maximum interaction with a single cavity mode. The atom and the cavity field coherently exchange an excitation at rate $2g$. Spontaneous atomic decay occurs at rate 2γ, and intracavity photons exit the cavity at rate 2κ. In practice, the atom may be localized in the cavity mode via an optical or radiofrequency trap, or it may be traversing the cavity with some velocity that limits its interaction time.

et al., 2001) and that many cavity QED experiments today incorporate more exotic resonators including stripline microwave resonators (Girvin *et al.*, 2009), photonic bandgap structures, and microtoroidal resonators (Vahala, 2003). The atom, meanwhile, may be a neutral atom or an ion but may also be an artificial structure, such as a quantum dot. It may even be an ensemble of atoms, each interacting with the cavity. In practice, one must account for the multiplicity of levels present in a realistic atom as well as for the possibility of coupling to multiple cavity modes.

This model allows us to identify relevant rates of the atom–cavity system. Specifically, interesting quantum dynamics occur via the coherent interaction (typically a dipole coupling) between the atom and the cavity mode. The rate of this interaction is identified as g and is compared with the incoherent channels present in the system, including spontaneous emission of the atom (at rate γ) and cavity field decay (at rate κ). The *strong coupling regime* is defined by the condition

$$g \gg \kappa, \tag{17.1}$$

$$g \gg \gamma, \tag{17.2}$$

where coherent interactions dominate the system dynamics. The single-atom cooperativity parameter,

$$C = \frac{g^2}{2\kappa\gamma}, \tag{17.3}$$

is thus much greater than 1, indicating that the presence of just one atom significantly modifies the intracavity field. If $C \gg 1$ but the cavity decay is now the dominant rate,

$$\kappa > \frac{g^2}{\kappa} > \gamma, \tag{17.4}$$

then the system is in the *bad cavity regime*. Here, the cavity field dynamics can be adiabatically eliminated, and atomic decay is now described by the cavity-enhanced rate $\gamma' = \gamma(1 + 2C)$.

Furthermore, in the laboratory, a (non-artificial) atom cannot simply be placed within a cavity. Rather, a neutral atom or ion may be confined between the cavity mirrors using electromagnetic fields, or one can think of the model as representing a snapshot in which the atom (which has some velocity due to its nonzero temperature or due to the influence of gravity) interacts with the cavity field for a brief interval of time as it transits the cavity mode.

Below, various realizations of laboratory cavity QED systems are summarized, with the intention to highlight advantages and limitations of specific approaches and to touch on recent developments in the field.

17.1.1 Neutral atoms in cavities

Early cavity QED experiments in the optical domain used a collimated beam from an atomic source which intersected the mode of the optical cavity (Rempe *et al.*, 1991; Thompson *et al.*, 1992). In order to approach the regime in which a single atom interacted with the cavity, the average atom number, $\bar{N}$, in the cavity could be tuned to approximately 1 via the temperature of the source. However, inherent Poissonian fluctuations in $\bar{N}$ obscured some features of the quantum interaction. In addition, due to the temperature of the atoms, the period of interaction with the cavity lasted only about 1 μs. This latter problem was addressed in the next generation of experiments, in which newly developed laser cooling and trapping techniques were incorporated in order to slow down the atoms. Cold atoms were trapped just above a cavity and released to fall under gravity through the mode (Hood *et al.*, 1998), or the cavity was placed at the turning point of an "atomic fountain" (Münstermann *et al.*, 1999).

More recent experiments have extended these interaction times by several orders of magnitude by trapping atoms within a cavity. The atoms are confined in an optical dipole trap which may be along the cavity axis at a secondary wavelength supported by the cavity (Ye *et al.*, 1999) or may be a standing wave perpendicular to the cavity (Sauer *et al.*, 2004; Nußmann *et al.*, 2005; Khudaverdyan *et al.*, 2009); the trap wavelength may be red-detuned from the atomic transition, in which case atoms are localized in the standing-wave maxima, or it may be blue-detuned, in which case atoms are located in the minima and multiple spatial modes may be necessary for three-dimensional confinement (Puppe *et al.*, 2007). The period of atom–cavity interaction may now be tens of seconds long, limited by background gas collisions (Hijlkema *et al.*, 2007). However, the complete Hamiltonian for the atom-cavity system must take into account the specific trap properties. For example, one must consider the relative positions of the trapping potential minima (where atoms are localized) and the cavity mode (which determines the strength of the atom–cavity interaction), as well as the temperature and thus the spatial extent of an atom in the trap. For a dipole trap supported by the cavity, an atom will be tightly confined along the cavity axis, but as the atomic wavelength and trap wavelength are incommensurate, the atom's coupling to the cavity depends on the standing wave minimum in which it is confined. In a perpendicular dipole trap, the trap waist typically extends across several half-wavelengths of the cavity. In both cases, data which is averaged over many atoms will also be averaged over various positions and thus various couplings. Furthermore, owing to AC Stark shifts from the trapping field, atomic transitions may be modified and thus spatially dependent (van Enk *et al.*, 2001).

Experiments with trapped atoms have also been able to address the problem of fluctuations in atom number, that is, to work with exactly one atom rather than just a very small value of $\bar{N}$. It is possible, for example, to transport exactly one atom into the cavity mode (Fortier *et al.*, 2007a), or to load an indeterminate number of atoms but to identify when exactly one is present (Weber *et al.*, 2009).

A separate class of experiments is concerned not with single-atom interactions but rather with the interaction of an ensemble of atoms with the cavity mode. This may be a cloud of cold atoms falling through a cavity mode (Chan *et al.*, 2003) or an ultracold gas or Bose–Einstein condensate which is prepared and then transported into the cavity (Brennecke *et al.*, 2007; Colombe *et al.*, 2007).

17.1.2 Trapped ions in cavities

We have seen that it is challenging to localize a single atom at a precise position with respect to the mode of an optical cavity. By working with ions rather than with neutral atoms, some of these challenges can be addressed. Individual ions can be stored for up to weeks at a time in a radiofrequency-Paul trap or ring trap. Trapped ions can be cooled to their motional ground state in three dimensions, resulting in a localization on the order of $\Delta x \sim 10$ nm, i.e., much smaller than the spatial structure of the cavity mode and thus of the atom–cavity coupling, which is defined along the cavity axis by the half-wavelength of the optical transition and in the perpendicular direction by the cavity waist. Furthermore, as a result of recent progress in ion-based quantum information processing, a sophisticated toolbox for coherent manipulation of ions' electronic and motional states has been developed (Wineland *et al.*, 1998; Häffner *et al.*, 2008).

Ion traps have been successfully integrated with high-finesse optical cavities in multiple experiments (Guthöhrlein *et al.*, 2001; Mundt *et al.*, 2002; Russo *et al.*, 2009; Leibrandt *et al.*, 2009; Herskind *et al.*, 2009). To date, however, only neutral atoms in cavities have reached the single-atom strong coupling regime (Equations 17.1 and 17.2). One limitation to achieving strong coupling with ions lies in the fact that ion traps, unlike optical or magneto-optical traps for neutral ions, are three-dimensional physical structures which must be integrated with the cavity mirrors. Moreover, the coupling rate g is inversely proportional to the cavity waist w_0 and to the square root of the cavity length L. The cavity waist w_0, in turn, is a function of cavity length L and mirror radius of curvature R through

$$w_0 = \left(\left(\frac{\lambda}{2\pi} \right)^2 L(2R - L) \right)^{1/4}. \tag{17.5}$$

Thus, we see that stronger coupling can be achieved with a short cavity where the mirrors have a small radius of curvature. However, the presence of a large dielectric surface close to an ion will alter the trapping potential experienced by the ion. In addition, stray charges on mirror surfaces may cause time-dependent shifts to this potential. Weighing these considerations, ion-trap cavity QED experiments to date have been constructed with relatively long cavities, settling for weaker coupling in order to ensure stable ion trapping and good optical access to the ion.

With trapped-ion systems, we again face the problem that we encountered with optical dipole traps for neutral atoms: how well does the trap minimum coincide with an antinode of the optical cavity? While optical traps can be steered from outside a vacuum chamber, the ability to adjust the position of an ion in a Paul trap is limited. Specifically, the trapping minimum lies along the axis of the Paul trap, perpendicular to its radiofrequency and ground electrodes. An ion can typically be translated for millimeters along this minimum by tuning the DC voltages that provide on-axis confinement. However, in the two orthogonal directions, micron-scale displacements will cause heating of the ion. One approach is to incorporate ultra-high-vacuum piezo stages within the chamber in order to adjust the relative positions of trap and cavity. This adjustment capability means that the ion-cavity coupling can be tuned *in situ*: a single ion can be brought from a place of minimum to maximum coupling (Mundt *et al.*, 2002). In-vacuum piezo stages can also be useful because ion-trap experiments may choose a near-concentric cavity geometry (near the edge of the stability diagram, discussed further in Chapter 13) in order to minimize the waist and maximize coupling. Given this choice of geometry, in which the spot size at the cavity mirrors is quite large, an ion must be centered exactly between the mirrors in order to experience maximum coupling. Figure 17.2 shows a cavity QED apparatus, including a linear Paul trap mounted from above and a piezo assembly beneath the cavity.

17.1.3 Emerging themes

The mirror coatings used in cavity QED experiments have typically been applied to BK7 or fused silica substrates, which may be machined afterwards on a lathe to smaller dimensions (Hood, 2000). As discussed in the previous section, it is advantageous to minimize both the mirror radius of curvature R and the cavity length L. For standard mirrors, the minimum R is constrained by the radius at which substrates can be superpolished to sub-Ångstrom roughness (a few centimeters). For a given R, the minimum L is then limited by the mirror diameter, since a nonzero gap between the two spherical cavity mirrors is necessary both for an atom to enter and for optical access. The mirror diameter is in practice constrained by the machining process to ~ 1 μm.

An interesting new development bypasses superpolished substrates entirely, enabling smaller values of both R and the mirror diameter: the demonstration of optical-fiber-based mirrors for cavity QED experiments (Trupke *et al.*, 2005; Steinmetz *et al.*, 2006). The highest-quality mirrors have been fabricated by first centering a cleaved optical fiber in

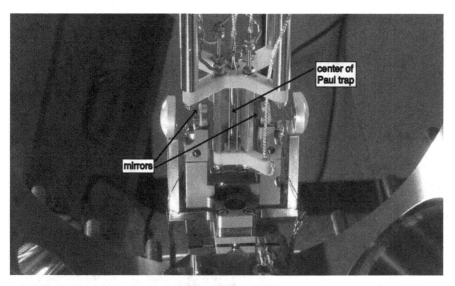

Figure 17.2 The current ion-trap cavity QED experiment in Innsbruck, an example of a high-finesse cavity integrated with a linear Paul trap. The Paul trap is suspended from above; sandwiched between two macor pieces are four blades at right angles for confinement in the xy-plane, two endcaps for confinement along the z-axis, and four thin wires to compensate micromotion of the ion. The cavity mirrors are 2 cm apart and are placed behind pairs of trap blades, which help to shield the ion from the dielectric mirror surface. As the cavity mode is near-concentric and thus has a short Rayleigh length, lenses for collimating the input and output beams are placed in vacuum. The cavity can be translated with respect to the ion trap via slip-stick and shear-mode piezos (Russo *et al.*, 2009).

the focus of a pulsed CO_2 laser. A short pulse ablates the fiber surface, which is then both extremely smooth and spherical over a certain diameter. The ablated fibers can afterwards be coated in a custom jig using ion-beam sputtering (see Chapter 2). Pre- and post-ablation fibers as well as a cavity assembled from coated fibers are shown in Figure 17.3. Uncoated fiber surface roughness has been measured at 2 Å, near the limit of atomic force microscope measurements (Hunger *et al.*, 2010). Recent finesse measurements comparing cavities built with fibers and substrates from the same coating run have demonstrated that additional scatter attributed to the fiber surface may be as low as 3 ppm. Radii of curvature as small as 40 μm have been demonstrated, enabling realistic possible waists as small as 1.3 μm (Hunger *et al.*, 2010). With such a steep value of R on a conventional substrate, it would not be possible to build short cavities before the mirrors came into contact, but here one has the advantage that the diameter of a standard fiber is only 125 μm, and the diameter of the spherical ablation is just 20–40 μm.

The small size of the fibers permits integration into a miniaturized device such as an atom chip, on which Colombe *et al.* (2007) have guided a Bose-Einstein condensate into the mode of a fiber cavity. Several projects for integration of fiber mirrors with ion traps are currently in development. A promising initial step has been the integration of uncoated fibers as light collection devices in surface ion traps; ions have been trapped as near as

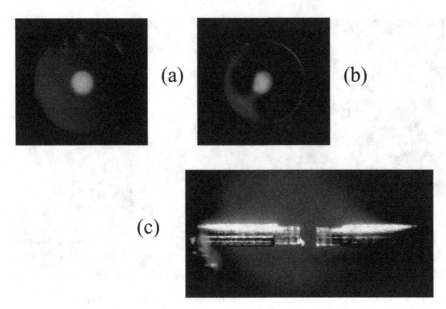

Figure 17.3 High-finesse, fiber-based mirrors, which are attractive for cavity QED experiments because of their high radius of curvature and potential for integration into small and perhaps scalable devices. (a) Microscope image of a cleaved fiber surface before laser ablation. (b) Concave fiber surface after laser ablation. (c) After dielectric mirror coatings have been sputtered onto the fibers, a cavity can be constructed from two fiber-based mirrors. The cavity length in this photograph is approximately 200 μm.

80 μm to the dielectric fiber surface (VanDevender *et al.*, 2010; Brady *et al.*, 2011). Fiber mirrors could also potentially be incorporated in cavity QED experiments with artificial atoms such as quantum dots, in which a distributed Bragg reflector fabricated beneath the dots would serve as the second cavity mirror (Muller *et al.*, 2010).

Another new theme in cavity QED research incorporates optical resonators with the quantized motion of nanomechanical oscillators, discussed in Chapter 16. Rather than coupling the electronic state of an atom to the optical mode of a cavity, the proposed experiments would use the cavity mode as a conduit to couple the atom and an oscillator, such as a thin membrane suspended in the cavity (Hammerer *et al.*, 2009). High-stress SiN films, for example, have been shown both to have excellent mechanical quality factors and to be compatible with high-finesse cavities (Wilson *et al.*, 2009).

17.2 Representative experiments

A central theme of cavity QED is the coherent exchange of individual quanta between atom and field in the presence of decoherent processes. Experiments in the field have thus focused on characterizing this light–matter interface, evaluating its nonclassical behavior, and demonstrating increasing control over various aspects of the quantum system.

The interaction of Figure 17.1 is described by the Jaynes–Cummings Hamiltonian in the rotating-wave approximation (Jaynes and Cummings, 1963),

$$H_{JC} = \hbar(\omega_c - \omega_p)\hat{a}^\dagger\hat{a} + \hbar(\omega_a - \omega_p)\hat{\sigma}^+\hat{\sigma}^- \qquad (17.6)$$
$$+ \hbar g(\hat{a}^\dagger\hat{\sigma}^- + \hat{a}\hat{\sigma}^+) + \hbar\epsilon\hat{a} + \hbar\epsilon^*\hat{a}^\dagger,$$

where $\hat{a}$ is the photon annihilation operator, $\hat{\sigma}^-$ is the atomic lowering operator, ω_a is the atomic transition frequency, ω_c is the cavity resonance frequency, ω_p is the frequency of a classical field driving the cavity, and ϵ is the Rabi frequency of this field. Additional terms can be included in order to account for multiple cavity modes or atomic levels, a classical field which drives the atom, or the presence of multiple atoms, in which case it is known as the Tavis–Cummings Hamiltonian (Tavis and Cummings, 1968).

In the absence of a driving field, the Jaynes–Cummings Hamiltonian can be diagonalized. When the cavity is resonant with the atomic transition, the first eigenstates above the ground state are symmetric and antisymmetric superpositions of one photon in the cavity and one atom in its excited state. The frequency splitting between the symmetric and antisymmetric states is given by $2g$ and is known as the vacuum Rabi splitting. It can be probed by introducing a weak cavity driving field and scanning the frequency of this field across the atom–cavity resonance. The ability to resolve the vacuum Rabi splitting spectroscopically is a hallmark of the strong coupling regime, and observation of this splitting was thus a milestone in early experiments (Thompson *et al.*, 1992). More recently, a complete vacuum-Rabi spectrum has been measured for a single atom confined in a dipole trap (Boca *et al.*, 2004).

In the strong coupling regime, saturation of the atomic transition occurs for photon numbers $\ll 1$, and it is possible to use just one intracavity atom as a laser gain medium (McKeever *et al.*, 2003). While the output of a conventional laser is a coherent state, the output state of this single-atom laser is nonclassical, as evinced by antibunched and sub-Poissonian photon statistics. By tuning the strength of the atom–field coupling via a Raman transition, the path from quantum to classical behavior in a single-atom laser has been mapped out (Dubin *et al.*, 2010). A single-atom laser can be operated in a pulsed regime as a single photon source, and sources based on both neutral atoms and ions have been shown to be deterministic, that is, to generate photons inside the cavity with near-unit efficiency (McKeever *et al.*, 2004; Keller *et al.*, 2004; Hijlkema *et al.*, 2007; Barros *et al.*, 2009).

Given a suitable atomic level structure, pairs of sequential single photons can be generated which are polarization-entangled, where the two polarizations correspond to the two orthogonal cavity modes. The entanglement is mediated by the intracavity atom (Wilk *et al.*, 2007). Separately, it has been shown that a photonic state can be mapped coherently onto a trapped atom in a cavity (Boozer *et al.*, 2007). Both experiments represent important steps towards quantum networks, where an entangled state stored in an intracavity atom could be mapped onto a photon, distributed over long distances, and transferred to a remote atom (Kimble, 2008). Such a network is depicted schematically in Figure 17.4.

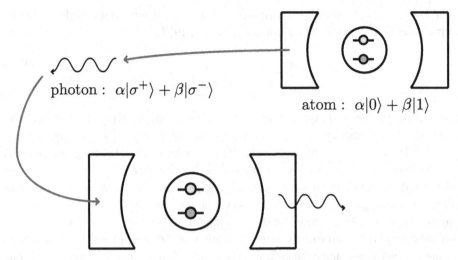

Figure 17.4 A model for a cavity-QED-based quantum network. Quantum information is stored and manipulated as a superposition of two long-lived atomic states. The information is then coherently mapped onto a photon polarization or number state for transport to a spatially separated atom. Such a network would enable distributed quantum computing, but for the mapping process to be efficient, transmission must dominate scatter and absorption losses in the cavity mirrors.

Recent work has achieved control not only over the atom–photon coupling process as above, but also over various other aspects of the system. Cavity-mediated cooling has been demonstrated, first for atomic ensembles (Chan *et al.*, 2003) and subsequently for individual atoms and ions (Maunz *et al.*, 2004; Leibrandt *et al.*, 2009). The trajectory of an atom within the cavity can be monitored via the cavity transmission and controlled in real time via a feedback laser (Kubanek *et al.*, 2009). Feedback is also an important theme in the collective coupling of an atomic ensemble to the cavity field, in which the density excitation of the atoms functions as a mechanical oscillator in a cavity optomechanics model (Brennecke *et al.*, 2008).

17.3 Optimizing cavity design

While researchers across the field share a common set of goals, there is no one ideal cavity design for cavity QED experiments, as the constraints of a particular experiment will dictate specific choices. The only constant is that for any experiment, it is advantageous to minimize scatter and absorption losses in the mirror coatings.

17.3.1 Transmission, absorption, and scatter losses

Advances in ion-beam-sputtering technology were critical in enabling access to the strong coupling regime. The demonstration of mirrors with total transmission, absorption, and scatter losses of only 1.6 ppm at 850 nm – corresponding to a cavity finesse $\mathcal{F} = 1.9 \times 10^6$, with inferred scatter and absorption losses of 1.1 ppm – remains unsurpassed 20 years later (Rempe et al., 1992). (See Table 10.1 for a comparison of these results with state-of-the-art mirrors at 1064 nm and Chapter 2 for a discussion of dielectric coating technology.) Such low optical loss meant that it was possible to build short cavities which still had long decay times, or equivalently, small decay rates κ. Short cavities, as we have seen, are necessary to minimize the cavity waist w_0 and thus to maximize the coupling g. If one can build a cavity with $g \gg \gamma$, the atomic decay rate, while still maintaining $\kappa \ll g$, then Equations 17.1 and 17.2 has been satisfied. In the first vacuum-Rabi splitting observation, for example, the system rates were given by $(g, \kappa, \gamma) = 2\pi \times (3.2, 0.9, 2.5)$ MHz (Thompson et al., 1992).

Minimizing total optical losses results in cavities with the highest finesse and, for a given length, the smallest κ. However, if the total optical loss is dominated by absorption and scatter, then it will be very difficult to extract any signal from the cavity.

The cavity finesse can be written as $\mathcal{F} = \frac{\pi\sqrt{R}}{1-R}$, where R is the fraction of intensity reflected by each cavity mirror, assuming that the mirrors are identical (see Equation 15.6). Similarly, scatter, absorption, and transmission at a single mirror can be described by coefficients S, A, and T, such that

$$R + S + A + T = 1. \tag{17.7}$$

(See Chapters 10 and 11 for more detailed discussions of absorption and scatter, respectively.) For a high-finesse cavity, where $R \gg \{S, A, T\}$, the finesse in the case of identical mirrors reduces to

$$\mathcal{F} = \frac{\pi}{S + A + T} = \frac{\pi}{\mathcal{L}}, \tag{17.8}$$

where $\mathcal{L}$ represents total optical loss. For nonidentical cavity mirrors, a more general expression is

$$\mathcal{F} = \frac{2\pi}{S_{\text{in}} + S_{\text{out}} + A_{\text{in}} + A_{\text{out}} + T_{\text{in}} + T_{\text{out}}}, \tag{17.9}$$

where the subscripts refer to input and output mirrors.

Given a photon in the cavity mode, the chance that it will exit the cavity through the output mirror is given by

$$P_{\text{out}} = \frac{T_{\text{out}}}{S_{\text{in}} + S_{\text{out}} + A_{\text{in}} + A_{\text{out}} + T_{\text{in}} + T_{\text{out}}}. \tag{17.10}$$

The same expression describes the time-reversed process of coherently mapping a single photon into a cavity (Boozer et al., 2007). Thus, applications which require mapping photonic states into and out of a cavity with high efficiency require the total optical loss $\mathcal{L}$ to be dominated by the transmission component. For example, proposals which rely on a cavity as an interface to transfer quantum information between photons and atoms in a quantum network assume that this transfer process will be nearly lossless. The first step is thus to decide what value of P_{out} is tolerable and then to assess what scatter and absorption values are likely to be at the wavelength of interest, where we have seen that in the infrared range, 1–2 ppm represents the current state of the art. Target values for $T_{in,out}$ can then be found which satisfy the requirements for P_{out}, and a coating run can be designed in order to meet these targets. As additional dielectric layers are added to the coating in order to reduce the transmission, the variation in transmission as a function of wavelength becomes increasingly steep. (This variation over wavelength is known as a "coating curve"; sample data and a theoretical model can be found in Hood *et al.* (2001).) It may be necessary to evaluate the coating properties at the desired wavelength in a test run before a final coating run, if possible with a laser that is tunable over a range of tens of nanometers.

The preceding discussion has focused on meeting the requirements of the strong coupling regime, where the intention is to maximize the time interval in which atoms and photons can interact coherently. Sometimes, however, it may be desirable for photons to exit the cavity as quickly as possible. For example, one can take advantage of the bad cavity regime defined in Equation 17.4 to collect spontaneous emission from the atom efficiently, or to generate a deterministic single-photon source with a high repetition rate. In this case, one can select a target transmission (thus adjusting κ) as well as a target cavity length (thus adjusting both κ and g) in order to satisfy both inequalities.

We have emphasized that absorption and scatter losses are constrained by the mirror coating process and the surface quality of the substrate, but that transmission is a controllable parameter of the coating. Meanwhile, the atomic decay rate γ (as well as g, which scales as $\sqrt{\gamma}$) is determined by the choice of atomic species and the particular atomic transition. This choice is constrained by several factors, including the availability of lasers for cooling, trapping, addressing, and possibly ionizing the atom; the suitability of the atomic structure, such as whether it possesses cycling transitions, magnetic-field-insensitive states, or long-lived transitions for quantum logic; and the quality of mirror coatings available at the atomic transition frequency. (Specifically, losses in optical coatings at blue and ultraviolet wavelengths are typically much higher than in the infrared.) Finally, the cavity length, and thus κ and g, may be limited by geometric constraints such as integration with an ion trap or the need for optical access between the mirrors.

As discussed in Chapters 10 and 11, particle contamination of the mirror surface increases the optical loss. In addition to the cleaning methods discussed in these chapters, a wafer spin cleaner or spin coater has been shown to be an effective method for cleaning mirrors from 7.75 mm to 1 inch diameter. The mirror is held in place via suction as it is spun, and spectrophotometric-grade solvents (first water, then acetone, then isopropyl alcohol) are applied with a squirt bottle. As each solvent is applied, a cleanroom cotton swab is swept gently from the center of the optic to the edge and rotated as it is swept. Ten to twenty seconds of spinning without solvent or swab is sufficient to dry the mirror at the end of the cycle.

Optical losses may also include clipping losses of the cavity mode at the edge of the mirror (see Chapter 13). Clipping losses are important to consider for mirrors applied to laser-ablated fibers, where the concave surface is only spherical over 20–40 μm. In

Innsbruck, we have recently succeeded in ablating structures with a diameter of 60–80 μm by using fibers with a 200 μm cladding diameter instead of the industry-standard 125 μm. It appears that the additional fiber surface enables better heat transport during the laser pulse. We intend to use these larger structures for fiber cavities integrated with ion traps, with a target ion–fiber distance of 100 μm.

Annealing is typically used as a final step to reduce absorption losses after ion-beam sputtering (see Section 10.1.2). Because we initially believed that annealing under air (as is standard practice) would cause oxidization problems with copper-coated fiber cavities, we have recently tested annealing Ta_2O_5/SiO_2 coatings under vacuum at 450 °C. This process, however, substantially *increased* the optical losses. The finesse could be recovered – and improved due to the desired annealing effects – after subsequent annealing under air. This behavior persisted during subsequent vacuum and air annealing cycles. We attribute the losses to depletion of oxygen from the surface layer of Ta_2O_5, a conjecture which is supported by comparative x-ray photoelectron spectroscopy (XPS) measurements of vacuum- and air-annealed mirrors. This finding suggests that although a high-temperature bakeout is typical for an ultrahigh-vacuum apparatus, it may be problematic for cavity QED experiments.

17.3.2 Asymmetric cavities

We have referred above to "input" and "output" cavity mirrors, implying a directionality in the process of coupling light into and out of the cavity. This simplification is, of course, artificial. In order to collect every possible photon emitted by a cavity, one should place detectors outside of both mirrors.

Nevertheless, it is often much simpler to designate one cavity mirror as the output port, as this avoids the technical challenge of combining the two output paths and also allows the other port to be used for locking, trapping, or cavity drive fields. In order to maximize the fraction of the intracavity field transmitted at the output port, given by Equation 17.10, one can choose $T_{in} \ll T_{out}$, i.e. an asymmetric or "single-sided" cavity. However, assuming that $T_{in} \gg \{S_{in,out}, A_{in,out}\}$ has been chosen in order to ensure a reasonable transmission efficiency, the increased value of T_{out} will compromise the cavity finesse and κ; thus the ratio $\frac{T_{in}}{T_{out}}$ is a compromise between detection efficiency and cavity coupling parameters.

17.3.3 Birefringence

In addition to the optical loss, other mirror properties play a role in atom–cavity interactions. While the simplest cavity QED model considers an atom interacting with just one cavity mode, a three-dimensional cavity supports two degenerate modes of orthogonal polarization. To the extent that the cavity mirrors are birefringent – that is, that the index of refraction in the mirror is anisotropic, so that different polarizations of light experience different effective cavity lengths – the mode frequencies will be nondegenerate. Some birefringence may be acquired during the mirror coating process (see Chapter 2), but the main contribution to the cavity birefringence occurs through stress during mounting, gluing, or baking. The birefringent splitting between mode frequencies $\Delta\nu$ relative to the cavity linewidth κ is given by

$$\frac{\Delta\nu}{\kappa} = \frac{\delta}{2\pi^2}\mathcal{F}, \tag{17.11}$$

where δ is the phase shift per round trip. Because the cavity amplifies the phase shift with each round trip, birefringence represents a particular challenge for high-finesse cavities (Ye and Lynn, 2003).

For cavity QED experiments, the simplest cases occur when no birefringence is present (two degenerate cavity modes), or when the birefringent mode splitting is much larger than the relevant rates $\{g, \kappa, \gamma\}$, so that an atom effectively interacts with just one cavity mode. (Note, however, that this second case precludes experiments in which entanglement is generated between atomic states and photon polarization, or in which a coherent atomic superposition is mapped onto a superposition of polarization states. A quantum state could be stored instead as a photon number state.) The most complicated situation lies somewhere in between, when the frequency of the birefringent splitting is on the order of $\{g, \kappa, \gamma\}$. In this case, the splitting must be carefully measured so that it can be incorporated into simulations of the system dynamics, in which photons can be transferred from one mode to another via an atom.

17.3.4 Co-resonant optical frequencies

Another consideration is whether multiple optical frequencies are present in the cavity, and if so, how the mirror coating curve varies over these frequencies. It is usually necessary to actively stabilize the cavity length via a Pound–Drever–Hall lock (see Section 15.2.1). Thus, the cavity is not only (near-)resonant with the atomic transition frequency but also must support a stabilization laser at a separate longitudinal mode. For cavities with length on the order of 10 µm, the wavelengths of neighboring longitudinal modes are separated by tens of nanometers.

Additional lasers at other longitudinal modes may be introduced for a red- or blue-detuned on-axis dipole trap or to drive Raman transitions between atomic states (Boozer et al., 2006). On one hand, the finesse of the mirrors at these wavelengths should be high enough to support a cavity mode. For dipole traps, for example, a higher finesse corresponds to a deeper trapping potential for a given input power. On the other hand, if the finesse is too high, thermal noise issues may introduce fluctuations within the system, see Section 17.4 below.

17.4 Thermal noise

Thermal noise is not an issue when cavity fields are on a single-photon level. In fact, some cavity QED experiments do not drive the cavity with any classical field near the atom–cavity resonance. Atoms may be driven directly from the side of the cavity at another transition frequency, and cavity-resonant photons are then generated from the vacuum.

However, thermal fluctuations may come into play at the higher field strengths used for actively stabilizing a cavity and for trapping atoms inside it. For example, the initial attempts of Ye *et al.* (1999) to confine atoms in an intracavity dipole trap were plagued by relatively short trapping lifetimes of 28 ms. These lifetimes could not be explained by background gas collisions (which should occur on a timescale of seconds) and were attributed to parametric heating at the trapping frequency; that is, frequency fluctuations in the trapping field were converted by the cavity to fluctuations in intensity, and the component at twice the trapping frequency heated atoms out of the trap. Brownian motion of the cavity mirrors was found to be a significant source of intensity fluctuations. In order to reduce this noise, which scales as $\mathcal{F}^2$, the trapping frequency was shifted by tens of nanometers to a frequency at which the cavity finesse was much lower; this shift was an important component in achieving $\sim$3 s trap lifetimes (Buck, 2003).

In order to diagnose Brownian motion in this experiment, the spectral density function for the cavity was measured and found to be in reasonable agreement with calculations in which finite element analysis was used to calculate the eigenmodes of the mirrors. The sensitivity of the mechanical system was found to be 1.24×10^{-19} m (Buck, 2003).

This finite element analysis has recently been refined, with the result that it now reproduces cavity displacement noise spectra with much higher accuracy (see Figure 17.5). A thorough understanding of these spectra is important for optomechanics experiments in which a cavity field is used to cool a membrane suspended between the cavity mirrors at room temperature. As the membrane approaches the thermal ground state, its displacement amplitude becomes comparable with displacement resulting from substrate thermal noise (Wilson *et al.*, 2009). One possible approach to reduce the thermal noise background consists of active feedback to the laser driving the cavity in order to counteract photothermoelastic noise (see Section 3.4). A second approach would be to choose membrane and cavity properties carefully to avoid resonant frequency overlap. Future cavity-QED optomechanics experiments (Hammerer *et al.*, 2009), in which the cavity field would enable coupling between the resonator and a trapped atom, would be confronted with the same challenges.

Thermal expansion in mirror coatings is also seen in frequency scans of high-finesse cavities, where in one scan direction, this expansion can lead to self-locking behavior as the change in effective cavity length maintains the resonance condition (Poirson *et al.*, 1997). Hunger *et al.* (2010) have used the frequency over which self-locking takes place to infer the mirror absorption in a high-finesse fiber cavity. This gives results which are consistent with ringdown, transmission, and surface roughness measurements. (See Chapter 10 for further discussion of absorption measurement techniques.) In addition, a power threshhold for this bistability effect is extracted.

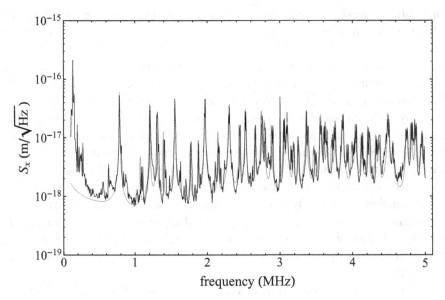

Figure 17.5 Measured and calculated thermal noise spectrum due to Brownian motion of the cavity mirror substrates used in Wilson *et al.* (2009). The BK7 substrates are 3 mm in diameter and 4 mm in length, with the front face chamfered to a mirror diameter of 1 mm. The model spectrum is obtained via finite element analysis (COMSOL 3.0) and treats the mirrors as free masses. Figure courtesy of D. J. Wilson, K. K. Ni, and H. J. Kimble.

17.5 Open questions

Several research groups are currently constructing ion traps with tightly integrated optical fibers or dielectric mirrors; for example, Herskind *et al.* (2011) have recently demonstrated a planar trap microfabricated on top of a high-finesse mirror, in which ions are confined <200 μm above the mirror. Thus, the next few years will bring important findings about the limits of this integration. One interesting question is whether it is possible to fabricate special mirror coatings which screen ions from the dielectric surface without compromising the mirror reflectivity. A candidate material for such coatings would be indium tin oxide, as thin layers are both electrically conductive and transparent at optical wavelengths. However, it is questionable whether an indium tin oxide layer that was thick enough to be conductive could be incorporated into a multilayer coating while maintaining high finesse.

A more general question concerns the extent to which quantum networking applications with atoms, ions, and cavities are in fact realistic. Cavity QED experiments to date have been proof-of-principle demonstrations and will continue to be so in the decade ahead. If this fundamental physics is to be implemented in long-distance networks connecting quantum computers via optical fiber, then quantum-interface schemes will have to be tolerant to the intrinsic scatter and absorption losses of cavities, and the process of constructing optical cavities will have to be streamlined from the complex and time-intensive activity which it is today. It may be that atomic physics experiments are precursors to solid-state implementations in which artificial atoms are fabricated within cavity structures.

Meanwhile, cavity QED experiments in the optical domain continue to provide a fascinating window into quantum mechanics at its most fundamental. On the surface, these experiments may seem quite distant from kilometer-long cavities built from kilogram-scale mirrors and driven by high-intensity laser fields. However, as we have seen, the technical challenges faced by precision measurement experiments often persist from the smallest scales to the largest.

References

Abbott, B. P., and The LIGO Scientific Collaboration. 2009. LIGO: The Laser Interferometer Gravitational-Wave Observatory. *Reports on Progress in Physics*, **72**(7), 076901.

Abbott, B., and The LIGO Scientific Collaboration. 2010. Searches for gravitational waves
from known pulsars with science run 5 LIGO data. *The Astrophysical Journal*, **713**(1),
671–685.

Abbott, B. P., and The LIGO and Virgo Collaborations. 2009. An upper limit on the
stochastic gravitational-wave background of cosmological origin. *Nature*, **460**, 990–
994.

Abbott, R., Adhikari, R., Allen, G. *et al.* 2002. Seismic isolation for Advanced LIGO.
Classical and Quantum Gravity, **19**(7), 1591–1597.

Abramovici, A., Althouse, W., Camp, J. *et al.* 1996. Improved sensitivity in a gravitational
wave interferometer and implications for LIGO. *Physics Letters A*, **218**(3–6), 157–163.

Abramovici, Alex, Althouse, William E., Drever, Ronald W. P. *et al.* 1992. LIGO: The laser
interferometer gravitational-wave observatory. *Science*, **256**, 325–333.

Accadia, T., and The Virgo Collaboration. 2010a. Commissioning status of the Virgo
interferometer. *Classical and Quantum Gravity*, **27**(8), 084002.

Accadia, T., and The Virgo Collaboration. 2010b. Virgo calibration and reconstruction of
the gravitational wave strain during VSR1. *Journal of Physics: Conference Series*,
228(1), 012015.

Acernese, F., and The Virgo Collaboration. 2006. The status of Virgo. *Classical and
Quantum Gravity*, **23**, S63–S69.

Acernese, F., and The Virgo Collaboration. 2010. Performances of the Virgo interferometer
longitudinal control system. *Astroparticle Physics*, **33**(2), 75–80.

Adler, F., Masłowski, P., Foltynowicz, A. *et al.* 2010. Mid-infrared Fourier transform
spectroscopy with a broadband frequency comb. *Optics Express*, **18**(21), 21 861–
21 872.

Advanced LIGO Team. 2007. *Advanced LIGO reference design*. M060056-08-M.

Ageev, Alexandr, Palmer, Belkis Cabrera, Felice, Antonio De, Penn, Steven D., and Saulson,
Peter R. 2004. Very high quality factor measured in annealed fused silica. *Classical
and Quantum Gravity*, **21**, 3887–3892.

Agresti, Juri. 2008. *Researches on non-standard optics for advanced gravitational wave
interferometers*. Ph.D. thesis, University of Pisa. LIGO-T040225-00-R.

Agresti, J., D'Ambrosio, E., DeSalvo, R. *et al.* 2006a. Design and construction of a prototype
of a flat top beam interferometer and initial tests. *Journal of Physics Conference Series*,
32(Mar.), 301–308.

Agresti, J., Castaldi, G., DeSalvo, R. *et al.* 2006b. Optimized multilayer dielectric mirror coatings for gravitational wave interferometers. Page 628608 of: *Proceedings of the SPIE*, vol. 6286.

Akhiezer, A. 1939. On the absorption of sound in solids. *Journal of Physics*, **1**, 277.

Alexandrovski, A., Route, R. K., and Fejer, M. M. 2001. *Absorption studies in sapphire*. G010152-00.

Alexandrovski, Alex, Markosyan, Ashot, Fejer, Martin, and Route, Roger. 2009. Photothermal common-path interferometry (PCI): New developments. Page 13 of: Clarkson, W. Andrew, Hodgson, Norman, and Shori, Ramesh K. (eds), *Proceedings of Solid State Lasers XVIII: Technology and Devices*, vol. 7193.

Allan, D. W. 1966. Statistics of atomic frequency standards. *Proceedings of the IEEE*, **54**(2), 221–230.

Alnis, J., Matveev, A., Kolachevsky, N. *et al.* 2008a. Stable diode lasers for hydrogen precision spectroscopy. *The European Physical Journal – Special Topics*, **163**, 89–94.

Alnis, J., Matveev, A., Kolachevsky, N., Udem, Th., and Hänsch, T. W. 2008b. Subhertz linewidth diode lasers by stabilization to vibrationally and thermally compensated ultralow-expansion glass Fabry–Pérot cavities. *Physical Review A*, **77**(5), 053809.

Anderson, O. L., and Bömmel, H. E. 1955. Ultrasonic absorption in fused silica at low temperatures and high frequencies. *Journal of the American Ceramic Society*, **38**, 125–131.

Anderson, O., and Ottermann, C. 1997. *Thin Films on Glass*. Springer-Verlag. Chap. Silicon dioxide.

Anderson, O., Bange, K., and Ottermann, C. 1997. *Thin Films on Glass*. Springer-Verlag. Chap. Titanium dioxide.

Anderson, P. W., Halperin, B. I., and Varma, C. M. 1972. Anomalous low-temperature thermal properties of glasses and spin glasses. *Philosophical Magazine*, **25**, 1–9.

Ando, Masaki, and The DECIGO Collaboration. 2010. DECIGO and DECIGO pathfinder. *Classical and Quantum Gravity*, **27**(8), 084010.

Ando, Masaki, and The TAMA Collaboration. 2001. Stable operation of a 300-m laser interferometer with sufficient sensitivity to detect gravitational-wave events within our galaxy. *Physical Review Letters*, **86**(18), 3950–3954.

Anetsberger, G., Gavartin, E., Arcizet, O. *et al.* 2010. Measuring nanomechanical motion with an imprecision below the standard quantum limit. *Physical Review A*, **82**, 061804:4.

Antonini, P. 2005. *Test of Lorentz invariance using sapphire optical resonators*. Ph.D. thesis, Heinrich-Heine-Universität Düsseldorf, Germany.

Antonini, P., Okhapkin, M., Göklü, E., and Schiller, S. 2005. Test of constancy of speed of light with rotating cryogenic optical resonators. *Physical Review A*, **71**(5), 050101.

Arai, K, and The TAMA Collaboration. 2008. Recent progress of TAMA300. *Journal of Physics: Conference Series*, **120**(3), 032010.

Arain, Muzammil A., Quetschke, Volker, Gleason, Joseph *et al.* 2007. Adaptive beam shaping by controlled thermal lensing in optical elements. *Applied Optics*, **46**(12), 2153–2165.

Arcizet, O., Cohadon, P. F., Briant, T., Pinard, M., and Heidmann, A. 2006. Radiation-pressure cooling and optomechanical instability of a micromirror. *Nature*, **444**(7115), 71–74.

Arkwright, J. W. 2006. Fabrication of optical elements with better than $\lambda/1000$ thickness uniformity by thin-film deposition through a multi-aperture mask. *Thin Solid Films*, **515**, 854.

Armani, D. K., Kippenberg, T. J., Spillane, S. M., and Vahala, K. J. 2003. Ultra-high-Q toroid microcavity on a chip. *Nature*, **421**(6926), 925–928.

Arndt, M., Aspelmeyer, M., and Zeilinger, A. 2009. How to extend quantum experiments. *Fortschritte der Physik*, **57**, 1153–1162.

Aspelmeyer, Markus, and Schwab, Keith. 2008. Focus on mechanical systems at the quantum limit. *New Journal of Physics*, **10**(9), 095001.

Aspelmeyer, M., Gröblacher, S., Hammerer, K., and Kiesel, N. 2010. Quantum optomechanics – throwing a glance. *Journal of the Optical Society of America B*, **27**(6), A189–A197.

Astrath, N. G. C., Rohling, J. H., Medina, A. N. *et al.* 2005. Time-resolved thermal lens measurements of the thermo-optical properties of glasses at low temperature down to 20 K. *Physical Review B*, **71**, 214202.

Atanassova, E., Tyuliev, G., Paskaleva, A., Spassov, D., and Kostov, K. 2004. XPS study of N_2 annealing effect on thermal Ta_2O_5 layers on Si. *Applied Surface Science*, **225**(1–4), 86–99.

Audoin, C., Santarelli, G., Makdissi, A., and Clairon, A. 1998. Properties of an oscillator slaved to a periodically interrogated atomic resonator. *IEEE Transactions on Ultrasonics, Ferroelectrics and Frequency Control*, **45**(4), 877–886.

Azzam, R. M. A., and Bashara, N. M. 1987. *Ellipsometry and Polarized Light*. Elsevier.

Bach, H., and Neuroth, N. (eds). 1995. *The Properties of Optical Glass*. Springer-Verlag.

Bagdasarov, Kh. S., Braginsky, V. B., and Mitrofanov, V. P. 1974. Mechanical dissipation in single-crystal sapphire. *Kristallografiya*, **19**, 883.

Bagini, V., Borghi, R., Gori, F. *et al.* 1996. Propagation of axially symmetric flattened Gaussian beams. *Journal of the Optical Society of America A*, **13**(7), 1385–1394.

Baker, John G., McWilliams, Sean T., van Meter, James R. *et al.* 2007. Binary black hole late inspiral: Simulations for gravitational wave observations. *Physical Review D*, **75**(12), 124024.

Bange, K. 1997a. *Thin Films on Glass*. Springer-Verlag. Chap. Properties and characterization of dielectric thin films.

Bange, K. 1997b. *Thin Films on Glass*. Springer-Verlag. Chap. Tantulum oxide layers.

Barish, Barry C., and Weiss, Rainer. 1999. LIGO and the detection of gravitational waves. *Physics Today*, **52**(10), 44–50.

Barr, B. W., and Burmeister, O. 2009. *Review of all-reflective optics for the Einstein Telescope*. Einstein Telescope Document: ET-028-09.

Barros, H. G., Stute, A., Northup, T. E. *et al.* 2009. Deterministic single-photon source from a single ion. *New Journal of Physics*, **11**(10), 103004.

Bassiri, R., Borisenko, K. B., Cockayne, D. J. H. *et al.* 2010. Probing the atomic structure of amorphous Ta_2O_5 mirror coatings for advanced gravitational wave detectors using transmission electron microscopy. *Journal of Physics: Conference Series*, **241**, 012070.

Bassiri, R., Borisenko, K. B., Cockayne, D. J. H. *et al.* 2011. Probing the atomic structure of amorphous Ta_2O_5 coatings. *Applied Physics Letters*, **98**, 031904.

Baumeister, P. W. 2004a. *Optical Coating Technology*. SPIE Press. Chap. How coatings are used and integrated into optical systems.

Baumeister, P. W. 2004b. *Optical Coating Technology*. SPIE Press. Chap. Collection of the evaporant upon the substrates.

Baumeister, P. W. 2004c. *Optical Coating Technology*. SPIE Press. Chap. Thin films, the building blocks of multilayers.

Bava, E., Galzerano, G., and Svelto, C. 2006. Amplitude and frequency noise sensitivities of optical frequency discriminators based on Fabry–Perot interferometers and the frequency modulation technique. *Review of Scientific Instruments*, **77**(12).

Beauville, F., and The Virgo Collaboration. 2004. The Virgo large mirrors: a challenge for low loss coatings. *Classical and Quantum Gravity*, **21**(5), S935–S945.

Becker, Jurgen, and Scheuer, Volker. 1990. Coatings for optical applications produced by ion beam sputter deposition. *Applied Optics*, **29**(28), 4303–4309.

Bélanger, P.-A., and Paré, C. 1991. Optical resonators using graded-phase mirrors. *Optics Letters*, **16**(July), 1057–1059.

Bennett, Jean M., Pelletier, Emile, Albrand, G. *et al.* 1989. Comparison of the properties of titanium dioxide films prepared by various techniques. *Applied Optics*, **28**(16), 3303–3317.

Benthem, Bruin, and Levin, Yuri. 2009. Thermorefractive and thermochemical noise in the beamsplitter of the GEO600 gravitational-wave interferometer. *Physical Review D*, **80**(6), 062004.

Berry, B. S., and Pritchet, W. C. 1975. Vibrating reed internal friction apparatus for films and foils. *IBM Journal of Research and Development*, **19**, 334–343.

Berthold, J. W., and Jacobs, S. F. 1976. Ultraprecise thermal expansion measurements of seven low expansion materials. *Applied Optics*, **15**, 2344–2347.

Betzweiser, J., Kawabe, K., Rakhmanov, M., and Savage, R. 2005. *Summary of recent measurements of g factor changes induced by thermal loading in the H1 interferometer.* LIGO-G050111-00-W.

Beyersdorf, Peter. 2001. *The polarization Sagnac interferometer for gravitational wave detection.* Ph.D. thesis, Stanford University.

Bignotto, M., Bonaldi, M., Cerdonio, M. *et al.* 2008. Low temperature mechanical dissipation measurements of silicon and silicon carbide as candidate material for DUAL detector. *Journal of Physics: Conference Series*, **122**(1), 012030.

Binh, L. N., Netterfield, R. P., and Martin, P. J. 1985. Low-loss waveguiding in ion-assisted-deposited thin films. *Applications of Surface Science*, **22–23**(Part 2), 656–662.

Bize, S., Laurent, P., Abgrall, M. *et al.* 2005. Cold atom clocks and applications. *Journal of Physics B: Atomic, Molecular and Optical Physics*, **38**, S449–S468.

Bjorlin, E. S., Kimura, T., Chen, Q., Wang, C., and Bowers, J. E. 2004. High output power 1540 nm vertical cavity semiconductor optical amplifiers. *Electronics Letters*, **40**(2), 121–123.

Black, E. 2001. An introduction to Pound–Drever–Hall laser frequency stabilization. *American Journal of Physics*, **69**, 79–87.

Black, Eric D., Villar, Akira, Barbary, Kyle *et al.* 2004a. Direct observation of broadband coating thermal noise in a suspended interferometer. *Physics Letters A*, **328**, 1–5.

Black, Eric D., Grudinin, Ivan S., Rao, Shanti R., and Libbrecht, Kenneth G. 2004b. Enhanced photothermal displacement spectroscopy for thin-film characterization using a Fabry-Perot resonator. *Journal of Applied Physics*, **95**(12), 7655–7659.

Black, Eric D., Villar, Akira, and Libbrecht, Kenneth G. 2004c. Thermoelastic-damping noise from sapphire mirrors in a fundamental-noise-limited interferometer. *Physical Review Letters*, **93**(Dec), 241101.

Blair, David, and Munch, Jesper. 2009. The Australian international gravitational observatory. *Australian Physics*, **46**(4).

Blair, D., Cleva, F., and Man, C. N. 1997. Optical absorption measurements in monocrystalline sapphire at 1 μm. *Optical Materials*, **8**, 233–236.

Blanchet, Luc, Iyer, Bala R., Will, Clifford M., and Wiseman, Alan G. 1996. Gravitational waveforms from inspiralling compact binaries to second-post-Newtonian order. *Classical and Quantum Gravity*, **13**(4), 575.

Blatt, S., Ludlow, A. D., Campbell, G. K. *et al.* 2008. New limits on coupling of fundamental constants to gravity using ^{87}Sr optical lattice clocks. *Physical Review Letters*, **100**(14), 140801.

Boca, A., Miller, R., Birnbaum, K. M. *et al.* 2004. Observation of the vacuum Rabi spectrum for one trapped atom. *Physical Review Letters*, **93**(23), 233603.

Boggess, T., Smirl, A., Moss, S., Boyd, I., and Van Stryland, E. 1985. Optical limiting in GaAs. *IEEE Journal of Quantum Electronics*, **21**(5), 488–494.

Böhm, H. R. B, Gigan, S. A., Blaser, F. B. *et al.* 2006. High reflectivity high-Q micromechanical Bragg mirror. *Applied Physics Letters*, **89**(22), 223101.

Bömmel, H. E., Mason, W. P., and Warner, A. W. 1956. Dislocations, relaxations, and anelasticity of crystal quartz. *Physical Review*, **102**, 64–71.

Bondarescu, Mihai, and Thorne, Kip S. 2006. New family of light beams and mirror shapes for future LIGO interferometers. *Physical Review D*, **74**, 082003.

Bondarescu, M., Kogan, O., and Chen, Y. 2008. Optimal light beams and mirror shapes for future LIGO interferometers. *Physical Review D*, **78**(8), 082002.

Bondu, François, Hello, Patrice, and Vinet, Jean-Yves. 1998. Thermal noise in mirrors of interferometric gravitational wave antennas. *Physics Letters A*, **246**(3–4), 227–236.

Bondu, R., Fritschel, R., Man, C. N., and Brillet, A. 1996. Ultrahigh-spectral-purity laser for the Virgo experiment. *Optics Letters*, **21**, 582–584.

Bongs, K., Burger, S., Dettmer, S. *et al.* 2001. Waveguide for Bose–Einstein condensates. *Physical Review A*, **63**(3), 031602.

Boozer, A. D., Boca, A., Miller, R., Northup, T. E., and Kimble, H. J. 2006. Cooling to the ground state of axial motion for one atom strongly coupled to an optical cavity. *Physical Review Letters*, **97**, 083602.

Boozer, A. D., Boca, A., Miller, R., Northup, T. E., and Kimble, H. J. 2007. Reversible state transfer between light and a single trapped atom. *Physical Review Letters*, **98**, 193601.

Borisenko, K. B., Chen, Y., Song, S. A., Nguyen-Manh, D., and Cockayne, D. J. H. 2009a. A concerted rational crystallization/amorphization mechanism of $Ge_2Sb_2Te_5$. *Journal of Non-Crystalline Solids*, **355**(43–44), 2122–2126.

Borisenko, Konstantin B., Chen, Yixin, Song, Se Ahn, and Cockayne, David J. H. 2009b. Nanoscale phase separation and building blocks of $Ge_2Sb_2Te_5N$ and $Ge_2Sb_2Te_5N_2$ thin films. *Chemistry of Materials*, **21**, 5244–5251.

Born, Max, and Wolf, Emil. 1999. *Principles of Optics: Electromagnetic Theory of Propagation, Interference and Diffraction of Light (7th Edition)*. Cambridge University Press.

Braccini, S., and The Virgo Collaboration. 2005. Measurement of the seismic attenuation performance of the Virgo Superattenuator. *Astroparticle Physics*, **23**(6), 557–565.

Brady, Gregory R., Ellis, A. Robert, Moehring, David L. *et al.* 2011. Integration of fluorescence collection optics with a microfabricated surface electrode ion trap. *Applied Physics B*, **103**(4), 801–808.

Braginsky, V. B., and Khalili, F. Ya. 1992. *Quantum Measurement*. Cambridge University Press.

Braginsky, V. B., and Vyatchanin, S. P. 2002. Low quantum noise tranquilizer for Fabry–Perot interferometer. *Physics Letters A*, **293**, 228–234.

Braginsky, V. B., and Vyatchanin, S. P. 2003a. Thermodynamical fluctuations in optical mirror coatings. *Physics Letters A*, **312**(3–4), 244–255.

Braginsky, V. B., and Vyatchanin, S. P. 2003b. Thermodynamical fluctuations in optical mirror coatings. ArXiv:cond-mat/0302617 v5.

Braginsky, V. B., and Vyatchanin, S. P. 2004. Corner reflectors and quantum-non-demolition measurements in gravitational wave antennae. *Physics Letters A*, **324**(1), 345–360.

Braginsky, V. B., Manukin, A. B., and Tikhonov, M. Y. 1970. Investigation of dissipative ponderomotive effects of electromagnetic radiation. *Soviet Physics JETP*, **31**, 829.

Braginsky, V. B., Vyatchanin, S. P., and Panov, V. I. 1979. On the ultimate stability of frequency in self-oscillators. *Soviet Physics-Doklady*, **247**, 583–586.

Braginsky, V. B., Mitrofanov, V. P., and Panov, V. I. 1985. *Systems with Small Dissipation*. University of Chicago Press.

Braginsky, V. B., Gorodetsky, M. L., and Vyatchanin, S. P. 1999. Thermodynamical fluctuations and photo-thermal shot noise in gravitational wave antennae. *Physics Letters A*, **264**, 1–10.

Braginsky, V. B., Gorodetsky, M. L., and Vyatchanin, S. P. 2000. Thermo-refractive noise in gravitational wave antennae. *Physics Letters A*, **271**(5–6), 303–307.

Braginsky, V. B., Khalili, F. Ya., and Volikov, P. S. 2001. The analysis of table-top quantum measurement with macroscopic masses. *Physics Letters A*, **287**(1–2), 31–38.

Braginsky, V. B., Gorodetsky, M. L., Khalili, F. Ya. *et al.* 2003. The noise in gravitational-wave detectors and other classical-force measurements is not influenced by test-mass quantization. *Physical Review D*, **67**, 082001.

Braginsky, V. B., Ryazhskaya, O. G., and Vyatchanin, S. P. 2006a. Limitations in quantum measurements resolution created by cosmic rays. *Physics Letters A*, **359**, 86–89.

Braginsky, V. B., Ryazhskaya, O. G., and Vyatchanin, S. P. 2006b. Notes about noise in gravitational wave antennas created by cosmic rays. *Physics Letters A*, **350**(1–2), 1–4.

Braxmaier, C., Müller, H., Pradl, O. *et al.* 2002. Tests of relativity using a cryogenic optical resonator. *Physical Review Letters*, **88**, 010401.

Brennecke, Ferdinand, Donner, Tobias, Ritter, Stephan. *et al.* 2007. Cavity QED with a Bose-Einstein condensate. *Nature*, **450**(7167), 268–271.

Brennecke, Ferdinand, Ritter, Stephan, Donner, Tobias, and Esslinger, Tilman. 2008. Cavity optomechanics with a Bose–Einstein condensate. *Science*, **322**(5899), 235–238.

Brillet, A., and Hall, J. L. 1979. Improved laser test of the isotropy of space. *Physical Review Letters*, **42**(9), 549–552.

Brodoceanu, D., Cole, G. D., Kiesel, N., Aspelmeyer, M., and Baeuerle, D. 2010. Femtosecond laser fabrication of high reflectivity micromirrors. *Applied Physics Letters*, **97**(4), 041104.

Brooks, Aidan F., Hosken, David, Munch, Jesper *et al.* 2009. Direct measurement of absorption-induced wavefront distortion in high optical power systems. *Applied Optics*, **48**(2), 355–364.

Brown, R. 1828. A brief account of microscopical observations made in the months of June, July and August, 1827, on the particles contained in the pollen of plants; and on the general existence of active molecules in organic and inorganic bodies. *Philosophical Magazine*, **4**, 161–173.

Brown, R. 1970. *Handbook of Thin Film Technology*. McGraw Hill. Chap. The nature of physical sputtering.

Brückner, Frank, Clausnitzer, Tina, Burmeister, Oliver *et al.* 2008. Monolithic dielectric surfaces as new low-loss light-matter interfaces. *Optics Letters*, **33**(3), 264–266.

Brückner, Frank, Friedrich, Daniel, Clausnitzer, Tina *et al.* 2009. Demonstration of a cavity coupler based on a resonant waveguide grating. *Optics Express*, **17**(1), 163–169.

Brückner, Frank, Friedrich, Daniel, Clausnitzer, Tina *et al.* 2010. Realization of a monolithic high-reflectivity cavity mirror from a single silicon crystal. *Physical Review Letters*, **104**(16), 163903.

Buck, Joseph R. 2003. *Cavity QED in microsphere and Fabry–Perot cavities*. Ph.D. thesis, California Institute of Technology, Pasadena, CA.

Bunkowski, A., Burmeister, O., Beyersdorf, P. *et al.* 2004. Low-loss grating for coupling to a high-finesse cavity. *Optics Letters*, **29**(20), 2342–2344.

Bunkowski, A., Burmeister, O., Friedrich, D., Danzmann, K., and Schnabel, R. 2006. High reflectivity grating waveguide coatings for 1064 nm. *Classical and Quantum Gravity*, **23**(24), 7297–7303.

Buonanno, Alessandra and Chen, Yanbei. 2001. Quantum noise in second generation, signal-recycled laser interferometric gravitational-wave detectors. *Physical Review D*, **64**(4), 042006.

Burmeister, Oliver, Britzger, Michael, Thüring, André *et al.* 2010. All-reflective coupling of two optical cavities with 3-port diffraction gratings. *Optics Express*, **18**(9), 9119–9132.

Buzea, Cristina, and Robbie, Kevin. 2005. State of the art in thin film thickness and deposition rate monitoring sensors. *Reports on Progress in Physics*, **68**(2), 385–409.

Callen, Herbert B., and Greene, Richard F. 1952. On a theorem of irreversible thermodynamics. *Physical Review*, **86**, 702–710.

Callen, Herbert B., and Welton, Theodore A. 1951. Irreversibility and generalized noise. *Physical Review*, **83**(1), 34–40.

Camp, Jordan, Billingsley, Garilynn, Kells, William P. *et al.* 2002. LIGO optics: initial and advanced. Page 1 of: *Proceedings of the SPIE*, vol. 4679.

Campbell, G. K., Ludlow, A. D., Blatt, S. *et al.* 2008. The absolute frequency of the ^{87}Sr optical clock transition. *Metrologia*, **45**, 539–548.

Campbell, G. K., Boyd, M. M., Thomsen, J. W. *et al.* 2009. Probing interactions between ultracold fermions. *Science*, **324**(5925), 360–363.

Caparrelli, S., Majorana, E., Moscatelli, V. *et al.* 2006. Vibration-free cryostat for low-noise applications of a pulse tube cryocooler. *Review of Scientific Instruments*, **77**(9), 095102.

Carmon, Tal, Kippenberg, Tobias, Yang, Lan *et al.* 2005. Feedback control of ultra-high-Q microcavities: Application to micro-Raman lasers and microparametric oscillators. *Optics Express*, **13**(9), 3558–3566.

Caves, Carlton M. 1981. Quantum-mechanical noise in an interferometer. *Physical Review D*, **23**(8), 1693–1708.

Caves, Carlton M., Thorne, Kip S., Drever, Ronald W. P., Sandberg, Vernon D., and Zimmermann, Mark. 1980. On the measurement of a weak classical force coupled to a quantum-mechanical oscillator. I. Issues of principle. *Reviews of Modern Physics*, **52**(2), 341–392.

Cerdonio, M., Conti, L., Heidmann, A., and Pinard, M. 2001. Thermoelastic effects at low temperatures and quantum limits in displacement measurements. *Physical Review D*, **63**, 082003.

Chan, Hilton W., Black, Adam T., and Vuletić, Vladan. 2003. Observation of collective-emission-induced cooling of atoms in an optical cavity. *Physical Review Letters*, **90**(6), 063003.

Chan, J., Mayer, Alegre, T. P., Satvi-Naeini, A. H. *et al.* 2011. Laser cooling of a nano mechanical oscillator into its quantum ground state. an XIV:1106.3614.

Chaneliere, C., Autran, J. L., Devine, R. A. B., and Balland, B. 1998. Tantalum pentoxide (Ta_2O_5) thin films for advanced dielectric applications. *Materials Science and Engineering: R: Reports*, **22**(6), 269–322.

Chao, Shiuh, Chang, Cheng-Kuel, and Chen, Jyh-Shin. 1991. TiO_2–SiO_2 mixed films prepared by the fast alternating sputter method. *Applied Optics*, **30**(22), 3233–3237.

Chao, Shiuh, Wang, Wen-Hsiang, Hsu, Min-Yu, and Wang, Liang-Chu. 1999. Characteristics of ion-beam-sputtered high-refractive-index TiO_2–SiO_2 mixed films. *Journal of the Optical Society of America A*, **16**(6), 1477–1483.

Chao, Shiuh, Wang, Wen-Hsiang, and Lee, Cheng-Chung. 2001. Low-loss dielectric mirror with ion-beam-sputtered TiO_2–SiO_2 mixed films. *Applied Optics*, **40**(13), 2177–2182.

Charbonneau, P. 2002. *An Introduction to Genetic Algorithms for Numerical Optimization.* NCAR Technical Note TN-450+IA.

Chelkowski, Simon, Hild, Stefan, and Freise, Andreas. 2009. Prospects of higher-order Laguerre–Gauss modes in future gravitational wave detectors. *Physical Review D*, **79**(12), 122002.

Chen, Jyh-Shin, Chao, Shiuh, Kao, Jiann-Shiun, Niu, Huan, and Chen, Chih-Hsin. 1996. Mixed films of TiO_2–SiO_2 deposited by double electron-beam coevaporation. *Applied Optics*, **35**(1), 90–96.

Chen, L., Hall, J. L., Ye, J. *et al.* 2006a. Vibration-induced elastic deformation of Fabry-Perot cavities. *Physical Review A*, **74**, 053801.

Chen, Yanbei, and Kawamura, Seiji. 2006. Displacement- and timing-noise-free gravitational-wave detection. *Physical Review Letters*, **96**(23), 231102.

Chen, Yanbei, Pai, Archana, Somiya, Kentaro *et al.* 2006b. Interferometers for displacement-noise-free gravitational-wave detection. *Physical Review Letters*, **97**(15), 151103.

Chou, C. W., Hume, D. B., Koelemeij, J. C. J., Wineland, D. J., and Rosenband, T. 2010. Frequency comparison of two high-accuracy Al^+ optical clocks. *Physical Review Letters*, **104**(7), 070802.

Chu, A., Lin, H., and Cheng, W. 1997. Temperature dependence of refractive index of Ta_2O_5 dielectric films. *Journal of Electronic Materials*, **26**, 889–892. 10.1007/s11664-997-0269-3.

Church, Eugene L. 1988. Fractal surface finish. *Applied Optics*, **27**(8), 1518–1526.

Cimma, B., Forest, D., Ganau, P. *et al.* 2006. Ion beam sputtering coatings on large substrates: Toward an improvement of the mechanical and optical performances. *Applied Optics*, **45**(Mar.), 1436–1439.

Clausnitzer, Tina, Kley, E.-B., Tünnermann, A. *et al.* 2005. Ultra low-loss low-efficiency diffraction gratings. *Optics Express*, **13**(12), 4370–4378.

Cleland, A. N., and Geller, M. R. 2004. Superconducting qubit storage and entanglement with nanomechanical resonators. *Physical Review Letters*, **93**, 070501.

Cockayne, D. J. H. 2009. The study of nanovolumes of amorphous materials using electron scattering. *Annual Review of Materials Research*, **37**, 159–187.

Cohadon, P. F., Heidmann, A., and Pinard, M. 1999. Cooling of a mirror by radiation pressure. *Physical Review Letters*, **83**(16), 3174–3177.

Cole, G. D., Bjorlin, E. S., Chen, Qi *et al.* 2005. MEMS-tunable vertical-cavity SOAs. *IEEE Journal of Quantum Electronics*, **41**(3), 390–407.

Cole, G. D., Groblacher, S., Gugler, K., Gigan, S., and Aspelmeyer, M. 2008. Monocrystalline $Al_x Ga_{1-x}$As heterostructures for high-reflectivity high-Q micromechnical resonators in the megahertz regime. *Applied Physics Letters*, **92**, 261108.

Cole, G. D., Wilson-Rae, I., Werbach, K., Vanner, M. R., and Aspelmeyer, M. 2010a. Minimization of phonon tunneling dissipation in mechanical resonators. arXiv:1007.4948.

Cole, Garrett D., Bai, Yu, Aspelmeyer, Markus, and Fitzgerald, Eugene A. 2010b. Freestanding $Al_x Ga_{1-x}$As heterostructures by gas-phase etching of germanium. *Applied Physics Letters*, **96**(26), 261102.

Cole, G. D., Wilson-Rae, I., Vanner, M. R. *et al.* 2010c (Jan.). Megahertz monocrystalline optomechanical resonators with minimal dissipation. Pages 847–850 of: *2010 IEEE 23rd International Conference on Micro Electro Mechanical Systems (MEMS)*.

Cole, Garrett D., Wilson-Rae, Ignacio, Werbach, Katharina, Vanner, Michael R., and Aspelmeyer, Markus. 2011. Phonon-tunnelling dissipation in mechanical resonators. *Nature Communications*, **2**, 231.

Colombe, Y., Steinmetz, T., Dubois, G. *et al.* 2007. Strong atom–field coupling for Bose–Einstein condensates in an optical cavity on a chip. *Nature*, **450**, 272–276.

Commandre, M., and Roche, P. 1996. Characterization of optical coatings by photothermal deflection. *Applied Optics*, **35**(25), 5021–5034.

Conti, L., Rosa, M. D., and Marin, F. 2003. High-spectral-purity laser system for the AURIGA detector optical readout. *Journal of the Optical Society of America*, **20**, 462–468.

Corbitt, Thomas, Wipf, Christopher, Bodiya, Timothy *et al.* 2007. Optical dilution and feedback cooling of a gram-scale oscillator to 6.9 mK. *Physical Review Letters*, **99**(16), 160801.

Crooks, D. R. M., Sneddon, P, Cagnoli, G. *et al.* 2002. Excess mechanical loss associated with dielectric mirror coatings on test masses in interferometric gravitational wave detectors. *Classical and Quantum Gravity*, **19**(5), 883–896.

Crooks, D. R. M., Cagnoli, G., Fejer, M. M. *et al.* 2004. Experimental measurements of coating mechanical loss factors. *Classical and Quantum Gravity*, **21**(5), S1059–S1065.

Crooks, D. R. M., Cagnoli, G., Fejer, M. M. *et al.* 2006. Experimental measurements of mechanical dissipation associated with dielectric coatings formed using SiO_2, Ta_2O_5 and Al_2O_3. *Classical and Quantum Gravity*, **23**(15), 4953–4965.

Cundiff, S. T., and Ye, J. 2003. Colloquium: Femtosecond optical frequency combs. *Reviews of Modern Physics*, **75**(1), 325–342.

Cunningham, L., Murray, P. G., Cumming, A. *et al.* 2010. Re-evaluation of the mechanical loss factor of hydroxide-catalysis bonds and its significance for the next generation of gravitational wave detectors. *Physics Letters A*, **374**(39), 3993–3998.

Cuthbertson, B. D., Tobar, M. E., Fvanov, E. N., and Blair, D. G. 1996. Parametric backaction effects in a high-Q cryogenic sapphire transducer. *Review of Scientific Instruments*, **67**, 2435–2442.

Cutler, Curt, Apostolatos, Theocharis A., Bildsten, Lars *et al.* 1993. The last three minutes: Issues in gravitational-wave measurements of coalescing compact binaries. *Physical Review Letters*, **70**(20), 2984–2987.

D'Ambrosio, E. 2003. Nonspherical mirrors to reduce thermoelastic noise in advanced gravitational wave interferometers. *Physical Review D*, **67**(10), 102004.

D'Ambrosio, E., O'Shaughnessy, R., Thorne, K. *et al.* 2004a. Advanced LIGO: non-Gaussian beams. *Classical and Quantum Gravity*, **21**(Mar.), 867–873.

D'Ambrosio, E., O'Shaughnessy, R., Strigin, S., Thorne, K. S., and Vyatchanin, S. 2004b (Sept.). *Reducing thermoelastic noise in gravitational-wave interferometers by flattening the light beams.* arXiv:gr-qc/0409075.

Davies, John H. 1998. *The Physics of Low-dimensional Semiconductors.* Cambridge University Press.

Day, T., Gustafson, E. K., and Byer, R. L. 1992. Sub-Hertz relative frequency stabilization of two-diode laser-pumped Nd:YAG lasers locked to a Fabry–Perot interferometer. *IEEE Journal of Quantum Electronics*, **28**(4), 1106–1117.

De Rosa, M., Conti, L., Cerdonio, M., Pinard, M., and Marin, F. 2002. Experimental measurement of the dynamic photothermal effect in Fabry–Perot cavities for gravitational wave detectors. *Physical Review Letters*, **89**(23), 237402.

de Silvestri, S., Laporta, P., Magni, V., Svelto, O., and Majocchi, B. 1988. Unstable laser resonators with super-Gaussian mirrors. *Optics Letters*, **13**(Mar.), 201–203.

Demiryont, H., Sites, James R., and Geib, Kent. 1985. Effects of oxygen content on the optical properties of tantalum oxide films deposited by ion-beam sputtering. *Applied Optics*, **24**(4), 490–495.

Demiryont, Hulya. 1985. Optical properties of SiO_2–TiO_2 composite films. *Applied Optics*, **24**(16), 2647–2650.

Diddams, S. A., Udem, T., Bergquist, J. C. *et al.* 2001. An optical clock based on a single trapped $^{199}Hg^+$ ion. *Science*, **293**(5531), 825–828.

Dobkin, D. M., and Zuraw, M. K. 2003. *Principles of Chemical Vapor Deposition: What's Going on Inside the Reactor.* Kluwer Academic.

Doremus, R. H. 1979. *Treatise on Material Science and Technology.* Academic.

Dorsel, A., McCullen, J. D., Meystre, P., Vignes, E., and Walther, H. 1983. Optical bistability and mirror confinement induced by radiation pressure. *Physical Review Letters*, **51**(17), 1550–1553.

Drever, R. W. P, Hall, J. L., Kowalski, F. V. *et al.* 1983. Laser phase and frequency stabilization using an optical resonator. *Applied Physics B*, **31**, 97–105.

Dubin, F., Russo, C., Barros, H. G. *et al.* 2010. Quantum to classical transition in a single-ion laser. *Nature Physics*, **6**(5), 350–353.

Duwel, Amy, Candler, Rob N., Kenny, Thomas W., and Varghese, Mathew. 2006. Engineering MEMS resonators with low thermoelastic damping. *Journal of Microelectromechanical Systems*, **15**(6), 1437–1445.

Edgar, M. P., Barr, B. W., Nelson, J. *et al.* 2010. Experimental demonstration of a suspended, diffractively coupled Fabry–Perot cavity. *Classical and Quantum Gravity*, **27**(8), 084029.

Eichenfield, Matt, Chan, Jasper, Camacho, Ryan M., Vahala, Kerry J., and Painter, Oskar. 2009. Optomechanical crystals. *Nature*, **462**(7269), 78–82.

Einstein, A. 1905. On the movement of small particles suspended in a stationary liquid demanded by the molecular-kinetic theory of heat. *Annalen der Physik*, **17**, 549–560.

Einstein, A. 1915. Zur allgemeinen Relativitätstheorie. *Preussische Akademie der Wissenschaften, Sitzungsberichte*, 778–786.

Einstein, A. 1916. Grundlage der allgemeinen Relativitätstheorie. *Annalen der Physik*, **49**, 769–822.

Eisele, C., Nevsky, A. Y., and Schiller, S. 2009. Laboratory test of the isotropy of light propagation at the 10^{-17} level. *Physical Review Letters*, **103**(9), 090401.

Eisele, Ch., Okhapkin, M., Nevsky, A.Yu., and Schiller, S. 2008. A crossed optical cavities apparatus for a precision test of the isotropy of light propagation. *Optics Communications*, **281**(5), 1189–1196.

Elson, J. M., and Bennett, J. M. 1979. Relation between the angular dependence of scattering and the statistical properties of optical surfaces. *Journal of the Optical Society of America*, **69**(1), 31–47.

Ernsting, I. 2009. *Entwicklung und Anwendung eines Frequenzkamm-basierten Lasersystems für die Präzisions-Spektroskopie an ultrakalten Molekülen und Atomen.* Ph.D. thesis, Heinrich-Heine-Universität Düsseldorf, Germany.

Evans, M., Ballmer, S., Fejer, M. *et al.* 2008. Thermo-optic noise in coated mirrors for high-precision optical measurements. *Physical Review D*, **78**(10), 102003.

Exner, F. M. 1900. Notiz zu Brown's molecularbewegung. *Annalen der Physik*, **2**, 843.

Fabre, C., Pinard, M., Bourzeix, S. *et al.* 1994. Quantum-noise reduction using a cavity with a movable mirror. *Physical Review A*, **49**(2), 1337–1343.

Favero, Ivan, and Karrai, Khaled. 2009. Optomechanics of deformable optical cavities. *Nature Photonics*, **3**, 201–205.

Fejer, M. M., Rowan, S., Cagnoli, G. *et al.* 2004. Thermoelastic dissipation in inhomogeneous media: Loss measurements and displacement noise in coated test masses for interferometric gravitational wave detectors. *Physical Review D*, **70**(8), 082003.

Ferreirinho, J. 1991. Internal friction in high Q materials. Pages 116–168 of: *The Detection of Gravitational Waves*. Cambridge University Press.

Fine, M. E., van Duyne, H., and Kenney, Nancy T. 1954. Low-temperature internal friction and elasticity effects in vitreous silica. *Journal of Applied Physics*, **25**, 402–405.

Flaminio, R., Franc, J., Michel, C. *et al.* 2010. A study of coating mechanical and optical losses in view of reducing mirror thermal noise in gravitational wave detectors. *Classical and Quantum Gravity*, **27**, 084030.

Flanagan, Eanna, and Thorne, Kip. 1995. *Scattered-light noise for LIGO*. T950102-00-R.

Forbes, A., Du Preez, N. C., Belyi, V., and Botha, L. R. 2009 (Aug.). Paint stripping with high power flattened Gaussian beams. In: *Society of Photo-Optical Instrumentation Engineers (SPIE) Conference Series*. Presented at the Society of Photo-Optical Instrumentation Engineers (SPIE) Conference, vol. 7430.

Foreman, Seth M., Ludlow, Andrew D., de Miranda, Marcio H. G. *et al.* 2007. Coherent optical phase transfer over a 32 km fiber with 1 s instability at 10^{-17}. *Physical Review Letters*, **99**(15), 153601.

Fortier, Kevin M., Kim, Soo Y., Gibbons, Michael J., Ahmadi, Peyman, and Chapman, Michael S. 2007a. Deterministic loading of individual atoms to a high-finesse optical cavity. *Physical Review Letters*, **98**(23), 233601.

Fortier, T. M., Ashby, N., Bergquist, J. C. *et al.* 2007b. Precision atomic spectroscopy for improved limits on variation of the fine structure constant and local position invariance. *Physical Review Letters*, **98**(7), 070801.

Fox, R. W. 2009. Temperature analysis of low-expansion Fabry-Perot cavities. *Optics Express*, **17**(17), 15 023–15 031.

Franc, J., Morgado, N., Flaminio, R. *et al.* 2009. Mirror thermal noise in laser interferometer gravitational wave detectors operating at room and cryogenic temperature. *ArXiv 0912.0107*, Dec.

Franc, Janyce, Galimberti, Massimo, Flaminio, Raffaele *et al.* 2010. *Role of high-order Laguerre-Gauss modes on mirror thermal noise in gravitational wave detectors*. ET note ET-0002A-09. Einstein Telescope.

Freise, A., and Strain, K. 2010. Interferometer techniques for gravitational-wave detection. *Living Reviews in Relativity*, **13**(Feb.).

Freise, A., Bunkowski, A., and Schnabel, R. 2007. Phase and alignment noise in grating interferometers. *New Journal of Physics*, **9**(12), 433.

Friedrich, Daniel, Burmeister, Oliver, Bunkowski, Alexander *et al.* 2008. Diffractive beam splitter characterization via a power-recycled interferometer. *Optics Letters*, **33**(2), 101–103.

Fritschel, Peter. 2006. *Backscattering from the AS port: Enhanced and Advanced LIGO.* LIGO-T060303-01.

Fritschel, P., and Zucker, M. E. 2010. *Wide-angle scatter from LIGO arm cavities.* LIGO-T070089.

Friz, M., and Waibel, F. 2003. *Optical Interference Coatings.* Springer-Verlag. Chap. Coating materials.

Fujiwara, Hiroyuki. 2007. *Spectroscopic Ellipsometry: Principles and Applications.* John Wiley and Sons.

Fulda, P., Kokeyama, K., Chelkowski, S., and Freise, A. 2010. Experimental demonstration of higher-order Laguerre-Gauss mode interferometry. *arXiv:1005.2990F*, May.

Galdi, V., Castaldi, G., Pierro, V. *et al.* 2006. Analytic structure of a family of hyperboloidal beams of potential interest for advanced LIGO. *Physical Review D,* **73**(12), 127101.

Genes, C., Vitali, D., Tombesi, P., Gigan, S., and Aspelmeyer, M. 2008a. Ground-state cooling of a micromechanical oscillator: Comparing cold damping and cavity-assisted cooling schemes. *Physical Review A,* **77**(3), 033804.

Genes, C., Mari, A., Tombesi, P., and Vitali, D. 2008b. Robust entanglement of a micromechanical resonator with output optical fields. *Physical Review A,* **78**(3), 032316.

Genes, C., Mari, A., Vitali, D., and Tombesi, P. 2009. Quantum effects in optomechanical systems. *Advances in Atomic, Molecular, and Optical Physics,* **57**, 33–86.

Gibson, Graham, Courtial, Johannes, Padgett, Miles *et al.* 2004. Free-space information transfer using light beams carrying orbital angular momentum. *Optics Express,* **12**(22), 5448–5456.

Gigan, S., Boehm, H. R., Paternostro, M. *et al.* 2006. Self-cooling of a micromirror by radiation pressure. *Nature,* **444**(7115), 67–70.

Gillespie, A., and Raab, F. 1995. Thermally excited vibrations of the mirrors of laser interferometer gravitational-wave detectors. *Physical Review D,* **52**(2), 577–585.

Gilroy, K. S., and Phillips, W. A. 1981. An asymmetric double-well potential model for structural relaxation processes in amorphous materials. *Philosophical Magazine B,* **43**, 735–746.

Girvin, S. M., Devoret, M. H., and Schoelkopf, R. J. 2009. Circuit QED and engineering charge-based superconducting qubits. *Physica Scripta,* **2009**(T137), 014012.

Goda, K., Miyakawa, O., Mikhailov, E. E. *et al.* 2008. A quantum-enhanced prototype gravitational-wave detector. *Nature Physics,* **4**(6), 472–476.

Gohle, C., Udem, T., Herrmann, M. *et al.* 2005. A frequency comb in the extreme ultraviolet. *Nature,* **436**, 234–237.

Gonzalez, Gabriela I., and Saulson, Peter R. 1994. Brownian motion of a mass suspended by an anelastic wire. *The Journal of the Acoustical Society of America,* **96**(1), 207–212.

Gonzalez, Gabriela I., and Saulson, Peter R. 1995. Brownian motion of a torsion pendulum with internal friction. *Physics Letters A,* **201**(1), 12–18.

Gori, F. 1994. Flattened gaussian beams. *Optics Communications,* **107**(May), 335–341.

Gorodetsky, Michael L. 2008. Thermal noises and noise compensation in high-reflection multilayer coating. *Physics Letters A,* **372**(46), 6813–6822.

Gorodetsky, M. L., and Grudinin, I. S. 2004. Fundamental thermal fluctuations in microspheres. *Journal of the Optical Society of America B,* **21**, 697–705.

Goßler, S., Bertolini, A., Born, M. *et al.* 2010. The AEI 10 m prototype interferometer. *Classical and Quantum Gravity,* **27**(8), 084023.

Gouy, M. 1888. Note sur le mouvement brownien. *Journal de Physique,* **7**, 561.

Green, J. E., Barnett, S. A., Sundgren, J. E., and Rockett, A. 1989. *Ion Beam Assisted Film Growth*. Elsevier. Chap. Low-energy ion/surface interaction during film growth from the vapor phase.

Green M. A., and Keevers, M. J. 1995. Optical properties of intrinsic silicon at 300 K. *Progress in Photovoltaics: Research and Applications*, **3**(3), 189–192.

Greene, Richard F., and Callen, Herbert B. 1951. On the formalism of thermodynamic fluctuation theory. *Physical Review*, **83**(Sep.), 1231–1235.

Greenhall, C. A. 1997 (May). Does Allan variance determine the spectrum? Pages 358–365 of: *Proceedings of the 1997 IEEE International Frequency Control Symposium*.

Gretarsson, Andri. 2008. *Thermo-optic noise from doped tantala/silica coatings*. LIGO-G080151-00-Z.

Gretarsson, A., and Harry, G. 1999. Dissipation of mechanical energy in used silica fibres. *Review of Scientific Instruments*, **70**(10), 4081–4087.

Gröblacher, S., Gigan, S., Böhm, H. R., Zeilinger, A., and Aspelmeyer, M. 2008. Radiation-pressure self-cooling of a micromirror in a cryogenic environment. *Europhysics Letters*, **81**, 54003.

Gröblacher, Simon, Hertzberg, J. B., Vanner, M. R. *et al.* 2009a. Demonstration of an ultracold micro-optomechanical oscillator in a cryogenic cavity. *Nature Physics*, **5**, 485–488.

Gröblacher, Simon, Hammerer, Klemens, Vanner, Michael R., and Aspelmeyer, Markus. 2009b. Observation of strong coupling between a micromechanical resonator and an optical cavity field. *Nature*, **460**(7256), 724–727.

Gurkovsky, A., and Vyatchanin, S. 2010. The thermal noise in multilayer coating. *Physics Letters A*, **374**, 3267–3274.

Guthöhrlein, G. R., Keller, M., Hayasaka, K., Lange, W., and Walther, H. 2001. A single ion as a nanoscopic probe of an optical field. *Nature*, **414**, 49–51.

Hadjar, Y., Cohadon, P. F., Aminoff, C. G., Pinard, M., and Heidmann, A. 1999. High-sensitivity optical measurement of mechanical Brownian motion. *Europhysics Letters*, **47**(5), 545–551.

Häffner, H., Roos, C. F., and Blatt, R. 2008. Quantum computing with trapped ions. *Physics Reports*, **469**(4), 155–203.

Hallam, J., Chelkowski, S., Freise, A. *et al.* 2009. Coupling of lateral grating displacement to the output ports of a diffractive FabryPerot cavity. *Journal of Optics A: Pure and Applied Optics*, **11**(8), 085502.

Hammerer, K., Wallquist, M., Genes, C. *et al.* 2009. Strong coupling of a mechanical oscillator and a single atom. *Physical Review Letters*, **103**(6), 063005.

Hao, Honggang, and Li, Bincheng. 2008. Photothermal detuning for absorption measurement of optical coatings. *Applied Optics*, **47**(2), 188–194.

Harper, J. M. E. 1984. *Sputter Deposition and Ion Beam Process*. American Vacuum Society. Chap. Ion beam application to thin films.

Harper, J. M. E., Cuomo, J. J., and Kaufman, H. R. 1982. Technology and applications of broad-beam ion sources used in sputtering. Part II. Applications. *Journal of Vacuum Science and Technology*, **21**(3), 737–756.

Harry, G. 2004. *Optical Coatings for Gravitational Wave Detection*. LIGO-G040434-00-R.

Harry, G. M., Gretarsson, A. M., Saulson, P. R. *et al.* 2002. Thermal noise in interferometric gravitational wave detectors due to dielectric optical coatings. *Classical and Quantum Gravity*, **19**, 897–917.

Harry, G. M., Crooks, D. R. M., Cagnoli, G. *et al.* 2006a. Thermal noise from optical coatings in gravitational wave detectors. *Applied Optics*, **45**, 1569–1574.

Harry, Gregory M., Fritschel, Peter, Shaddock, Daniel A., Folkner, William, and Phinney, E. Sterl. 2006b. Laser interferometry for the Big Bang Observer. *Classical and Quantum Gravity*, **23**(15), 4887–4894.

Harry, G. M., Abernathy, M. R., Becerra-Toledo, A. *et al.* 2007. Titania-doped tantala/silica coatings for gravitational-wave detection. *Classical and Quantum Gravity*, **24**, 405–415.

Harry, Gregory, and The LIGO Scientific Collaboration. 2010. Advanced LIGO: The next generation of gravitational wave detectors. *Classical and Quantum Gravity*, **27**(Apr.), 084006.

Hartle, James. 2003. *Gravity: An Introduction to Einstein's General Relativity*. Boston: Addison-Wesley.

Hartmann, J. 1900. Bemerkungen uber den Bau und die Justirung von Spektrographen. *Zeitschrift für Instrumentenkunde*, **20**, 47–58.

Heavner, T. P., Jefferts, S. R., Donley, E. A., Shirley, J. H., and Parker, T. E. 2005. NIST-F1: recent improvements and accuracy evaluations. *Metrologia*, **42**, 411–422.

Hello, P., and Vinet, J. Y. 1990a. Analytical models of thermal aberrations in massive mirrors heated by high-power laser-beams. *Journal de Physique*, **51**, 1267–1282.

Hello, P., and Vinet, J. Y. 1990b. Analytical models of transient thermoelastic deformations of mirrors heated by high-power cw laser-beams. *Journal de Physique*, **51**, 1243–1261.

Heptonstall, A., Cagnoli, G., Hough, J., and Rowan, S. 2006. Characterisation of mechanical loss in synthetic fused silica ribbons. *Physics Letters A*, **354**, 353–359.

Herman, M. A., and Sitter, H. 1989. *Molecular Beam Epitaxy, Fundamentals and Current Status*. Springer-Verlag, Berlin.

Herrmann, S., Senger, A., Möhle, K. *et al.* 2009. Rotating optical cavity experiment testing Lorentz invariance at the 10^{-17} level. *Physical Review D*, **80**(10), 105011.

Herskind, P. F., Dantan, A., Marler, J. P., Albert, M., and Drewsen, M. 2009. Realization of collective strong coupling with ion Coulomb crystals in an optical cavity. *Nature Physics*, **5**(7), 494–498.

Herskind, P. F., Wang, S. X., Shi, M. 2011. Microfabricated surface trap for scalable ion–photon interfaces. *Optics Letters*, **36**(16), 3045–3047.

Hijlkema, M., Weber, B., Specht, H. P. *et al.* 2007. A single-photon server with just one atom. *Nature Physics*, **3**, 253–255.

Hild, Stefan. 2007. *Beyond the first generation: extending the science range of the gravitational wave detector GEO 600*. Ph.D. thesis, University of Hannover.

Hillion, P. 1994. Gaussian beam at a dielectric interface. *Journal of Optics*, **25**, 155–164.

Hils, D., and Hall, J. L. 1987. Response of a Fabry–Pérot cavity to phase modulated light. *Review of Scientific Instruments*, **58**, 1406–1412.

Hils, D., and Hall, J. L. 1990. Improved Kennedy–Thorndike experiment to test special relativity. *Physical Review Letters*, **64**(15), 1697–1700.

Hirakawa, Hiromasa, and Narihara, Kazumichi. 1975. Search for gravitational radiation at 145 Hz. *Physical Review Letters*, **35**(6), 330–334.

Hirota, Hidenobu, Itoh, Mikitaka, Oguma, Manabu, and Hibino, Yoshinori. 2005. Temperature coefficients of refractive indices of TiO_2-SiO_2 films. *Japanese Journal of Applied Physics*, **44**, 1009–1010.

Ho, C. Y., Powell, R. W., and Liley, P. E. 1972. Thermal conductivity of the elements. *Journal of Physical and Chemical Reference Data*, **1**, 279–421.

Hollberg, L., Diddams, S., Bartels, A., Fortier, T., and Kim, K. 2005a. The measurement of optical frequencies. *Metrologia*, **42**(3), S105–S124.

Hollberg, L., Oates, C. W., Wilpers, G. *et al.* 2005b. Optical frequency/wavelength references. *Journal of Physics B: Atomic, Molecular and Optical Physics*, **38**(9), S469–S495.

Hood, Christina J. 2000. *Real-time measurement and trapping of single atoms by single photons*. Ph.D. thesis, California Institute of Technology, Pasadena, CA.

Hood, C. J., Chapman, M. S., Lynn, T. W., and Kimble, H. J. 1998. Real-time cavity QED with single atoms. *Physical Review Letters*, **80**(19), 4157–4160.

Hood, Christina J., Kimble, H. J., and Ye, Jun. 2001. Characterization of high-finesse mirrors: Loss, phase shifts, and mode structure in an optical cavity. *Physical Review A*, **64**(3), 033804.

Hunger, D., Steinmetz, T., Colombe, Y. *et al.* 2010. A fiber Fabry–Perot cavity with high finesse. *New Journal of Physics*, **12**(6), 065038.

Iga, Kenichi. 2008. Vertical-cavity surface-emitting laser: Its conception and evolution. *Japanese Journal of Applied Physics*, **47**(1), 1–10.

Ignatchenko, V. A., and Laletin, O. N. 2004. Waves in a superlattice with arbitrary interlayer boundary thickness. *Physics of the Solid State*, **46**, 2292–2300.

Ikushima, Y., Li, R., Tomaru, T. *et al.* 2008. Ultra-low-vibration pulse-tube cryocooler system – cooling capacity and vibration. *Cryogenics*, **48**, 406–412.

Inci, M. N. 2004. Simultaneous measurements of the thermal optical and linear thermal expansion coefficients of a thin film etalon from the reflection spectra of a super-luminescent diode. *Journal of Physics D*, **37**, 3151–3154.

Inci, M. Naci, and Yoshino, T. 2000. A fiber optic wavelength modulation sensor based on tantalum pentoxide coatings for absolute temperature measurement. *Optical Review*, **7**, 205–208.

Itano, W. M., Bergquist, J. C., Bollinger, J. J. *et al.* 1993. Quantum projection noise: Population fluctuations in two-level systems. *Physical Review A*, **47**(5), 3554–3570.

Jackson, W. B., Amer, N. M., Boccara, A. C., and Fournier, D. 1981. Photothermal deflection spectroscopy and detection. *Applied Optics*, **20**(8), 1333–1344.

Jacobs, S. F. 1986. Dimensional stability of materials useful in optical engineering. *Optica Acta*, **11**, 1377–1388.

Jafry, Y, and Sumner, T J. 1997. Electrostatic charging of the LISA proof masses. *Classical and Quantum Gravity*, **14**(6), 1567–1574.

Jafry, Y. R., Cornelisse, J., and Reinhard, R. 1994. LISA – A laser interferometer space antenna for gravitational-wave measurements. *European Space Agency Journal*, **18**, 219–228.

JAHM software, Inc. 1998. *Material Property Database (MPDB software)*.

Jähne, K., Genes, C., Hammerer, K. *et al.* 2009. Cavity-assisted squeezing of a mechanical oscillator. *Physical Review A*, **79**, 063819.

Jaynes, E. T., and Cummings, F. W. 1963. Comparison of quantum and semiclassical radiation theories with application to the beam maser. *Proceedings of the IEEE*, **51**, 89–109.

Jefferts, S. R., Monroe, C., Bell, E. W., and Wineland, D. J. 1995. Coaxial-resonator-driven rf (Paul) trap for strong confinement. *Physical Review A*, **51**(4), 3112–3116.

Jellison, G. E., and Modine, F. A. 1996. Parameterization of the optical functions of amorphous materials in the interband region. *Applied Physics Letters*, **69**, 371–373.

Jennrich, O. 2009. LISA technology and instrumentation. *Classical and Quantum Gravity*, **26**(15), 153001.

Jewell, J. L., Scherer, A., McCall, S. L. *et al.* 1989. Low-threshold electrically pumped vertical-cavity surface-emitting microlasers. *Electronics Letters*, **25**(17), 1123–1124.

Jiang, Y., Fang, S., Bi, Z., Xu, X., and Ma, L. 2010. Nd:YAG lasers at 1064 nm with 1 Hz linewidth. *Applied Physics B: Lasers and Optics*, **98**, 61–67.

Jiang, Y. Y., Ludlow, A. D., Lemke, N. D. *et al.* 2011. Making optical atomic clocks more stable with 10^{-16} level laser stabilization. *Nature Photonics*, doi:10,1038/nphdcon.2010. 313.

Joe, M., Kim, J.-H., Choi, C., Kahng, B., and Kim, J.-S. 2009. Nanopatterning by multiple-ion-beam sputtering. *Journal of Physics: Condensed Matter*, **21**(22), 224011.

Jones, David J., Diddams, Scott A., Ranka, Jinendra K. *et al.* 2000. Carrier-envelope phase control of femtosecond mode-locked lasers and direct optical frequency synthesis. *Science*, **288**(5466), 635–639.

Jonscher, A. K. 1964. Semiconductors at cryogenic temperatures. *Proceedings of the IEEE*, **52**, 1092–1104.

Kajima, Mariko, Kusumi, Nobuhiro, Moriwaki, Shigenori, and Mio, Norikatsu. 1999. Wideband measurement of mechanical thermal noise using a laser interferometer. *Physics Letters A*, **264**(4), 251–256.

Kalb, Austin. 1986. Neutral ion beam sputter deposition of high-quality optical films. *Optics News*, **12**(8), 13–17.

Kalb, A., Mildebrath, M., and Sanders, V. 1986. Neutral ion beam deposition of high reflectance coatings for use in ring laser gyroscopes. *Journal of Vacuum Science and Technology A*, **4**, 436–437.

Kamp, Carl Justin, Kawamura, Hinata, Passaquieti, Roberto, and DeSalvo, Riccardo. 2009. Directional radiative cooling thermal compensation for gravitational wave interferometer mirrors. *Nuclear Instruments and Methods in Physics Research Section A: Accelerators, Spectrometers, Detectors and Associated Equipment*, **607**(3), 530–537.

Karow, H. H. 2004. *Fabrication Methods for Precision Optics*. Wiley-Interscience.

Katori, H., Takamoto, M., Pal'chikov, V.G., and Ovsiannikov, V.D. 2003. Ultrastable optical clock with neutral atoms in an engineered light shift trap. *Physical Review Letters*, **91**(17), 173005.

Kaufman, H. R., Cuomo, J. J., and Harper, J. M. E. 1982. Technology and applications of broad-beam ion sources used in sputtering. Part I. ion source technology. *Journal of Vacuum Science and Technology*, **21**(3), 725–736.

Kawamura, Seiji. 2010. Ground-based interferometers and their science reach. *Classical and Quantum Gravity*, **27**(8), 084001.

Kawamura, Seiji, and Chen, Yanbei. 2004. Displacement-noise-free gravitational-wave detection. *Physical Review Letters*, **93**(21), 211103.

Keller, M., Lange, B., Hayasaka, K., Lange, W., and Walther, H. 2004. Continuous generation of single photons with controlled waveform in an ion-trap cavity system. *Nature*, **431**, 1075–1078.

Keller, U., Weingarten, K. J., Kartner, F. X. *et al.* 1996. Semiconductor saturable absorber mirrors (SESAM's) for femtosecond to nanosecond pulse generation in solid-state lasers. *IEEE Journal of Selected Topics in Quantum Electronics*, **2**(3), 435–453.

Khalili, F. Ya. 2001. Frequency-dependent rigidity in large-scale interferometric gravitational-wave detectors. *Physics Letters A*, **288**(5–6), 251–256.

Khalili, F. 2005. Reducing the mirrors coating noise in laser gravitational-wave antennae by means of double mirrors. *Physics Letters A*, **334**, 67–72.

Khazanov, E., Andreev, N. F., Mal'shakov, A. *et al.* 2004. Compensation of thermally induced modal distortions in Faraday isolators. *IEEE Journal of Quantum Electronics*, **40**, 1500–1510.

Khudaverdyan, M., Alt, W., Kampschulte, T. *et al.* 2009. Quantum jumps and spin dynamics of interacting atoms in a strongly coupled atom-cavity system. *Physical Review Letters*, **103**(12), 123006.

Kimble, H. J. 2008. The quantum internet. *Nature*, **453**(7198), 1023–1030.

Kimble, H. J., and Levin, Yuri, and Matsko, Andrey B., and Thorne, Kip S., and Vyatchanin, Sergey P. 2001. Conversion of conventional gravitational-wave interferometers into quantum nondemolition interferometers by modifying their input and/or output optics. *Physical Review D*, **65**(2), 022002.

Kimble, H. J., Lev, Benjamin L., and Ye, Jun. 2008. Optical interferometers with reduced sensitivity to thermal noise. *Physical Review Letters*, **101**(26), 260602.

Kippenberg, T. J., and Vahala, K. J. 2008. Cavity optomechanics: Back action at the mesoscale. *Science*, **321**(5893), 1172–1176.

Kittel, C. 1995. *Introduction to Solid State Physics*. 7th edn. Wiley.

Kleckner, D., and Bouwmeester, D. 2006. Sub-kelvin optical cooling of a micromechanical resonator. *Nature*, **444**, 75–78.

Kleckner, Dustin, Marshall, William, de Dood, Michiel J. A. *et al.* 2006. High finesse optomechanical cavity with a movable thirty-micron-size mirror. *Physical Review Letters*, **96**, 173901.

Kleinman, D. A., Miller, R. C., and Nordland, W. A. 1973. Two-photon absorption of Nd laser radiation in GaAs. *Applied Physics Letters*, **23**(5), 243–244.

Knudsen, S., Tveten, A. B., and Dandridge, A. 1995. Measurements of fundamental thermal induced phase fluctuations in the fiber of a Sagnac interferometer. *IEEE Photonics Technology Letters*, **7**, 90–92.

Kogelnik, H., and Li, T. 1966. Laser beams and resonators. *Applied Optics*, **5**(10), 1550–1567.

Kolachevsky, N., Matveev, A., Alnis, J. *et al.* 2009. Measurement of the $2S$ hyperfine interval in atomic hydrogen. *Physical Review Letters*, **102**(21), 213002.

Konagai, Makoto, Sugimoto, Mitsunori, and Takahashi, Kiyoshi. 1978. High efficiency GaAs thin film solar cells by peeled film technology. *Journal of Crystal Growth*, **45**, 277–280.

Kondratiev, N. M., Gurkovsky, A. G., and Gorodetsky, M. L. 2011. Thermal noise and coating optimization in multilayer dielectric mirrors. *Physical Review D*, **84**, 022001.

Kopperschmidt, P., Kästner, G., Senz, S., Hesse, D., and Gösele, U. 1997. Wafer bonding of gallium arsenide on sapphire. *Applied Physics A: Materials Science & Processing*, **64**, 533–537.

Kordonski, W. I., and Golini, D. 2000. Fundamentals of magnetorheological fluid utilization in high precision finishing. Page 682 of: *Proceedings of the 7th International Conference on Electro-rheological Fluids and Magneto-rheological Suspensions*. World Scientific, Singapore.

Kovalik, J., and Saulson, P. R. 1993. Mechanical loss in fibers for low pendulums. *Review of Scientific Instruments*, **64**, 2942–2946.

Kubanek, A., Koch, M., Sames, C. *et al.* 2009. Photon-by-photon feedback control of a single-atom trajectory. *Nature*, **462**(7275), 898–901.

Kuga, Takahiro, Torii, Yoshio, Shiokawa, Noritsugu *et al.* 1997. Novel optical trap of atoms with a doughnut beam. *Physical Review Letters*, **78**(25), 4713–4716.

Kuroda, K., and The LCGT Collaboration. 2010. Status of LCGT. *Classical and Quantum Gravity*, **27**(8), 084004.

Kuznetsov, M., Hakimi, F., Sprague, R., and Mooradian, A. 1999. Design and characteristics of high-power (> 0.5 W CW) diode-pumped vertical-external-cavity surface-emitting semiconductor lasers with circular TEM00 beams. *IEEE Journal of Selected Topics in Quantum Electronics*, **5**(3), 561–573.

Kwee, Patrick, Willke, Benno, and Danzmann, Karste. 2009. Shot-noise-limited laser power stabilization with a high-power photodiode array. *Optics Letters*, **34**, 2912–2914.

Lakes, S. R. 2009. *Viscoelastic Materials*. Cambridge University Press.

Lam, C. C., and Douglass, D. H. 1981. Internal friction measurements in boron-doped single-crystal silicon. *Physics Letters*, **85A**(1), 41–42.

Lawrence, Ryan. 2003. *Active wavefront correction in laser interferometric gravitational wave detectors*. Ph.D. thesis, Massachusetts Institute of Technology.

Lawrence, Ryan, Ottaway, David, Zucker, Michael, and Fritschel, Peter. 2004. Active correction of thermal lensing through external radiative thermal actuation. *Optics Letters*, **29**(22), 2635–2637.

Le Targat, R., Baillard, X., Fouché, M. *et al.* 2006. Accurate optical lattice clock with [87]Sr atoms. *Physical Review Letters*, **97**(13), 130801.

Lebedev, P. 1901. Untersuchungen über die Druckkräfte des Lichtes. *Annalen der Physik*, **311**, 433–458.

Lee, Cheng-Chung, and Tang, Chien-Jen. 2006. TiO_2–Ta_2O_5 composite thin films deposited by radio frequency ion-beam sputtering. *Applied Optics*, **45**(36), 9125–9131.

Lee, Cheng-Chung, Tang, Chien-Jen, and Wu, Jean-Yee. 2006. Rugate filter made with composite thin films by ion-beam sputtering. *Applied Optics*, **45**(7), 1333–1337.

Legero, T., Kessler, T, and Sterr, U. 2010. Tuning the thermal expansion properties of optical reference cavities with fused silica mirrors. *Journal of the Optical Society of America B*, **27**(5), 914–919.

Leggett, A. J. 2002. Testing the limits of quantum mechanics: Motivation, state of play, prospects. *Journal of Physics: Condensed Matter*, **14**, R415–R451.

Leibfried, D., Blatt, R., Monroe, C., and Wineland, D. 2003. Quantum dynamics of single trapped ions. *Reviews of Modern Physics*, **75**, 281–324.

Leibrandt, David R., Labaziewicz, Jaroslaw, Vuletić, Vladan, and Chuang, Isaac L. 2009. Cavity sideband cooling of a single trapped ion. *Physical Review Letters*, **103**(10), 103001.

Lemke, N. D., Ludlow, A. D., Barber, Z. W. *et al.* 2009. Spin-1/2 optical lattice clock. *Physical Review Letters*, **103**(6), 063001.

Lemonde, P., Laurent, P., Santarelli, G. *et al.* 2001. Cold-atom clocks on Earth and in space. Pages 131–153 of: Luiten, Andre (ed.), *Frequency Measurement and Control*. Topics in Applied Physics, vol. 79. Springer Berlin/Heidelberg.

Lequime, Michel, Zerrad, Myriam, Deumie, Carole, and Amra, Claude. 2009. A goniometric light scattering instrument with high-resolution imaging. *Optics Communications*, 1265–1273.

Levin, Yu. 1998. Internal thermal noise in the LIGO test masses: A direct approach. *Physical Review D*, **57**(2), 659–663.

Levin, Yuri. 2008. Fluctuation–dissipation theorem for thermo-refractive noise. *Physics Letters A*, **372**(12), 1941–1944.

Li, M., Pernice, W. H. P., Xiong, C. *et al.* 2008. Harnessing optical forces in integrated photonic circuits. *Nature*, **456**(7221), 480–484.

Lienerth, C., Thummes, G., and Heiden, C. 2001. Progress in low noise cooling performance of a pulse-tube cooler for HT-SQUID operation. *IEEE Transactions on Applied Superconductivity*, **11**, 812–815.

Liu, X., and Pohl, R. O. 1998. Low-energy excitations in amorphous films of silicon and germanium. *Physical Review B*, **58**, 9067–9081.

Liu, Xiao, Vignola, J. F., Simpson, H. J. *et al.* 2005. A loss mechanism study of a very high Q silicon micromechanical oscillator. *Journal of Applied Physics*, **97**(2), 023524.

Liu, Yuk Tung, and Thorne, Kip S. 2000. Thermoelastic noise and homogeneous thermal noise in finite sized gravitational-wave test masses. *Physical Review D*, **62**(12), 122002.

Lodewyck, J., Westergaard, P. G., and Lemonde, P. 2009. Nondestructive measurement of the transition probability in a Sr optical lattice clock. *Physical Review A*, **79**(6), 061401.

Lück, H., and The GEO 600 Collaboration. 2006. Status of the GEO 600 detector. *Classical and Quantum Gravity*, **23**, S71–S78.

Lück, H., Freise, A., Goßler, S. *et al.* 2007. Thermal correction of the radii of curvature of mirrors for GEO 600. *Classical and Quantum Gravity*, **21**, S985–S989.

Lück, H., Degallaix, J., Grote, H. *et al.* 2008. Opto-mechanical frequency shifting of scattered light. *Journal of Optics A: Pure and Applied Optics*, 085004 (6pp).

Ludlow, Andrew, D., Boyd, Martin, M., Zelevinsky, Tanya *et al.* 2006. Systematic study of the ^{87}Sr clock transition in an optical lattice. *Physical Review Letters*, **96**, 033003.

Ludlow, A. D., Huang, X., Notcutt, M. *et al.* 2007. Compact, thermal-noise-limited optical cavity for diode laser stabilization at 1×10^{-15}. *Optics Letters*, **32**(Mar.), 641–643.

Ludlow, A. D., Zelevinsky, T., Campbell, G. K. *et al.* 2008. Sr lattice clock at 1×10^{-16} fractional uncertainty by remote optical evaluation with a Ca clock. *Science*, **319**, 1805–1808.

Ludlow, A. D., Huang, X., Notcutt, M. *et al.* 2009. A narrow linewidth and frequency-stable probe laser source for the ^{88}Sr$^+$ single ion optical frequency standard. *Applied Physics B*, **95**, 45–54.

Ludowise, M. J. 1985. Metalorganic chemical vapor deposition of III-V semiconductors. *Journal of Applied Physics*, **58**(8), R31–R55.

Lunin, B. S. 2005. *Physical and Chemical Bases for the Development of Hemispherical Resonators for Solid-State Gyroscopes*. Moscow: Moscow Aviation Institute.

Lyon, K. G., Salinger, G. L., Swenson, C. A., and White, G. K. 1977. Linear thermal expansion measurements on silicon from 6 to 340 K. *Journal of Applied Physics*, **48**, 865–868.

Macfarlane, G. G., McLean, T. P., Quarrington, J. E., and Roberts, V. 1958. Fine structure in the absorption-edge spectrum of Si. *Physical Review*, **111**, 1245–1254.

Macfarlane, G. G., McLean, T. P., Quarrington, J. E., and Roberts, V. 1959. Exciton and phonon effects in the absorption spectra of germanium and silicon. *Journal of Physics and Chemistry of Solids*, **8**, 388–392.

Macleod, A. H. 1981. Monitoring of optical coatings. *Applied Optics*, **20**, 82–89.

Macleod, A. H. 2010. *Thin Film Optical Filters*. 4th edn. Taylor & Francis Group.

Madej, A. A., Bernard, J. E., Dubé, P., Marmet, L., and Windeler, R. S. 2004. Absolute frequency of the ^{88}Sr$^+$ $5s^2$S$_{1/2}$–$4d^2$D$_{5/2}$ reference transition at 445 THz and evaluation of systematic shifts. *Physical Review A*, **70**(1), 012507.

Majorana, E., and Ogawa, Y. 1997. Mechanical thermal noise in coupled oscillators. *Physics Letters A*, **233**(3), 162–168.

Mancini, S., and Tombesi, P. 1994. Quantum noise reduction by radiation pressure. *Physical Review A*, **49**(5), 4055–4065.

Margolis, H. S., Barwood, G. P., Huang, G. *et al.* 2004. Hertz-level measurement of the optical clock frequency in a single ^{88}Sr$^+$ ion. *Science*, **306**(5700), 1355–1358.

Mari, A., and Eisert, J. 2009. Gently modulating optomechanical systems. *Physical Review Letters*, **103**, 213603.

Markosyan, Ashot, Armandula, Helena, Fejer, Martin M., and Route, Roger. 2008. *PCI technique for thermal absorption measurements*. LIGO-G080315-00.

Marquardt, F., and Girvin, S. M. 2009. Optomechanics. *Physics*, **2**, 40.

Marquardt, Florian, Chen, Joe P., Clerk, A. A., and Girvin, S. M. 2007. Quantum theory of cavity-assisted sideband cooling of mechanical motion. *Physical Review Letters*, **99**(9), 093902.

Marshall, W., Simon, C., Penrose, R., and Bouwmeester, D. 2003. Towards quantum superposition of a mirror. *Physical Review Letters*, **91**, 130401.

Martin, I. 2009. *Studies of materials for use in future interferometric gravitational wave detectors*. Ph.D. thesis, University of Glasgow.

Martin, I., Armandula, H., Comtet, C. *et al.* 2008. Measurements of a low-temperature mechanical dissipation peak in a single layer of Ta_2O_5 doped with TiO_2. *Classical and Quantum Gravity*, **25**, 055005.

Martin, I. W., Chalkley, E., Nawrodt, R. *et al.* 2009. Comparison of the temperature dependence of the mechanical dissipation in thin films of Ta_2O_5 and Ta_2O_5 doped with TiO_2. *Classical and Quantum Gravity*, **26**(15), 155012.

Martin, I. W., Bassiri, R., Nawrodt, R. *et al.* 2010. Effect of heat treatment on mechanical dissipation in Ta_2O_5 coatings. *Classical and Quantum Gravity*, **27**(22), 225020.

Martin, P. J., and Netterfield, R. P. 1989. *Handbook of Ion Beam Processing Technology*. Noyes. Chap. Ion-assisted dielectric and optical coatings.

Matsko, A. B., Savchenkov, A. A., Yu, N., and Maleki, L. 2007. Whispering-gallery-mode resonators as frequency references. I. Fundamental limitations. *Journal of the Optical Society of America B*, **24**, 1324–1334.

Mauceli, E., Geng, Z. K., Hamilton, W. O. *et al.* 1996. The Allegro gravitational wave detector: Data acquisition and analysis. *Physical Review D*, **54**(2), 1264–1275.

Maunz, P., Puppe, T., Schuster, I. *et al.* 2004. Cavity cooling of a single atom. *Nature*, **428**, 50–52.

McClelland, D. E., Camp, J. B., Mason, J., Kells, W., and Whitcomb, S. E. 1999. Arm cavity resonant sideband control for laser interferometric gravitational wave detectors. *Optics Letters*, **24**(15), 1014–1016.

McGuigan, D. H., Lam, C. C., Gram, R. Q. *et al.* 1978. Measurements of the mechanical Q of single-crystal silicon at low temperatures. *Journal of Low Temperature Physics*, **30**, 621–629.

McIvor, G., Waldman, S., and Willems, P. 2007. *Analysis of LIGO test mass internal modes as a measure of coating absorption*. LIGO-G070636-00.

McKeever, J., Boca, A., Boozer, A. D., Buck, J. R., and Kimble, H. J. 2003. Experimental realization of a one-atom laser in the regime of strong coupling. *Nature*, **425**, 268–271.

McKeever, J., Boca, A., Boozer, A. D. *et al.* 2004. Deterministic generation of single photons from one atom trapped in a cavity. *Science*, **303**, 1992–1994.

McLachlan, D. Jr., and Chamberlain, L. L. 1964. Atomic vibrations and melting point in metals. *Acta Metallurgica*, **12**, 571–576.

McSkimin, H. J. 1953. Measurement of elastic constants at low temperatures by means of ultrasonic waves – data for silicon and germanium single crystals, and for fused silica. *Journal of Applied Physics*, **24**, 988–997.

Meers, Brian J. 1988. Recycling in laser-interferometric gravitational-wave detectors. *Physical Review D*, **38**(8), 2317–2326.

Melliar-Smith, C. M., and Mogab, C. J. 1978. *Thin Film Processes*. Academic Press. Chap. Plasma-assisted etching techniques for pattern delineation.

Melninkaitis, Andrius, Tolenis, Tomas, Mažulė, Lina *et al.* 2011. Characterization of zirconia– and niobia–silica mixture coatings produced by ion-beam sputtering. *Applied Optics*, **50**(9), C188–C196.

Metzger, C. H., and Karrai, K. 2004. Cavity cooling of a microlever. *Nature*, **432**(7020), 1002–1005.

Michael, C. P., Srinivasan, K., Johnson, T. J. *et al.* 2007. Wavelength- and material-dependent absorption in GaAs and AlGaAs microcavities. *Applied Physics Letters*, **90**(5), 051108–051108–3.

Mie, G. 1908. Beiträge zur Optik Trüber Medien, speziell Kolloidaler Metallösungen. *Annalen der Physik*, **25**, 377–452.

Milam, D., Lowdermilk, W. H., Rainer, F. *et al.* 1982. Influence of deposition parameters on laser-damage threshold of silica-tantala AR coatings. *Applied Optics*, **21**(20), 3689–3694.

Milatz, J. M. W., Van Zolingen, J., and Van Iperen, B. B. 1953. The reduction in the brownian motion of electrometers. *Physica*, **19**, 195–202.

Milburn, G. J., Jacobs, K., and Walls, D. F. 1994. Quantum-limited measurements with the atomic force microscope. *Physical Review A*, **50**(6), 5256–5263.

Miller, John. 2010. *On non-Gaussian beams and optomechanical parametric instabilities in interferometric gravitational wave detectors.* Ph.D. thesis, University of Glasgow.

Miller, R., Northup, T. E., Birnbaum, K. M. *et al.* 2005. Trapped atoms in cavity QED: coupling quantized light and matter. *Journal of Physics B: Atomic, Molecular and Optical Physics*, **38**(9), S551–S565.

Millo, J., Magalhães, D. V., Mandache, C. *et al.* 2009. Ultrastable lasers based on vibration insensitive cavities. *Physical Review A*, **79**(5), 053829.

Misner, Charles W., Thorne, Kip S., and Wheeler, John A. 1973. *Gravitation (Physics Series).* 2nd edn. W. H. Freeman.

Mitrofanov, V. P., and Tokmakov, K. V. 2003. Effect of heating on dissipation of mechanical energy in fused silica fibers. *Physics Letters A*, **308**(2–3), 212–218.

Mitrofanov, V. P., Prokhorov, L. G., and Tokmakov, K. V. 2002. Variation of electric charge on prototype of fused silica test mass of gravitational wave antenna. *Physics Letters A*, **300**, 370–374.

Miyoki, S., and The CLIO and LCGT Collaboration. 2010. Underground cryogenic laser interferometer CLIO. *Journal of Physics: Conference Series*, **203**, 012075.

Miyoki, S., Tomaru, T., Ishitsuka, H. *et al.* 2001. Cryogenic contamination speed for cryogenic laser interferometric gravitational wave detector. *Cryogenics*, **41**, 415–420.

Mizuno, J., Strain, K. A., Nelson, P. G. *et al.* 1993. Resonant sideband extraction: A new configuration for interferometric gravitational wave detectors. *Physics Letters A*, **175**(5), 273–276.

Mohanty, P., Harrington, D. A., Ekinci, K. L. *et al.* 2002. Intrinsic dissipation in high-frequency micromechanical resonators. *Physical Review B*, **66**(8), 085416.

Mor, O., and Arie, A. 1997. Performance analysis of Drever–Hall laser frequency stabilization using a proportional+integral servo. *IEEE Journal of Quantum Electronics*, **33**(4), 532–540.

Mortonson, M. J., Vassiliou, C. C., Ottaway, D. J., Shoemaker, D. H., and Harry, G. M. 2003. Effects of electrical charging on the mechanical Q of a fused silica disk. *Review of Scientific Instruments*, **74**, 4840–4845.

Mours, B., Tournefier, E., and Vinet, J.-Y. 2006. Thermal noise reduction in interferometric gravitational wave antennas: Using high order TEM modes. *Classical and Quantum Gravity*, **23**(Oct.), 5777–5784.

Muller, Andreas, Flagg, Edward B., Lawall, John R., and Solomon, Glenn S. 2010. Ultrahigh-finesse, low-mode-volume Fabry–Perot microcavity. *Optics Letters*, **35**(13), 2293–2295.

Müller, H., Braxmaier, C., Herrmann, S. *et al.* 2002. Testing the foundations of relativity using cryogenic optical resonators. *International Journal of Modern Physics D*, **11**, 1101–1108.

Müller, H., Herrmann, S., Braxmaier, C., Schiller, S., and Peters, A. 2003a. Modern Michelson–Morley experiment using cryogenic optical resonators. *Physical Review Letters*, **91**(2), 020401.

Müller, H., Herrmann, S., Braxmaier, C., Schiller, S., and Peters, A. 2003b. Precision test of the isotropy of light propagation. *Applied Physics B*, **77**, 719–731.

Mundt, A. B., Kreuter, A., Becher, C. *et al.* 2002. Coupling a single atomic quantum bit to a high finesse optical cavity. *Physical Review Letters*, **89**(10), 103001.

Münstermann, P., Fischer, T., Pinkse, P. W. H., and Rempe, G. 1999. Single slow atoms from an atomic fountain observed in a high-finesse optical cavity. *Optics Communications*, **159**, 63–67.

Murray, Peter. 2008. *Measurement of the mechanical loss of test mass materials for advanced gravitational wave detectors.* Ph.D. thesis, University of Glasgow.

Nakagawa, N., Auld, B. A., Gustasfson, E., and Fejer, M. M. 1997. Estimation of thermal noise in the mirrors of laser interferometric gravitational wave detectors: Two point correlation function. *Review of Scientific Instruments*, **68**, 3553–3556.

Nakagawa, N., Gustafson, E. K., Beyersdorf, Peter T., and Fejer, M. M. 2002a. Estimating the off resonance thermal noise in mirrors, Fabry-Perot interferometers, and delay lines: The half infinite mirror with uniform loss. *Physical Review D*, **65**(Mar.), 082002.

Nakagawa, N., Gretarsson, A. M., Gustafson, E. K., and Fejer, M. M. 2002b. Thermal noise in half-infinite mirrors with nonuniform loss: A slab of excess loss in a half-infinite mirror. *Physical Review D*, **65**(Apr), 102001.

Narayan, R., Paczynski, B., and Piran, T. 1992. Gamma-ray bursts as the death throes of massive binary stars. *Astrophysical Journal, Part 2 – Letters*, **395**(Aug.), L83–L86.

Nawrodt, R., Zimmer, A., Koettig, T. *et al.* 2007a. High mechanical Q-factor measurements on calcium fluoride at cryogenic temperatures. *The European Physical Journal – Applied Physics*, **38**, 53–59.

Nawrodt, R., Zimmer, A., Koettig, T. *et al.* 2007b. Mechanical Q-factor measurements on a test mass with a structured surface. *New Journal of Physics*, **9**, 225.

Nawrodt, R., Zimmer, A., Koettig, T. *et al.* 2008. High mechanical Q-factor measurements on silicon bulk samples. *Journal of Physics: Conference Series*, **122**(1), 012008.

Netterfield, R. P., and Gross, M. 2007. Investigation of ion beam sputtered silica-titania mixtures for use in gravitational-wave detectors. Page ThD2 of: *Proceedings of Optical Interference Coatings (CD)*. Optical Society of America.

Netterfield, Roger P., Gross, Mark, Baynes, Fred N. *et al.* 2005. Low mechanical loss coatings for LIGO optics: Progress report. Page 58700 of: Fulton, Michael L., and Kruschwitz, Jennifer D. T. (eds), *Proceedings of SPIE, Advances in Thin-Film Coatings for Optical Applications II*, vol. **5870**.

Neugebauer, P. A. 1970. *Handbook of Thin Film Technology.* McGraw Hill. Chap. Condensation, nucleation, and growth of thin films.

Neuroth, N. 1995. *The Properties of Optical Glass.* Springer-Verlag. Chap. Transmission and reflection.

Nichols, E. F., and Hull, G. F. 1901. A preliminary communication on the pressure of heat and light radiation. *Physical Review (Series I)*, **13**(5), 307–320.

Notcutt, M., Taylor, C. T., Mann, A. G., and Blair, D. G. 1995. Temperature compensation for cryogenic cavity stabilized lasers. *Journal of Physics D: Applied Physics*, **28**, 1807–1810.

Notcutt, M., Taylor, C. T., Mann, A. G., Gummer, R., and Blair, D. G. 1996. Cryogenic system for a sapphire Fabry–Perot optical frequency standard. *Cryogenics*, **36**, 13–16.

Notcutt, Mark, Ma, Long-Sheng, Ye, Jun, and Hall, John L. 2005. Simple and compact 1 Hz laser system via an improved mounting configuration of a reference cavity. *Optics Letters*, **30**(14), 1815–1817.

Notcutt, Mark, Ma, Long-Sheng, Ludlow, Andrew D. *et al.* 2006. Contribution of thermal noise to frequency stability of rigid optical cavity via Hertz-linewidth lasers. *Physical Review A*, **73**(3), 031804.

Nowick, A. S., and Berry, B. S. 1972. *Anelastic Relaxation in Crystalline Solids*. Academic Press.

Numata, Kenji. 2003 (March). *Direct measurement of mirror thermal noise*. Ph.D. thesis, University of Tokyo, Tokyo, Japan.

Numata, Kenji, Bianc, Giuseppe Bertolotto, Tanaka, Mitsuru *et al.* 2001. Measurement of the mechanical loss of crystalline samples using a nodal support. *Physics Letters A*, **284**(4–5), 162–171.

Numata, Kenji, Ando, Masaki, Yamamoto, Kazuhiro, Otsuka, Shigemi, and Tsubono, Kimio. 2003. Wide-band direct measurement of thermal fluctuations in an interferometer. *Physical Review Letters*, **91**(26), 260602.

Numata, K., Kemery, A., and Camp, J. 2004. Thermal-noise limit in the frequency stabilization of lasers with rigid cavities. *Physical Review Letters*, **93**(25), 250602.

Nußmann, Stefan, Hijlkema, Markus, Weber, Bernhard *et al.* 2005. Submicron positioning of single atoms in a microcavity. *Physical Review Letters*, **95**(17), 173602.

Obeidat, Amjad, Khurgin, Jacob, and Knox, Wayne. 1997. Effects of two-photon absorption in saturable Bragg reflectors used in femtosecond solid state lasers. *Optics Express*, **1**(3), 68–72.

Ohring, M. 2002. *The Materials Science of Thin Films: Deposition and Structure*. Academic Press.

Okaji, M., Yamada, N., Nara, K., and Kato, H. 1995. Laser interferometric dilatometer at low temperatures: Application to fused silica SRM 739. *Cryogenics*, **35**, 887–891.

Ono, Takahito, Wang, Dong F., and Esashi, Masayoshi. 2003. Time dependence of energy dissipation in resonating silicon cantilevers in ultrahigh vacuum. *Applied Physics Letters*, **83**(10), 1950–1952.

O'Shaughnessy, R., Strigin, S., and Vyatchanin, S. 2004. *The implications of Mexican-hat mirrors: Calculations of thermoelastic noise and interferometer sensitivity to perturbation for the Mexican-hat-mirror proposal for advanced LIGO*. arXiv:gr-qc/0409050.

O'Shaughnessy, R., Kalogera, V., and Belczynski, Krzysztof. 2010. Binary compact object coalescence rates: The role of elliptical galaxies. *The Astrophysical Journal*, **716**(1), 615–633.

Oskay, W. H., Diddams, S. A., Donley, E. A. *et al.* 2006. Single-atom optical clock with high accuracy. *Physical Review Letters*, **97**(2), 020801.

Ottaway, David, Betzwieser, Joseph, Ballmer, Stefan, Waldman, Sam, and Kells, William. 2006. In situ measurement of absorption in high-power interferometers by using beam diameter measurements. *Optics Letters*, **31**(4), 450–452.

Park, Young-Shin, and Wang, Hailin. 2009. Resolved-sideband and cryogenic cooling of an optomechanical resonator. *Nature Physics*, **5**, 489–493.

Parker, E. H. C. (ed). 1985. *The Technology and Physics of Molecular Beam Epitaxy*. Plenum Press.

Paternostro, M., Vitali, D., Gigan, S. *et al.* 2007. Creating and probing multipartite macroscopic entanglement with light. *Physical Review Letters*, **99**, 250401.

Penn, S. D., Harry, G. M., Gretarsson, A. M. *et al.* 2001. High quality factor measured in fused silica. *Review of Scientific Instruments*, **72**(9), 3670–3673.

Penn, Steven D., Sneddon, Peter H., Armandula, Helena *et al.* 2003. Mechanical loss in tantala/silica dielectric mirror coatings. *Classical and Quantum Gravity*, **20**(13), 2917–2928.

Penn, Steven D., Ageev, Alexander, Busby, Dan *et al.* 2006. Frequency and surface dependence of the mechanical loss in fused silica. *Physics Letters A*, **352**, 3–6.

Phillips, W. A. 1972. Tunneling states in amorphous solids. *Journal of Low Temperature Physics*, **7**, 351–360.

Pierro, V., Galdi, V., Castaldi, G. *et al.* 2007. Perspectives on beam-shaping optimization for thermal-noise reduction in advanced gravitational-wave interferometric detectors: Bounds, profiles, and critical parameters. *Physical Review D*, **76**(12), 122003.

Pinard, L., and The Virgo Collaboration. 2004. Low loss coatings for the Virgo large mirrors. Pages 483–492 of: Amra, C., Kaiser, N., and Macleod, H. A. (eds.), *Proceedings of SPIE, Advances in Optical Thin Films*, vol. **5250**.

Pinard, M., Fabre, C., and Heidmann, A. 1995. Quantum-nondemolition measurement of light by a piezoelectric crystal. *Physical Review A*, **51**(3), 2443–2449.

Pinto, Innocenzo M., Piero, Vincenzo, Principe, Maria, and DeSalvo, Riccardo. 2010. *Mixture theory approach to coating materials optimization.* LIGO-G1000537.

Plissi, M. V., Torrie, C. I., Husman, M. E. *et al.* 2000. GEO 600 triple pendulum suspension system: Seismic isolation and control. *Review of Scientific Instruments*, **71**, 2539–2545.

Plissi, M. V., Torrie, C. I., Barton, M. *et al.* 2004. An investigation of eddy-current damping of multi-stage pendulum suspensions for use in interferometric gravitational wave detectors. *Review of Scientific Instruments*, **75**, 4516–4522.

Poirson, Jérôme, Bretenaker, Fabien, Vallet, Marc, and Floch, Albert Le. 1997. Analytical and experimental study of ringing effects in a Fabry–Perot cavity. Application to the measurement of high finesses. *Journal of the Optical Society of America B*, **14**(11), 2811–2817.

Pollack, S. E., Turner, M. D., Schlamminger, S., Hagedorn, C. A., and Gundlach, J. H. 2010. Charge management for gravitational-wave observatories using UV LEDs. *Physical Review D*, **81**(2), 021101.

Pond, B. J., DeBar, J. I., Carniglia, C. K., and Raj, T. 1989. Stress reduction in ion beam sputtered mixed oxide films. *Applied Optics*, **28**(14), 2800–2805.

Principe, M., Pinto, I. M., and Galdi, V. 2007. *A general formula for the thermorefractive noise coefficient of stacked-doublet mirror coatings.* LIGO-T070159.

Principe, Maria, DeSalvo, Riccardo, Pinto, Innocenzo, and Galdi, Vincenzo. 2008. *Minimum Brownian noise dichroic dielectric mirror coatings for AdLIGO.* LIGO-T080337.

Pulker, H. K. 1984a. *Coatings on Glass.* Elsevier. Chap. Glass and thin films.

Pulker, H. K. 1984b. *Coatings on Glass.* Elsevier. Chap. Cleaning of substrate surfaces.

Pulker, H. K. 1984c. *Coatings on Glass.* Elsevier. Chap. Film formation methods.

Punturo, M., and The Einstein Telescope Collaboration. 2007. *Einstein gravitational wave telescope, proposal to the European Commission, Framework Programme 7.* http://www.ego-gw.it/ILIAS-GW/FP7-DS/fp7-DS.htm.

Punturo, M, and The Einstein Telescope Collaboration. 2010. The third generation of gravitational wave observatories and their science reach. *Classical and Quantum Gravity*, **27**(8), 084007.

Puppe, T., Schuster, I., Grothe, A. *et al.* 2007. Trapping and observing single atoms in a blue-detuned intracavity dipole trap. *Physical Review Letters*, **99**(1), 013002.

Quessada, A., Kovacich, R. P., Courtillot, I. *et al.* 2003. The Dick effect for an optical frequency standard. *Journal of Optics B: Quantum and Semiclassical Optics*, **5**, S150–S154.

Quetschke, V., Gleason, J., Rakhmanov, M. *et al.* 2006. Adaptive control of laser modal properties. *Optics Letters*, **31**(2), 217–219.

Quinn, T. J., Speake, C. C., and Brown, L. M. 1997. Materials problems in the construction of long-period pendulums. *Philosophical Magazine A*, **65**, 261–276.

Rabl, P., Genes, C., Hammerer, K., and Aspelmeyer, M. 2009. Phase-noise induced limitations on resolved-sideband cavity cooling of mechanical resonators. *Physical Review A*, **80**, 063819.

Rabl, P., Kolkowitz, S. J., Koppens, F. H. *et al.* 2010. A quantum spin transducer based on nano electro-mechanical resonator arrays. *Nature Physics*, **6**(8), 602–608.

Radebaugh, R. 2009. Cryocoolers: the state of the art and recent developments. *Journal of Physics: Condensed Matter*, **21**, 164219.

Rafac, R. J., Young, B. C., Beall, J. A. *et al.* 2000. Sub-dekahertz ultraviolet spectroscopy of ^{199}Hg$^+$. *Physical Review Letters*, **85**(Sep.), 2462–2465.

Raimond, J. M., Brune, M., and Haroche, S. 2001. Manipulating quantum entanglement with atoms and photons in a cavity. *Reviews of Modern Physics*, **73**(3), 565–582.

Rao, S. 2003. *Mirror thermal noise in interferometric gravitational wave detectors*. Ph.D. thesis, California Institute of Technology.

Reid, S., Cagnoli, G., Crooks, D. R. M. *et al.* 2006. Mechanical dissipation in silicon flexures. *Physics Letters A*, **351**, 205–211.

Reinisch, J., and Heuer, A. 2005. What is moving in silica at 1 K? A computer study of the low-tempreature anomalies. *Physical Review Letters*, **95**, 155502.

Reitzenstein, S., Hofmann, C., Gorbunov, A. *et al.* 2007. AlAs/GaAs micropillar cavities with quality factors exceeding 150,000. *Applied Physics Letters*, **90**(25), 251109–251109–3.

Rempe, G., Thompson, R. J., Brecha, R. J., Lee, W. D., and Kimble, H. J. 1991. Optical bistability and photon statistics in cavity quantum electrodynamics. *Physical Review Letters*, **67**(13), 1727–1730.

Rempe, G., Thompson, J., Kimble, H. J., and Lalezari, R. 1992. Measurement of ultralow losses in an optical interferometer. *Optics Letters*, **17**(5), 363–365.

Richard, J.-P. 1992. Approaching the quantum limit with optically instrumented multimode gravitational-wave bar detectors. *Physical Review D*, **46**(6), 2309–2317.

Richard, J.-P., and Hamilton, J. J. 1991. Cryogenic monocrystalline silicon Fabry–Perot cavity for the stabilization of laser frequency. *Review of Scientific Instruments*, **62**(10), 2375–2378.

Richard, J.-P., Hamilton, J. J., and Pang, Y. 1990. Fabry–Perot optical resonator at low temperatures. *Journal of Low Temperature Physics*, **81**, 189–198.

Rocheleau, T., Ndukum, T., Macklin, C. *et al.* 2009. Preparation and detection of a mechanical resonator near the ground state of motion. *Nature*, **463**, 72–75.

Romero-Isart, O., Pflanzer, A. C., Blaser, F. *et al.* 2011. Large quantum superpositions and interference of massive nano-objects. *Physical Review Letters*, **107**, 020405.

Rosenband, T., Hume, D. B., Schmidt, P. O. *et al.* 2008. Frequency ratio of Al$^+$ and Hg$^+$ single-ion optical clocks; metrology at the 17th decimal place. *Science*, **319**(5871), 1808–1812.

Rowan, S. 2000. *Implications of thermo-elastic damping for cooled detectors*. Aspen 2000 Winter Conference on Gravitational Waves and Their Detection.

Rowan, S., Twyford, S., Hutchins, R., and Hough, J. 1997. Investigations into the effects of electrostatic charge on the Q factor of a prototype fused silica suspension for use in gravitational wave detectors. *Classical and Quantum Gravity*, **14**, 1537–1541.

Rowan, S., Byer, R. L., Fejer, M. M. *et al.* 2003. Test mass materials for a new generation of gravitational wave detectors. *Proceedings of SPIE*, **4856**, 292–297.

Russo, C., Barros, H., Stute, A. *et al.* 2009. Raman spectroscopy of a single ion coupled to a high-finesse cavity. *Applied Physics B: Lasers and Optics*, **95**(2), 205–212.

Rutman, J., and Walls, F. L. 1991. Characterization of frequency stability in precision frequency sources. *Proceedings of the IEEE*, **79**(7), 952–960.

Ryazhskaya, O. G. 1996. Muons and neutrinos in the cosmic radiation. *Il Nuovo Cimento*, **19C**, 655–670.

Salomon, C., Hils, D., and Hall, J. L. 1988. Laser stabilization at the millihertz level. *Journal of the Optical Society of America B*, **5**, 1576–1587.

Sankur, Haluk, Gunning, William J., and DeNatale, Jeffrey F. 1988. Intrinsic stress and structural properties of mixed composition thin films. *Applied Optics*, **27**(8), 1564–1567.

Santarelli, G., Audoin, C., Makdissi, A. *et al.* 1998. Frequency stability degradation of an oscillator slaved to a periodically interrogated atomic resonator. *IEEE Transactions on Ultrasonics, Ferroelectrics and Frequency Control*, **45**(4), 887–894.

Sassolas, B., Flaminio, R., Franc, J. *et al.* 2009. Masking technique for coating thickness control on large and strongly curved aspherical optics. *Applied Optics*, **48**, 3760–3765.

Sauer, J. A., Fortier, K. M., Chang, M. S., Hamley, C. D., and Chapman, M. S. 2004. Cavity QED with optically transported atoms. *Physical Review A*, **69**(5), 051804.

Saulson, Peter R. 1990. Thermal noise in mechanical experiments. *Physical Review D*, **42**(8), 2437–2445.

Saulson, Peter R., Stebbins, Robin T., Dumont, Frank D., and Mock, Scott E. 1994. The inverted pendulum as a probe of anelasticity. *Review of Scientific Instruments*, **65**, 182–191.

Savchenkov, A. A., Matsko, A. B., Yu, N., Ilchenko, V. S., and Maleki, L. 2007. Whispering-gallery-mode resonators as frequency references. II. Stabilization. *Journal of the Optical Society of America B*, **24**, 2988–2998.

Schermer, J. J., Bauhuis, G. J., Mulder, P. *et al.* 2006. Photon confinement in high-efficiency, thin-film III-V solar cells obtained by epitaxial lift-off. *Thin Solid Films*, **511–512**, 645–653.

Schiller, S., Lämmerzahl, C., Müller, H. *et al.* 2004. Experimental limits for low-frequency space-time fluctuations from ultrastable optical resonators. *Physical Review D*, **69**, 027504.

Schiller, S., Antonini, P., and Okhapkin, M. 2006. *Lecture Notes in Physics*. Springer. Chap. A precision test of the isotropy of the speed of light using rotating cryogenic optical cavities, pages 401–415.

Schliesser, A., Del'Haye, P., Nooshi, N., Vahala, K. J., and Kippenberg, T. J. 2006. Radiation pressure cooling of a micromechanical oscillator using dynamical backaction. *Physical Review Letters*, **97**(24), 243905.

Schmidt-Kaler, F., Gulde, S., Riebe, M. *et al.* 2003. The coherence of qubits based on single Ca^+ ions. *Journal of Physics B: Atomic, Molecular and Optical Physics*, **36**, 623–636.

Schnabel, R., Britzger, M., Brückner, F. *et al.* 2010. Building blocks for future detectors: Silicon test masses and 1550 nm laser light. *Journal of Physics: Conference Series*, **228**, 012029.

Schubert, E. F. 2003. *Light Emitting Diodes*. Cambridge University Press.

Schutz, Bernard. 2009. *A First Course in General Relativity*. 2nd edn. Cambridge University Press.

Scott, W. W., and MacCrone, R. K. 1968. Apparatus for mechanical loss measurements at audio frequencies and low temperatures. *Review of Scientific Instruments*, **39**, 821–823.

Seeber, B., and White, G. K. 1998. *Handbook of Applied Superconductivity, Volume 2*. 2nd edn. Taylor & Francis.

Seel, S., Storz, R., Ruoso, G., Mlynek, J., and Schiller, S. 1997. Cryogenic optical resonators: A new tool for laser frequency stabilization at the 1 Hz level. *Physical Review Letters*, **78**, 4741–4744.

Selhofer, Hubert, and Müller, René. 1999. Comparison of pure and mixed coating materials for AR coatings for use by reactive evaporation on glass and plastic lenses. *Thin Solid Films*, **351**(1–2), 180–183.

Seshan, Krishna (ed). 2002. *Handbook of Thin-Film Deposition Processes and Techniques – Principles, Methods, Equipment and Applications*. William Andrew Publishing/Noyes.

Sheard, Benjamin S., Gray, Malcolm B., Mow-Lowry, Conor M., McClelland, David E., and Whitcomb, Stanley E. 2004. Observation and characterization of an optical spring. *Physical Review A*, **69**(5), 051801.

Sheppard, C. J. R., and Saghafi, S. 1996. Flattened light beams. *Optics Communications*, **132**(Feb.), 144–152.

Sherstov, I., Okhapkin, M., Lipphardt, B., Tamm, C., and Peik, E. 2010. Diode-laser system for high-resolution spectroscopy of the $^{2}S_{1/2} \rightarrow {}^{2}F_{7/2}$ octupole transition in $^{171}Yb^{+}$. *Physical Review A*, **81**(2), 021805.

Shoemaker, D., Schilling, R., Schnupp, L. *et al.* 1988. Noise behavior of the Garching 30-meter prototype gravitational-wave detector. *Physical Review D*, **38**(2), 423–432.

Siegman, A. E. 1986. *Lasers*. University Science Books. See also: Errata List for LASERS, http://www.stanford.edu/~siegman/lasers_book_errata.pdf.

Smith, G. L., Hoyle, C. D., Gundlach, J. H. *et al.* 1999. Short-range tests of the equivalence principle. *Physical Review D*, **61**(2), 022001.

Soloviev, A. A., Kozhevatov, I. E., Palashov, O. V., and Khazanov, E. A. 2006. Compensation for thermally induced aberrations in optical elements by means of additional heating by CO_2 laser radiation. *Quantum Electronics*, **36**, 939–945.

Somiya, K. 2009a. *Discussion about losses in the perpendicular and parallel directions*. LIGO-T0900033.

Somiya, Kentaro. 2009b. Reduction and possible elimination of coating thermal noise using a rigidly controlled cavity with a quantum-nondemolition technique. *Physical Review Letters*, **102**(23), 230801.

Somiya, Kentaro, and Yamamoto, Kazuhiro. 2009. Coating thermal noise of a finite-size cylindrical mirror. *Physical Review D*, **79**(10), 102004.

Somiya, Kentaro, Kokeyama, Keiko, and Nawrodt, Ronny. 2010. Remarks on thermoelastic effects at low temperatures and quantum limits in displacement measurements. *Physical Review D*, **82**, 127101.

Spitzer, W. G., and Whelan, J. M. 1959. Infrared absorption and electron effective mass in n-type gallium arsenide. *Physical Review*, **114**(1), 59–63.

Stannigel, K., Rabl, P., Sørensen, A. S. *et al.* 2010. Opto-mechanical transducers for long-distance quantum communication. *Physical Review Letters*, **105**, 220501.

Steinmetz, T., Colombe, Y., Hunger, D. *et al.* 2006. Stable fiber-based Fabry–Pérot cavity. *Applied Physics Letters*, **89**, 111110.

Stenzel, Olaf, Wilbrandt, Steffen, Schürmann, Mark *et al.* 2011. Mixed oxide coatings for optics. *Applied Optics*, **50**(9), C69–C74.

Sterr, U., Degenhardt, C., Stoehr, H. *et al.* 2004. The optical calcium frequency standards of PTB and NIST. *Comptes Rendus Physique*, **5**, 845–855.

Sterr, U., Legero, T., Kessler, T. *et al.* 2009 (Aug.). Ultrastable lasers: New developments and applications. In: *Society of Photo-Optical Instrumentation Engineers (SPIE) Conference Series*. Presented at the Society of Photo-Optical Instrumentation Engineers (SPIE) Conference, vol. 7431.

Stolz, C. J., and Taylor, J. R. 1992. Damage threshold study of ion beam sputtered coatings for a visible high-repetition laser at LLNL. *Proceedings of SPIE*, **1848**, 182–191.

Stone, J. 1988. Stress-optic effects, birefringence, and reduction of birefringence by annealing in fiber Fabry–Perot interferometers. *Journal of Lightwave Technology*, **6**(7), 1245–1248.

Storz, R., Braxmaier, C., Jäack, K., Pradl, O., and Schiller, S. 1998. Ultrahigh long-term dimensional stability of a sapphire cryogenic optical resonator. *Optics Letters*, **23**, 1031–1033.

Stover, John C. 1995. *Optical scattering: measurement and analysis*. 2nd edn. SPIE – The International Society for Optical Engineering.

Strakna, R. E. 1961. Investigation of low temperature ultrasonic absorption in fast-neutron irradiated SiO_2 glass. *Physical Review*, **123**, 2020–2026.

Stringfellow, G. B. 1989. *Organometallic Vapor Phase Epitaxy: Theory and Practice*. Academic Press.

Stuart, R. V., and Wehner, G. K. 1964. Angular distribution of sputtered Cu atoms. *Journal of Applied Physics*, **35**(6), 1819.

Sullivan, B. T., and Dobrowski, J. A. 1992a. Deposition error compensation for optical multilayer coatings. I. Theoretical description. *Applied Optics*, **31**, 3821–3835.

Sullivan, B. T., and Dobrowski, J. A. 1992b. Deposition error compensation for optical multilayer coatings. II. Experimental results-sputtering system. *Applied Optics*, **32**, 2351–2360.

Sun, Ke-Xun, and Byer, Robert L. 1998. All-reflective Michelson, Sagnac, and Fabry–Perot interferometers based on grating beam splitters. *Optics Letters*, **23**(8), 567–569.

Sun, Ke-Xun, Allard, Brett, Buchman, Saps, Williams, Scott, and Byer, Robert L. 2006. LED deep UV source for charge management of gravitational reference sensors. *Classical and Quantum Gravity*, **23**(8), S141–S150.

Sun, Ke-Xun, Leindecker, Nick, Markosyan, Ashot *et al.* 2008. *Effects of ultraviolet irradiation to LIGO mirror coatings*. LIGO-G080150-00.

Swenson, C. A. 1983. Recommended values for the thermal expansivity of silicon from 0 to 1000 K. *Journal of Physical and Chemical Reference Data*, **12**(2), 179–182.

Takahashi, Ryutaro, and the TAMA Collaboration. 2004. Status of TAMA300. *Classical and Quantum Gravity*, **21**, S403–S408.

Takamoto, M., Hong, F., Higashi, R., and Katori, H. 2005. An optical lattice clock. *Nature*, **435**, 321–324.

Takashashi, Haruo. 1995. Temperature stability of thin-film narrow-bandpass filters produced by ion-assisted deposition. *Applied Optics*, **34**(4), 667–675.

Tarallo, Marco G., Miller, John, Agresti, J. *et al.* 2007. Generation of a flat-top laser beam for gravitational wave detectors by means of a nonspherical Fabry–Perot resonator. *Applied Optics*, **46**(26), 6648–6654.

Tavis, Michael, and Cummings, Frederick W. 1968. Exact solution for an N-molecule-radiation-field Hamiltonian. *Physical Review*, **170**(2), 379–384.

Taylor, C. T., Notcutt, M., Wong, E. K., Mann, A. G., and Blair, D. G. 1996. Measurement of the coefficient of thermal expansion of a cryogenic, all-sapphire, Fabry-Perot optical cavity. *Optics Communications*, **131**, 311–314.

Taylor, C. T., Notcutt, M., Wong, Eng Kiong, Mann, A. G., and Blair, D. G. 1997. Measurement of the thermal expansion coefficient of an all-sapphire optical cavity. *IEEE Transactions on Instrumentation and Measurement*, **46**(2), 183–185.

Tellier, C. R. 1982. Some results on chemical etching of AT-cut quartz wafers in ammonium bifluoride solutions. *Journal of Materials Science*, **17**, 1348–1354.

Teufel, J. D., Donner, T., Li, D., *et al.* 2011. Sideband cooling of micromechanical motion to the ground state. *Nature*, **475**, 359–363.

The LIGO Scientific Collaboration. 2010. *A gravitational wave observatory operating beyond the quantum limit: Squeezed light in application.* In press, Nature Physics.

Thompson, J. D., Zwickl, B. M., Jayich, A. M. *et al.* 2008. Strong dispersive coupling of a high-finesse cavity to a micromechanical membrane. *Nature*, **452**(7183), 72–45.

Thompson, R. J., Rempe, G., and Kimble, H. J. 1992. Observation of normal-mode splitting for an atom in an optical cavity. *Physical Review Letters*, **68**(8), 1132–1135.

Thorne, Kip S., O'Shaugnessy, Richard, and d'Ambrosio, Erika. 2000. *Beam reshaping to reduce thermoelastic noise.* LIGO-G000223-00-D.

Thorpe, M. J., Moll, K. D., Jones, R. J., Safdi, B., and Ye, J. 2006. Broadband cavity ring-down spectroscopy for sensitive and rapid molecular detection. *Science*, **311**(5767), 1595–1599.

Ting, S. M., and Fitzgerald, E. A. 2000. Metal-organic chemical vapor deposition of single domain GaAs on Ge/Ge$_x$Si$_{1-x}$/Si and Ge substrates. *Journal of Applied Physics*, **87**(5), 2618–2628.

Tittonen, I., Breitenbach, G., Kalkbrenner, T. *et al.* 1999. Interferometric measurements of the position of a macroscopic body: Towards observation of quantum limits. *Physical Review A*, **59**(2), 1038–1044.

Tomaru, Takayuki, Uchiyama, Takashi, Tatsumi, Daisuke *et al.* 2001. Cryogenic measurement of the optical absorption coefficient in sapphire crystals at 1.064 μm for the large-scale cryogenic gravitational wave telescope. *Physics Letters A*, **283**(1–2), 80–84.

Tomaru, T., Suzuki, T., Miyoki, S. *et al.* 2002. Thermal lensing in cryogenic sapphire substrates. *Classical and Quantum Gravity*, **19**(7), 2045–2049.

Tomaru, T., Suzuki, T., Haruyama, T. *et al.* 2004. Vibration analysis of cryocoolers. *Cryogenics*, **44**, 309–317.

Tomaru, T., Tokunari, M., Kuroda, K. *et al.* 2008a. Reduction of heat load of LCGT cryostat. *Journal of Physics: Conference Series*, **122**, 012009.

Tomaru, Takayuki, Tokunari, Masao, Kuroda, Kazuaki *et al.* 2008b. Conduction effect of thermal radiation in a metal shield pipe in a cryostat for a cryogenic interferometric gravitational wave detector. *Japanese Journal of Applied Physics*, **47**, 1771–1774.

Topp, K. A., and Cahill, David G. 1996. Elastic properties of several amorphous solids and disordered crystals below 100 K. *Zeitschrift für Physik B*, **101**, 235–245.

Touloukian, Y. S., and Ho, C. Y. 1970. *Thermophysical Properties of Matter: The TRPC Data Series.* Plenum Press.

Tovar, A. A. 2001. Propagation of flat-topped multi-Gaussian laser beams. *Journal of the Optical Society of America A*, **18**(Aug.), 1897–1904.

Trupke, M., Hinds, E. A., Eriksson, S. *et al.* 2005. Microfabricated high-finesse optical cavity with open access and small volume. *Applied Physics Letters*, **87**, 211106.

Tsao, J. Y. 1993. *Materials Fundament of Molecular Beam Epitaxy*. Academic Press.

Uchiyama, T., Tomaru, T., Tobar, M. E. *et al.* 1999. Mechanical quality factor of a cryogenic sapphire test mass for gravitational wave detectors. *Physics Letters A*, **261**(1–2), 5–11.

Uchiyama, T., Miyoki, S., Ohashi, M. *et al.* 2006. Cryogenic systems of the Cryogenic Laser Interferometer Observatory. *Journal of Physics: Conference Series*, **32**, 259–264.

Udem, T., Holzwarth, R., and Hänsch, T. W. 2002. Optical frequency metrology. *Nature*, **416**(6877), 233–237.

Ugolini, D., Girard, M., Harry, G. M., and Mitrofanov, V. P. 2008. Discharging fused silica test masses with ultraviolet light. *Physics Letters A*, **372**(36), 5741–5744.

Vahala, Kerry J. 2003. Optical microcavities. *Nature*, **424**(6950), 839–846.

Vahlbruch, Henning, Mehmet, Moritz, Chelkowski, Simon *et al.* 2008. Observation of squeezed light with 10-dB quantum-noise reduction. *Physical Review Letters*, **100**(3), 033602.

van Enk, S. J., McKeever, J., Kimble, H. J., and Ye, J. 2001. Cooling of a single atom in an optical trap inside a resonator. *Physical Review A*, **64**(1), 013407.

van Vliet, K. M., and Menta, H. 1981. Theory of transport noise in semiconductors. *Physica Status Solidi. B, Basic Research*, **106**, 11–30.

van Vliet, K. M., van der Ziel, A., and Schmidt, R. 1980. Temperature fluctuation noise of thin films supported by a substrate. *Journal of Applied Physics*, **51**, 2947–2956.

VanDevender, A. P., Colombe, Y., Amini, J., Leibfried, D., and Wineland, D. J. 2010. Efficient fiber optic detection of trapped ion fluorescence. *Physical Review Letters*, **105**(2), 023001.

Vanner, M. R., Dikovski, I., Cole, G. D. *et al.* 2011. Pulsed quantum optomechanics. *Proceedings of the National Academy of Sciences (USA)*, **108**.

Ventura, G., and Risegari, L. 2007. *The Art of Cryogenics: Low-Temperature Experimental Techniques*. Elsevier Science.

Villa, F., Martinez, A., and Regalado, F. E. 2000. Correction mask for thickness uniformity in large-area thin films. *Applied Optics*, **39**(10), 1602–1610.

Villar, A., Black, E., Ogin, G. *et al.* 2010a. *Loss angles from the direct measurement of Brownian noise in coatings*. LIGO-G1000937.

Villar, Akira E., Black, Eric D., DeSalvo, Riccardo *et al.* 2010b. Measurement of thermal noise in multilayer coatings with optimized layer thickness. *Physical Review D*, **81**(12), 122001.

Vinet, Jean-Yves. 2009. On special optical modes and thermal issues in advanced gravitational wave interferometric detectors. *Living Reviews in Relativity*, **12**(5).

Vinet, J.-Y., Hello, P., Man, C. N., and Brillet, A. 1992. A high accuracy method for the simulation of non-ideal optical cavities. *Journal de Physique I*, **2**(July), 1287–1303.

Vitali, D., Gigan, S., Ferreira, A. *et al.* 2007. Optomechanical entanglement between a movable mirror and a cavity field. *Physical Review Letters*, **98**, 030405.

Volpyan, O. D., and Yakovlev, P. P. 2002. The effect of heat treatment on the optical properties of Ta_2O_5 films. *Journal of Optical Technology*, **69**(5), 319–321.

Vossen, J. L., and Kern, W. (eds). 1978. *Thin Film Processes*. Academic Press.

Vukcevich, M. R. 1972. A new interpretation of the anomalous properties of vitreous silica. *Journal of Non-Crystalline Solids*, **11**, 25–63.

Walsh, Christopher J., Leistner, Achim J., Seckold, Jeffrey, Oreb, Bozenko F., and Farrant, David I. 1999. Fabrication and measurement of optics for the Laser Interferometer Gravitational Wave Observatory. *Applied Optics*, **38**(13), 2870–2879.

Wang, C., and Hartnett, J. G. 2010. A vibration free cryostat using pulse tube cryocooler. *Cryogenics*, **50**, 336–341.

Wang, C., Thummes, G., Heiden, C., Best, K.-J., and Oswald, B. 1999. Cryogen free operation of a niobium–tin magnet using a two-stage pulse tube cooler. *IEEE Transactions on Applied Superconductivity*, **9**, 402–405.

Wang, Wen-Hsiang, and Chao, Shiuh. 1998. Annealing effect on ion-beam-sputtered titanium dioxide film. *Optics Letters*, **23**(18), 1417–1419.

Wanser, K. H. 1992. Fundamental phase noise limit in optical fibres due to temeperature fluctuations. *Electronics Letters*, **28**, 53–54.

Weber, B., Specht, H. P., Müller, T. *et al.* 2009. Photon-photon entanglement with a single trapped atom. *Physical Review Letters*, **102**(3), 030501.

Weber, J. 1960. Detection and generation of gravitational waves. *Physical Review*, **117**(1), 306–313.

Webster, Stephen A., Oxborrow, Mark, and Gill, Patrick. 2004. Subhertz-linewidth Nd:YAG laser. *Optics Letters*, **29**, 1497–1499.

Webster, S. A., Oxborrow, M., and Gill, P. 2007. Vibration insensitive optical cavity. *Physical Review A*, **75**(1), 011801.

Webster, S. A., Oxborrow, M., Pugla, S., Millo, J., and Gill, P. 2008. Thermal-noise-limited optical cavity. *Physical Review A*, **77**(3), 033847.

Wehner, G. K., and Anderson, G. S. 1970. *Handbook of Thin Film Technology*. McGraw Hill. Chap. The nature of physical sputtering.

Wehner, G. K., and Rosenberg, D. 1960. Angular distribution of sputtered material. *Journal of Applied Physics*, **31**(1), 177–179.

Wei, D. T., and Louderback, A. W. 1979. *US Patent 4,142,958: Method for fabricating multi-layer optical films*.

Weisberg, J. M., Nice, D. J., and Taylor, J. H. 2010. Timing measurements of the relativistic binary pulsar PSR B1913+16. *The Astrophysical Journal*, **722**(Oct.), 1030–1034.

Weiss, Rai. 1972. *Electromagnetically coupled broad-band gravitational wave antenna*. Tech. rept. Massachusetts Institute of Technology. LIGO-P720002-00.

Wen, Linqing, and Chen, Yanbei. 2010. Geometrical expression for the angular resolution of a network of gravitational-wave detectors. *Physical Review*, **81**(8), 082001.

White, G. K. 1993. Reference materials for thermal expansion: Certified or not? *Thermochimica Acta*, **218**, 83–99.

Wiedersich, J., Adichtchev, S. V., and Rössler, E. 2000. Spectral shape of relaxations in silica glass. *Physical Review Letters*, **84**, 2718–2721.

Wilk, T., Webster, S. C., Kuhn, A., and Rempe, G. 2007. Single-atom single-photon quantum interface. *Science*, **317**, 488–490.

Willke, B., and The GEO-HF Collaboration. 2006. The GEO-HF project. *Classical and Quantum Gravity*, **23**(8), S207–S214.

Willke, B., Danzmann, K., Frede, M. *et al.* 2008. Stabilized lasers for advanced gravitational wave detectors. *Classical and Quantum Gravity*, **25**(11), 114040.

Wilmsen, C. W., Temkin, H., and Coldren, L. A. (eds). 1999. *Vertical-Cavity Surface-Emitting Lasers: Design, Fabrication, Characterization, and Applications*. Cambridge University Press.

Wilson, D. J., Regal, C. A., Papp, S. B., and Kimble, H. J. 2009. Cavity optomechanics with stoichiometric SiN films. *Physical Review Letters*, **103**(20), 207204.

Wilson-Rae, I. 2008. Intrinsic dissipation in nanomechanical resonators due to phonon tunneling. *Physical Review B*, **77**, 245418.

Wilson-Rae, I., Nooshi, N., Zwerger, W., and Kippenberg, T. J. 2007. Theory of ground state cooling of a mechanical oscillator using dynamical backaction. *Physical Review Letters*, **99**(9), 093901.

Wineland, D. J., Monroe, C., Itano, W. M. *et al.* 1998. Experimental issues in coherent quantum-state manipulation of trapped atomic ions. *Journal of Research of the National Institute of Standards and Technology*, **103**, 259–328.

Winkler, W., Danzmann, K., Rüdiger, A., and Schilling, R. 1991. Heating by optical absorption and the performance of interferometric gravitational-wave detectors. *Physical Review A*, **44**(Dec.), 7022–7036.

Wong, N. C., and Hall, J. L. 1985. Servo control of amplitude modulation in frequency-modulation spectroscopy: Demonstration of shot-noise-limited detection. *Journal of the Optical Society of America B*, **2**(9), 1527–1533.

Wortman, J. J., and Evans, R. A. 1965. Young's modulus, shear modulus, and Poisson's ratio in silicon and germanium. *Journal of Applied Physics*, **36**, 153–156.

Wu, S. C., Wan, Z. Z., Li, H., and Z., Liu Z. 2006. Photo-thermal shot noise in end mirrors of LIGO due to correlation of power fluctuations. *Chinese Physics Letters*, **23**, 3173–3176.

Xie, H., Zeng, X. T., and Yeo, W. K. 2008. Temperature dependent properties of titanium oxide thin films by spectroscopic ellipsometry. *SimTech Reports*, **9**, 29.

Yablonovitch, Eli, Gmitter, T., Harbison, J. P., and Bhat, R. 1987. Extreme selectivity in the lift-off of epitaxial GaAs films. *Applied Physics Letters*, **51**(26), 2222–2224.

Yamamoto, Hiro. 2007. *LIGO I mirror scattering loss by microroughness*. LIGO-T070082-03-E.

Yamamoto, K. 2000. *Study of the thermal noise caused by inhomogeneously distributed loss*. Ph.D. thesis, University of Tokyo.

Yamamoto, Kazuhiro, Otsuka, Shigemi, Ando, Masaki, Kawabe, Keita, and Tsubono, Kimio. 2001. Experimental study of thermal noise caused by an inhomogeneously distributed loss. *Physics Letters A*, **280**(5–6), 289–296.

Yamamoto, Kazuhiro, Otsuka, Shigemi, Ando, Masaki, Kawabe, Keita, and Tsubono, Kimio. 2002. Study of the thermal noise caused by inhomogeneously distributed loss. *Classical and Quantum Gravity*, **19**(7), 1689–1696.

Yamamoto, K., Miyoki, S., Uchiyama, T. *et al.* 2004. Mechanical loss of the reflective coating and fluorite at low temperature. *Classical and Quantum Gravity*, **21**(5), S1075–S1081.

Yamamoto, K., Miyoki, S., Uchiyama, T. *et al.* 2006a. Measurement of the mechanical loss of a cooled refractive coating for gravitational wave detection. *Physical Review D*, **74**, 022002.

Yamamoto, K., Uchiyama, T., Miyoki, S. *et al.* 2006b. Measurement of vibration of the top of the suspension in a cryogenic interferometer with operating cryocoolers. *Journal of Physics: Conference Series*, **32**, 418–423.

Yamamoto, K., Uchiyama, T., Miyoki, S. *et al.* 2008. Current status of the CLIO project. *Journal of Physics: Conference Series*, **122**(1), 012002.

Yang, Jinling, Ono, Takahito, and Esashi, Masayoshi. 2000. Surface effects and high quality factors in ultrathin single-crystal silicon cantilevers. *Applied Physics Letters*, **77**(23), 3860–3862.

Yang, Jinling, Ono, T., and Esashi, M. 2002. Energy dissipation in submicrometer thick single-crystal silicon cantilevers. *Journal of Microelectromechanical Systems*, **11**(6), 775–783.

Yasamura, K. Y., Stowe, T. D., Chow, E. M. *et al.* 2000. Quality factors in micron and sub-micron thick cantilevers. *Journal of Microelectromechanical Systems*, **9**, 117–125.

Ye, J., and Lynn, T. W. 2003. Applications of optical cavities in modern atomic, molecular, and optical physics. *Advances in Atomic, Molecular, and Optical Physics*, **49**, 1–83.

Ye, J., Vernooy, D. W., and Kimble, H. J. 1999. Trapping of single atoms in cavity QED. *Physical Review Letters*, **83**(24), 4987–4990.

Ye, J., Kimble, H. J., and Katori, H. 2008. Quantum State engineering and precision metrology using state-insensitive light traps. *Science*, **320**(5884), 1734–1738.

Yoon, Jongseung, Jo, Sungjin, Chun, Ik Su *et al.* 2010. GaAs photovoltaics and optoelectronics using releasable multilayer epitaxial assemblies. *Nature*, **465**(7296), 329–333.

Yost, D. C., Schibli, T. R., Ye, J. *et al.* 2009. Vacuum-ultraviolet frequency combs from below-threshold harmonics. *Nature Physics*, **5**, 815–820.

Young, B. C., Cruz, F. C., Itano, W. M., and Bergquist, J. C. 1999. Visible lasers with subhertz linewidths. *Physical Review Letters*, **82**(19), 3799–3802.

Zelenogorsky, Victor V., Solovyov, Alexander A., Kozhevator, Ilya E. *et al.* 2006. High-precision methods and devices for in situ measurements of thermally induced aberrations in optical elements. *Applied Optics*, **45**(17), 4092–4101.

Zelevinsky, T., Blatt, S., Boyd, M. M. *et al.* 2008. Highly coherent spectroscopy of ultracold atoms and molecules in optical lattices. *ChemPhysChem*, **9**, 375–382.

Zeller, R. C., and Pohl, R. O. 1971. Thermal conductivity and specific heat of noncrystalline solids. *Physical Review B*, **4**(6), 2029–2041.

Zener, C. 1937. Internal friction in solids. I. Theory of internal friction in reeds. *Physical Review*, **52**(3), 230–235.

Zener, C. 1938. Internal friction in solids II. General theory of thermoelastic internal friction. *Physical Review*, **53**(1), 90–99.

Zener, C. 1948. *Elasticity and Anelasticity in Metals*. University of Chicago Press.

Zhang, L.-T., Armandula, Helena, Billingsley, Garilynn, Cardenas, Lee, and Kells, Bill. 2008. *The coating scattering and absorption measurements of LIGO mirrors at Caltech*. LIGO-G080162-00.

Zhao, C., Degallaix, J., Ju, L. *et al.* 2006. Compensation of strong thermal lensing in high-optical-power cavities. *Physical Review Letters*, **96**(23), 231101.

Zorn, M., Haberland, K., Knigge, A. *et al.* 2002. MOVPE process development for 650 nm VCSELS using optical in-situ techniques. *Journal of Crystal Growth*, **235**(1–4), 25–34.

Zwickl, B. M., Shanks, W. E., Jayich, A. M. *et al.* 2008. High quality mechnaical and optical properties of commercial silicon nitride membranes. *Applied Physics Letters*, **92**, 103125.

Note: LIGO Technical Documents having the form LIGO-axxxxxx, where "a" is any letter and "x" is any number, can be retrieved from the web at dcc.ligo.org.

Printed in the United States
by Baker & Taylor Publisher Services